Lecture Notes in Computer Science

# Lecture Notes in Artificial Intelligence 14992

Founding Editor

Jörg Siekmann

The series Lecture Notes in Artificial Intelligence (LNAI) was established in 1988 as a topical subseries of LNCS devoted to artificial intelligence.

The series publishes state-of-the-art research results at a high level. As with the LNCS mother series, the mission of the series is to serve the international R & D community by providing an invaluable service, mainly focused on the publication of conference and workshop proceedings and postproceedings.

Andreas Hotho · Sebastian Rudolph
Editors

# KI 2024: Advances in Artificial Intelligence

## 47th German Conference on AI
## Würzburg, Germany, September 25–27, 2024
## Proceedings

Springer

*Editors*
Andreas Hotho
Julius-Maximilians-Universität Würzburg
Würzburg, Germany

Sebastian Rudolph
TU Dresden
Dresden, Germany

ISSN 0302-9743 ISSN 1611-3349 (electronic)
Lecture Notes in Artificial Intelligence
ISBN 978-3-031-70892-3 ISBN 978-3-031-70893-0 (eBook)
https://doi.org/10.1007/978-3-031-70893-0

LNCS Sublibrary: SL7 – Artificial Intelligence



This Springer imprint is published by the registered company Springer Nature Switzerland AG
The registered company address is: Gewerbestrasse 11, 6330 Cham, Switzerland

# Preface

This proceedings volume contains the papers presented at the 47th German Conference on Artificial Intelligence (KI 2024), held in Würzburg, Germany, during September 25–27, 2024, and co-located with the LWDA 2024 conference ("Lernen, Wissen, Daten, Analysen", i.e., "Learning, Knowledge, Data, Analytics").

KI 2024 was the 47th German Conference on Artificial Intelligence (AI) organized in cooperation with the German Section on Artificial Intelligence of the German Informatics Society (GI-FBKI, Fachbereich Künstliche Intelligenz der Gesellschaft für Informatik (GI) e.V.). The German AI Conference basically started 49 years ago with the first meeting of the national special interest group on AI within the GI on October 7, 1975. KI is one of the major European AI conferences and traditionally brings together academic and industrial researchers from all areas of AI, providing an ideal place for exchanging news and research results on theory and applications. While KI is primarily attended by researchers from Germany and neighboring countries, it warmly welcomes international participation.

The technical program of KI 2024 featured paper presentations as well as two keynotes, two tutorials, a doctoral symposium, and eight workshops. Overall, KI 2024 received 67 submissions from authors in twelve countries, which were single-blind reviewed, in all but very few cases by three Program Committee members each. The Program Committee, comprising 43 experts from five countries, accepted 19 full papers (43% of the submissions in that category), seven technical communications (47% of the submissions in that category), and five extended abstracts reporting on work published at other major international AI venues (45% of the submissions in that category). As a highlight of this year's edition of the KI conference, the main program featured a session organized jointly with the Bavarian AI network baiosphere on the first day, which included a panel discussion with the network's speakers, a presentation of next-generation scientists, and a welcome reception. Further, for the first time, KI and LWDA were held together. We were honored that two prominent researchers kindly agreed to give very interesting keynote talks (alphabetical order, see also the abstracts below):

- Elisabeth André, University of Augsburg, Germany
- Christian Baukhage, University of Bonn, Germany

As Program Committee chairs, we would like to thank our speakers for their interesting and inspirational talks, the Workshop and Tutorial Chairs Janna Hastings and Ute Schmid, the Doctoral Symposium Chairs Ingo Scholtes and Petra Mutzel, the Local Chairs Anna Krause and Alison Wichmann, as well as the Publicity Chairs Martin Rackl and Albin Zehe. Our special gratitude goes to the Program Committee, whose sophisticated and conscientious judgement ensured the high quality of the KI conference. Without their substantial voluntary work, this conference would not have been possible.

The following tutorials and workshops took place at KI 2024:

- *Research Data Management in Data Science and AI – Avoiding a Replicability Crisis (RDM4AI)*, tutorial by Ricardo Usbeck, Angelie Kraft, and Leyla Jael Castro
- *Exploiting Structure in Decision Making under the Lens of Recent Advances in StaRAI*, tutorial by Marcel Gehrke, Florian Andreas Marwitz, Nazli Nur Karabulut, and Tanya Braun
- *38th Workshop on (Constraint) Logic Programming (WLP 2024)*, workshop organized by Ulrich John, Dietmar Seipel, and Sibylle Schwarz
- *Humanities-Centred AI (CHAI)*, workshop organized by Sylvia Melzer, Stefan Thiemann, Hagen Peukert, and Erik Radisch
- *2nd Workshop on Public Interest AI*, workshop organized by Hadi Ashgari and Theresa Züger
- *HuMaIn: Human-Machine Interaction Workshop*, workshop organized by Ivan Yamshchikov, Pascal Meißner, and Sharwin Rezagholi
- *10th Workshop on Formal and Cognitive Reasoning*, workshop organized by Özgür Özçep, Nele Rußwinkel, Kai Sauerwald, and Diedrich Wolter
- *AI in Production*, workshop organized by Torsten Munkelt, Martin Krockert, Nick Hartmann, Christoph Wehner, Lukas Bahr, Dominik Pietsch, Stephan Scheele, Marvin Matthes, Judith Knoblach, and Kirsten Hoffmann
- *Second Workshop on AI for AI Education (AI4AILearning)*, workshop organized by Michael Kohlhase, Jochen Leidner, Ute Schmid, and Diedrich Wolter
- *Joint Workshop FG WM + PUK*, workshop organized by Stefan Edelkamp, Jürgen Sauer, Andreas Korger, and Pascal Reuss

Last not least, we would like to thank our sponsors for their support:

- German Section on Artificial Intelligence of the German Informatics Society (https://fb-ki.gi.de)
- SECAI: The School of Embedded Composite Artificial Intelligence (https://secai.org)
- baiosphere – the Bavarian AI network (https://baiosphere.org)
- Vogel Stiftung Dr. Eckernkamp (https://www.vogel-stiftung.de/)
- CAIDAS: Center for Artificial Intelligence and Data Science (https://www.caidas.uni-wuerzburg.de/)
- Institute of Computer Science, Universität Würzburg (https://www.informatik.uni-wuerzburg.de/en/)
- Springer (https://www.springer.com)

September 2024

Andreas Hotho
Sebastian Rudolph

# Organization

## Program Committee Chairs

| | |
|---|---|
| Andreas Hotho | University of Würzburg, Germany |
| Sebastian Rudolph | TU Dresden, Germany |

## Workshop and Tutorial Chairs

| | |
|---|---|
| Janna Hastings | University of Zurich, Switzerland |
| Ute Schmid | Otto-Friedrich-Universität Bamberg, Germany |

## Doctoral Symposium Chairs

| | |
|---|---|
| Petra Mutzel | University of Bonn, Germany |
| Ingo Scholtes | University of Würzburg, Germany |

## Local Chairs

| | |
|---|---|
| Anna Krause | University of Würzburg, Germany |
| Alison Wichmann | University of Würzburg, Germany |

## Publicity Chairs

| | |
|---|---|
| Martin Rackl | University of Würzburg, Germany |
| Albin Zehe | University of Würzburg, Germany |

## Proceedings Chair

| | |
|---|---|
| Sebastian Rudolph | TU Dresden, Germany |

## Program Committee

| | |
|---|---|
| Martin Aleksandrov | FU Berlin, Germany |
| Martin Atzmueller | Osnabrück University, Germany |
| Franz Baader | TU Dresden, Germany |
| Joachim Baumeister | denkbares GmbH, Germany |
| Christoph Benzmüller | Otto-Friedrich-Universität Bamberg, Germany |
| Ralph Bergmann | University of Trier, Germany |
| Christian Bizer | University of Mannheim, Germany |
| Ulf Brefeld | Leuphana Universität Lüneburg, Germany |
| Philipp Cimiano | Bielefeld University, Germany |
| Stefan Dietze | GESIS, Germany |
| Stefan Edelkamp | CTU Prague, Czech Republic |
| Ulrich Furbach | University of Koblenz, Germany |
| Johannes Fähndrich | Hochschule für Polizei Baden-Württemberg, Germany |
| Johannes Fürnkranz | Johannes Kepler University Linz, Austria |
| Stephan Günnemann | Technical University of Munich, Germany |
| Tom Hanika | University of Hildesheim, Germany |
| Andreas Harth | Friedrich-Alexander-Universität Erlangen-Nürnberg and Fraunhofer IIS, Germany |
| Gabriele Kern-Isberner | Technische Universität Dortmund, Germany |
| Matthias Klusch | DFKI, Germany |
| Franziska Klügl | Örebro University, Sweden |
| Stefan Kopp | Bielefeld University, Germany |
| Markus Krötzsch | TU Dresden, Germany |
| Christoph Lange | Fraunhofer FIT and RWTH Aachen, Germany |
| Bernd Ludwig | University Regensburg, Germany |
| Lukas Malburg | University of Trier/DFKI, Germany |
| Mirjam Minor | Goethe University Frankfurt, Germany |
| Till Mossakowski | University of Magdeburg, Germany |
| Ralf Möller | University of Lübeck, Germany |
| Axel-Cyrille Ngonga Ngomo | Paderborn University, Germany |
| Falco Nogatz | SUSI&James GmbH, Germany |
| Heiko Paulheim | University of Mannheim, Germany |
| Axel Polleres | Vienna University of Economics and Business, Austria |
| Simon Razniewski | TU Dresden, Germany |
| Jürgen Sauer | University of Oldenburg, Germany |
| Ute Schmid | University of Bamberg, Germany |

| | |
|---|---|
| Lutz Schröder | Friedrich-Alexander-Universität Erlangen-Nürnberg, Germany |
| René Schumann | University of Applied Sciences Western Switzerland, Switzerland |
| Daniel Sonntag | DFKI, Germany |
| Myra Spiliopoulou | Otto-von-Guericke-University Magdeburg, Germany |
| Frieder Stolzenburg | Harz University of Applied Sciences, Germany |
| Matthias Thimm | FernUniversität in Hagen, Germany |
| Julian Tritscher | University of Würzburg, Germany |
| Ricardo Usbeck | Leuphana University Lüneburg, Germany |
| Stefan Wrobel | Fraunhofer IAIS/University of Bonn, Germany |
| Özgür Lütfü Özcep | University of Hamburg, Germany |

## Additional Reviewers

Mehdi Akbari Gurabi
Thomas Asselborn
Debayan Banerjee
Farouk Beddiaf
Johannes Bühl
Sisay Chala
Ömer Ibrahim Erduran
Björn Filter
Michael Freund
Lukas Gerlach
Martin Glauer
Anna-Christina Glock
Alexander Graß
Paul Heisterkamp
Wilfried Honekamp
Junbo Huang
Hannes Kath
Angelie Kraft
Stefanie Krause
Mirko Lenz
Soham Majumder
Carsten Maletzki
Florian Marwitz
Marcel Mauri
Sylvia Melzer
Rahel Moser
Bartosz Piotrowski
Jakob Piribauer
Yannick Rudolph
Kai Sauerwald
Simon Schiff
Alexander Schultheis
Veronika Solopova
Tim Strobel
Ilira Troshani
Moritz Wohlstein
Xi Yan
Christian Zeyen

## Sponsors

# Abstracts of Invited Talks

# Amplifying Human-Human and Human-Agent Interaction with AI

Elisabeth André

University of Augsburg, Germany

Advances in analytic and generative AI present promising opportunities for enhancing communication by augmenting individuals' perceptual and expressive capabilities. This talk will explore recent advancements and practical applications of conversation-enhancing technologies, with a focus on their potential benefits for individuals facing communication challenges. Collaborative efforts have resulted in innovative applications tailored to diverse user needs. Examples include signing avatars designed to assist the hearing impaired, as well as more expressive voices for users of AAC (Augmentative and Alternative Communication) devices. Additionally, virtual reality environments have been developed to facilitate social skill training for both children and adults, leveraging role play interactions with virtual agents. We will discuss how analytic and generative AI can serve as powerful tools to overcome communication barriers, foster authentic expression, and promote social inclusion. By analyzing the potential of these technologies to support users with varying communication abilities, we aim to address the challenges inherent in designing empathetic solutions that cater to the specific needs of diverse audiences, while mitigating the risk of reinforcing stereotypes.

# Quantum AI/ML – Hype or Hope?

Christian Baukhage

University of Bonn, Germany

Over the past decade, we have seen encouraging progress in technical realizations of quantum computers and there are now rising hopes for practical applications of quantum computing.

A noteworthy trend in this regard is that the quantum computing community has jumped on the machine learning bandwagon and is promising quantum advantages for artificial intelligence.

Our goals with this presentation are therefore threefold: First, we provide an ever so brief introduction to quantum computing and its expected benefits. Second, we point out the sobering fact that most promises as to quantum supremacy for ML and AI are and likely will remain severely exaggerated. Third, we emphasize that there still is hope for quantum AI as long as we are looking for appropriate use cases. As an example of such a use case, we discuss how quantum algorithms can accelerate Bayesian network inference.

# Contents

## Full Technical Papers

From Resolving Inconsistencies in Qualitative Constraints Networks to Identifying Robust Solutions: A Universal Encoding in ASP ............... 3
*Moritz Bayerkuhnlein, Tobias Schwartz, and Diedrich Wolter*

Efficiently Training Neural Networks for Imperfect Information Games by Sampling Information Sets ........................................ 17
*Timo Bertram, Johannes Fürnkranz, and Martin Müller*

A Note on Linear Time Series Prediction ................................ 30
*Christopher Bonenberger, Markus Schneider, Wolfgang Ertel, and Friedhelm Schwenker*

Data Augmentation in Latent Space with Variational Autoencoder and Pretrained Image Model for Visual Reinforcement Learning ............ 45
*Xuzhe Dang and Stefan Edelkamp*

Could the Declarer Have Discarded It? Refined Anticipation of Cards in Skat ....................................................................... 60
*Stefan Edelkamp*

A Framework for General Trick-Taking Card Games ...................... 73
*Stefan Edelkamp*

Mechanisms for Data Sharing in Collaborative Causal Inference ............ 86
*Björn Filter, Ralf Möller, and Özgür Lütfü Özçep*

SaVeWoT: Scripting and Verifying Web of Things Systems and Their Effects on the Physical World .......................................... 99
*Justus Fries, Michael Freund, and Andreas Harth*

Active Learning in Multi-label Classification of Bioacoustic Data ........... 114
*Hannes Kath, Thiago S. Gouvêa, and Daniel Sonntag*

Quantifying the Trade-Offs Between Dimensions of Trustworthy AI - An Empirical Study on Fairness, Explainability, Privacy, and Robustness ...... 128
*Nils Kemmerzell and Annika Schreiner*

Image Dataset Quality Assessment Through Descriptive Out-of-Distribution Detection ..... 147
*Sami Kharma and Jürgen Großmann*

Saxony-Anhalt is the Worst: Bias Towards German Federal States in Large Language Models ..... 160
*Anna Kruspe and Mila Stillman*

Towards Privacy-Preserving Relational Data Synthesis via Probabilistic Relational Models ..... 175
*Malte Luttermann, Ralf Möller, and Mattis Hartwig*

Evaluating AI-Based Components in Autonomous Railway Systems: A Methodology ..... 190
*Jan Roßbach, Oliver De Candido, Ahmed Hammam, and Michael Leuschel*

SocialCOP: Reusable Building Blocks for Collective Constraint Optimization ..... 204
*Julia Ruttmann and Alexander Schiendorfer*

Context-Specific Selection of Commonsense Knowledge Using Large Language Models ..... 218
*Oliver Jakobs and Claudia Schon*

Graph2RETA: Graph Neural Networks for Pick-up and Delivery Route Prediction and Arrival Time Estimation ..... 232
*Wilson Sentanoe, Sreyashi Saha, Yasas Diyanananda, Aqsa Manzoor, Buddhika Dasanayake, and Daniela Thyssens*

Data Generation for Explainable Occupational Fraud Detection ..... 246
*Julian Tritscher, Maximilian Wolf, Anna Krause, Andreas Hotho, and Daniel Schlör*

Leveraging Weakly Supervised and Multiple Instance Learning for Multi-label Classification of Passive Acoustic Monitoring Data ..... 260
*Ilira Troshani, Thiago S. Gouvêa, and Daniel Sonntag*

**Technical Communications**

Leveraging YOLO for Real-Time Video Analysis of Animal Welfare in Pig Slaughtering Processes ..... 275
*Christian Beecks, Anandraj Amalraj, Alexander Graß, Marc Jentsch, Felix Kitschke, Maximilian Norz, and Patric Schäffer*

Early Explorations of Lightweight Models for Wound Segmentation on Mobile Devices ..... 282
*Vanessa Borst, Timo Dittus, Konstantin Müller, and Samuel Kounev*

Automated Design in Hybrid Action Spaces by Reinforcement Learning and Differential Evolution ..... 292
*Quirin Göttl, Haris Asif, Alexander Mattick, Robert Marzilger, and Axel Plinge*

Instance Segmentation with a Novel Tree Log Detection Dataset ..... 300
*Julian Haasis, Christopher Bonenberger, and Markus Schneider*

LaFAM: Unsupervised Feature Attribution with Label-Free Activation Maps ..... 308
*Aray Karjauv and Sahin Albayrak*

Uli-RL: A Real-World Deep Reinforcement Learning Pedagogical Agent for Children ..... 316
*Anna Riedmann, Julia Götz, Carlo D'Eramo, and Birgit Lugrin*

Explanatory Interactive Machine Learning with Counterexamples from Constrained Large Language Models ..... 324
*Emanuel Slany, Stephan Scheele, and Ute Schmid*

**Extended Abstracts of Papers from other AI Conferences**

MIND Your Language: A Multilingual Dataset for Cross-Lingual News Recommendation (Extended Abstract) ..... 335
*Andreea Iana, Goran Glavaš, and Heiko Paulheim*

A Human-in-the-Loop Tool for Annotating Passive Acoustic Monitoring Datasets (Extended Abstract) ..... 341
*Hannes Kath, Thiago S. Gouvêa, and Daniel Sonntag*

Out-of-Distribution Detection with Logical Reasoning (Extended Abstract) ..... 346
*Konstantin Kirchheim, Tim Gonschorek, and Frank Ortmeier*

A Brief Systematization of Explanation-Aware Attacks ..... 350
*Maximilian Noppel and Christian Wressnegger*

Star-Shaped Denoising Diffusion Probabilistic Models (Extended Abstract) ..... 355
*Andrey Okhotin, Dmitry Molchanov, Vladimir Arkhipkin, Grigory Bartosh, Viktor Ohanesian, Aibek Alanov, and Dmitry Vetrov*

**Author Index** ..... 361

# Full Technical Papers

# From Resolving Inconsistencies in Qualitative Constraints Networks to Identifying Robust Solutions: A Universal Encoding in ASP

Moritz Bayerkuhnlein[1,2(✉)], Tobias Schwartz[2], and Diedrich Wolter[2]

[1] University of Bamberg, Bamberg, Germany
moritz.bayerkuhnlein@uni-bamberg.de
[2] University of Lübeck, Lübeck, Germany
{tobias.schwartz,diedrich.wolter}@uni-luebeck.de
https://www.uni-luebeck.de

**Abstract.** Qualitative Constraint Networks (QCNs) are foundational to Qualitative Spatial and Temporal Reasoning (QSTR) for modelling real-world entity relations, facilitating decision-making and planning. However, perturbations or even inconsistencies in inputs may arise in various application contexts, such as from unforeseen circumstances or by merging data from different sources. Therefore, it is crucial for qualitative reasoning systems to address these challenges, e.g., not just identify any solution, but to optimize solutions for resilience against perturbations. Likewise, the ability is needed to resolve any inconsistencies with minimal repairs. Both tasks are challenging optimization problems on their own, as determining network consistency is already NP-hard for most qualitative constraint languages. In this paper, we present a universal encoding in Answer-Set Programming (ASP) to address these challenges. Our encoding allows for efficient resolution of both the robustness and repair problem by exploiting ASP optimization techniques. We demonstrate the effectiveness of our encoding in an experimental evaluation. Our results show that our encoding can match the state-of-the-art on some qualitative calculi in terms of computational efficiency. On top, our encoding offers a flexible and powerful framework for tackling optimization problems in qualitative reasoning systems.

**Keywords:** Answer Set Programming · Qualitative Spatial and Temporal Reasoning · Robustness · Repair

## 1 Introduction

The portrayal of the world through the laws of nature is precise but fragile, as it is impractical to comprehensively monitor and control every element. From a computational perspective, it is thus sensible to employ qualitative representations that abstract from numerical details to enhance reasoning efficiency. Reasoning


A. Hotho and S. Rudolph (Eds.): KI 2024, LNAI 14992, pp. 3–16, 2024.
https://doi.org/10.1007/978-3-031-70893-0_1

about fundamental cognitive concepts such as space and time through qualitative information falls within the domain of Qualitative Spatial and Temporal Reasoning (QSTR) [8,13]. Like natural language, QSTR leverages qualitative relations, like *before*, *left of*, and *outside*, to depict the world's configuration and the relationships between entities. Many formalisms, known as qualitative calculi, have been proposed and used across multiple application domains, including robotics, geographic information systems, and language understanding [11,30].

In QSTR, knowledge bases are represented as Qualitative Constraint Networks (QCNs), where vertices correspond to entities, and edges indicate their relations, specified as disjunctive qualitative constraints. Given such a QCN, the literature focuses mainly on its satisfiability, that is, methods for determining whether there is an interpretation satisfying all constraints. For most widely-adopted qualitative calculi, the satisfiability problem is NP-complete [12].

Instead of finding a specific instantiation of entities, we may be interested in a qualitative solution, called *scenario*, where we refine the original QCN so that it yields a consistent assignment of constraints. Thus, a scenario may describe multiple specific instantiations. Typically, the exact solutions are not of particular interest, as the goal of solving the satisfiability problem is to only prove the existence of at least one valid solution.

Recent work in the field suggests that not all solutions are created equal [25,28,32]. Imagine, for example, that you have to schedule two independent meetings for work, an internal team meeting A and meeting B with some external guests. You reckon that you can have one meeting before and one after lunch break. Where any order of those meetings entails a consistent scenario, it may be of higher likelihood that a meeting with external guests may start delayed in case they arrive late. So, it might be advisable to have the more controllable meeting A scheduled before lunch break L and then have meeting B at last, where a delay cannot interfere with follow-up appointments. This ability of withstanding perturbations is called *robustness*; and we may call one scenario more *robust* than another if it can withstand more perturbations [32].

In practice, knowledge bases may contain contradicting information, e.g., as a result of merging different data sources, and consequently the representing QCN becomes inconsistent. In these cases, we may be inclined to *repair* the network with a minimal set of changes. Solving a QCN is generally a combinatorial problem, and state-of-the-art solvers implement practical algorithms and heuristics to speed up the solving process [29,31]. Repairing a solution, or finding the most robust one, are optimization problems that build on top of the solving process and are thus at least NP-hard themselves [9,28].

In this paper, we tackle both of these optimization problems declaratively using Answer Set Programming (ASP). ASP has been used successfully in QSTR to deal with the complexity of finding individual solutions [3,5,20], where every answer set corresponds to a qualitative scenario. We improve upon their work by providing a general and flexible, yet concise ASP encoding for solving QCNs and show how our encoding can be easily extended to also solve both described optimization problems.

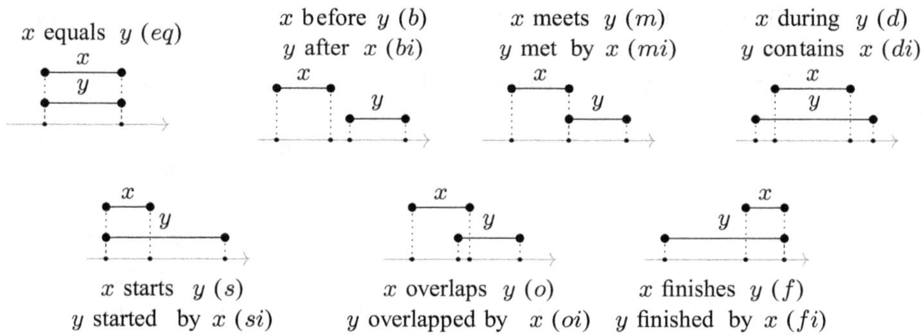


**Fig. 1.** All 13 base relations B of Allen's Interval Algebra.

Having access to a general and extensible framework based on logic programming may lower the barrier of entry for AI practitioners to integrate qualitative reasoning within their system. Thus, assisting the conception of correct but practical spatial and temporal reasoning agents.

We demonstrate the effectiveness of our encoding by comparing it with state-of-the-art solvers in an experimental evaluation.

## 2 Preliminaries

In qualitative reasoning, we abstract from an infinite domain D (e.g., space or time) using a finite set B of *jointly exhaustive and pairwise disjoint* binary relations, called the *base relations*. Entities in this domain are related pairwise, with base relations representing definite knowledge that constrains possible instantiations of D. Indefinite knowledge is represented by relations in $2^{\mathsf{B}}$ which support standard set operations [21].

For example, in Allen's Interval Algebra [1], the domain D is defined to be the set of intervals on the line of rational number, and any configuration of two intervals is described using the 13 base relations displayed in Fig. 1, derived from 7 relations and their inverse, denoted as $\cdot i$ (identity relation $eq = eqi$).

Spatial or temporal information in a qualitative calculus can be modeled as a Qualitative Constraint Network (QCN). A QCN is a graph where vertices represent spatial or temporal entities, and edges are labeled with corresponding qualitative relations. Formally, a QCN is defined as follows.

**Definition 1 (Qualitative Constraint Network).** *A Qualitative Constraint Network (QCN) is a tuple $\mathcal{N} = (V, C)$ where:*

***V*** *a finite set of variables $V = \{v_1, \ldots, v_n\}$ representing entities of domain* D
***C*** *the constraints as a mapping $C : V \times V \to 2^{\mathsf{B}}$ as edges between the variables*

We call a QCN $\mathcal{N}$ *atomic*, iff $\forall v, v' \in V$, $C(v, v') = \{b\}$ with $b \in \mathsf{B}$. Furthermore, a sub-QCN $\mathcal{N}'$ of $\mathcal{N}$, denoted by $\mathcal{N}' \subseteq \mathcal{N}$, is a QCN $(V, C')$ such that $C'(v, v') \subseteq$

$C(v, v')$ for all $v, v' \in V$. Then, a *solution* to a QCN $\mathcal{N}$, also called scenario $S$ of $\mathcal{N}$, is an atomic sub-QCN of $\mathcal{N}$.

We say that $\mathcal{N}$ is *trivially inconsistent* iff $\exists v, v' \in V$ such that $C(v, v') = \emptyset$, where the QCN is not *satisfiable* because no solution can be found. If $\mathcal{N}$ is satisfiable, i.e., it entails at least one consistent scenario $S$ of $\mathcal{N}$, we say that $\mathcal{N}$ is consistent.

The composition operation between two relations $b \circ b'$ that relate entities $i, j$ over an intermediate entity $k$ results in the relation between $i, j$, the weak composition ($\diamond$) holds when $b \diamond b' \supseteq b \circ b'$ and $b \diamond b'$ has minimal cardinality. For a qualitative calculus, the weak compositions are defined using a *composition table*; for further details on the specifics of qualitative calculi refer to [12].

### 2.1 Answer Set Programming

Answer Set Programming (ASP) is a declarative programming approach that is based on the stable model semantics of logic programs [6].

A program $\Pi$ in ASP consists of *rules*

$$a_1 \vee \cdots \vee a_l \leftarrow a_{l+1}, \ldots, a_m, \neg a_{m+1}, \ldots, \neg a_n.$$

with $l, m, n \in \mathbb{N}$ such that $l \leq m \leq n$, each $a_i$ is an *atom* or predicate $p(t_1, \ldots, t_n)$, with predicate name $p$ and terms $t_i$. During grounding, a version of $\Pi$ is computed in which all variables are replaced by constant symbols, which are referred to as ground($\Pi$).

A *rule* is a *fact* for some $a$ if $a \leftarrow \top$. In ASP, clauses of the form $\bot \leftarrow a_1, \ldots a_m, \neg a_{m+1}, \ldots, \neg a_n$ act as *constraints.* To avoid confusion with terminology from QSTR, in what follows, we will refer to them as *denials.* The program ground($\Pi$) can be seen as a propositional program that involves all ground atoms. A set of atoms is called an *interpretation* $I$. If $I \models \Pi$, i.e., $I$ satisfies all rules of $\Pi$, $I$ is called a model [6,17]. A logic program $\Pi$ can have multiple *stable* models (answer sets) that are minimal among the set of $I$, representing outcomes in non-monotonic reasoning where the truth of statements can depend on the absence of other statements. Stable models of ground($\Pi$) are also those of $\Pi$. Finally, Answer Set Optimization (ASO) is applied were interpretations are optimized towards a preferred solutions using a objective function.

The efficiency achieved by some encoding in ASP relies heavily on preferences and optimization towards preferred solutions. In our case, finding the most robust solution or the solution with the least number of repairs. Naturally, this raises the complexity from simply checking and generating solutions to additionally ensuring that no better alternative exists w.r.t. the specified preferences [14]. Preferences are indicated by numerical values that count and evaluate occurrences of predicates; depending on the purpose, solutions are sought that maximize or minimize these values.

**Listing 1.1.** Base relations $B$ and selected compositions for IA

```
baserel(eq; b; bi; m; mi; o; oi; s; si; f; fi; d; di).
compose(s,  s,   s).           compose(s,  b,  b).
compose(f,  o,   (d; s; o)).   compose(f,  m,  m).
compose(m,  mi,  (eq;f;fi)).   compose(m,  o,  b). ...
```

## 3 Encoding QCNs in ASP

For encoding a QCN $\mathcal{N} = (V, C)$ in ASP we represent $C$ as labelled edges. Labels indicate possible relations. By choosing one base relation from the set of relations, a scenario can be found that satisfies all integrity constraints (if such exists). We model this using *choice* rules as usual [20]. Embedding the calculus semantically benefits the reasoning process of finding a satisfiable scenario, as has been demonstrated for Allen's Interval Algebra [18]. We take inspiration from a general encoding for finding satisfying scenarios in a QCN proposed by Beryannis et al. [3], which uses algebraic properties of qualitative calculi to discern a consistent solution without embedding the specific semantics for each calculus [12].

A qualitative calculus may be specified in ASP using its base relations $\mathsf{B}$ `baserel/1` and its composition table, defined using predicate `compose/3`. Then, $\mathcal{N}$ is encoded using `node/1` for $V$ indexed by id in numerical order, and edges `edge/2` for $C$, referencing nodes by index, always connecting smaller to larger indices. The relations of $C(v, v')$ are stated as one or more `constraint/3`, relating edges to base relations. Any edge that is not explicitly labelled is linked by the universal relation $\mathsf{B}$, i.e., it contains all base relations. Figure 2 illustrates this encoding.

```
node(1..4).
edge(F,T) ← constraint(F,T,_).
constraint(1,2,(o;si)).
constraint(1,4,(di;b)).
constraint(2,3,(fi;si;o)).
constraint(2,4,(di;d;o)).
constraint(3,4,(b;bi)).
```

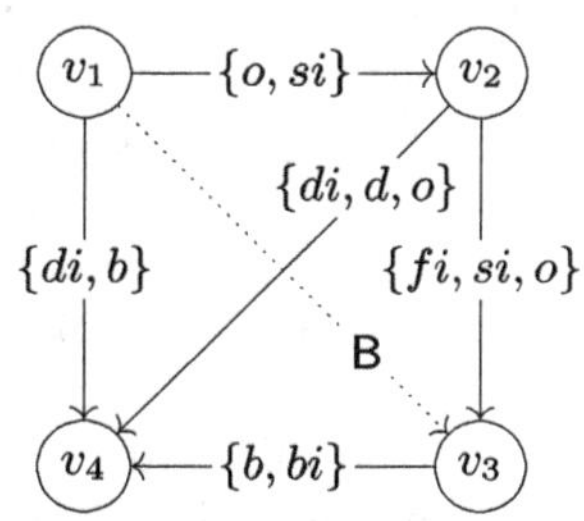


**Fig. 2.** QCN represented in Answer Set Programming.

Solving the QCN is a matter of constraint satisfaction. Here, the task is to select a suitable base relation as atomic refinement for every edge, and ensure their consistency using integrity constraints. Refining a constraint to a single

**Listing 1.2.** General encoding for finding satisfying solutions of QCN

```
1 {label(FROM,TO,REL):baserel(REL)} = 1 ← edge(FROM,TO).
2 ⊥ ← label(FROM,TO,REL), not constraint(FROM,TO,REL).
3 ⊥ ← label(I,K,R1), label(K,J,R2), label(I,J,COMP),
4     not compose(R1,R2,COMP).
```

atomic base relation makes the relation concrete. This selection propagates constraints to adjacent edges, causing impossible base relations to be eliminated from the options. This procedure is known as *refinement search* [23]. The selection of a base relation from all options is realized using a *choice* rule (Listing 1.2, line 1). Note that we do not directly select from the set of constraints i.e. `constraint(FROM,TO,R)`, for finding solutions, this will be crucial when searching for repairs. For now we choose from the set of all base relations $\mathsf{B}$ and the denial in line 2 will enforce that no base relation is chosen that is outside of the specified constraints.

Finally, the propagated integrity constraints are made general through the algebraic property of qualitative calculi and local consistency, known as path consistency ($\diamond$-consistency). A complete QCN $\mathcal{N} = (V, C)$ is $\diamond$-consistent iff $\forall (i,j)$, $(i,k)$, $(k,j) \in C$, defining triangles in the constraint network, we have $C(i,j) \subseteq C(i,k) \diamond C(k,j)$. Path consistency is translated into the denial in Listing 1.2 line 3, using the labeled edges and their compositions.

### 3.1 Chordal Networks

For verifying consistency in a QCN $\mathcal{N}$, the integrity constraint checks the local consistency of potentially all triangles $\triangle ijk$ with $i, j, k \in V$. Recall that, while typically not shown, when not otherwise constrained, the universal relation $\mathsf{B}$ holds between any two nodes in a given QCN, essentially forming a complete graph for consistency checking (Fig. 3a). However, verifying local consistency on all possible triangles is redundant. Instead, we may only add universal relations that are required to make the QCN a chordal graph. A graph is chordal if every node is part of at least one cycle of length 3, i.e. a triangle. Hence, this process of explicitly stating all required edges in a QCN is also known as *triangulation* [7].

Finding the minimal chordal graph (Fig. 3c), however, is itself an NP-hard problem. Therefore, in practice, approximate techniques are often used to generate chordal graphs for consistency checking (Fig. 3b) [4,26].

Listing 1.3 shows our encoding for obtaining a chordal graph, using a declarative program. Every `edge/2` that is part of a triangle has the `chord/2` property. In order to satisfy this required property for every edge, the solver may choose to add new edges between variables. We then optimize for a QCN with the least added edges.

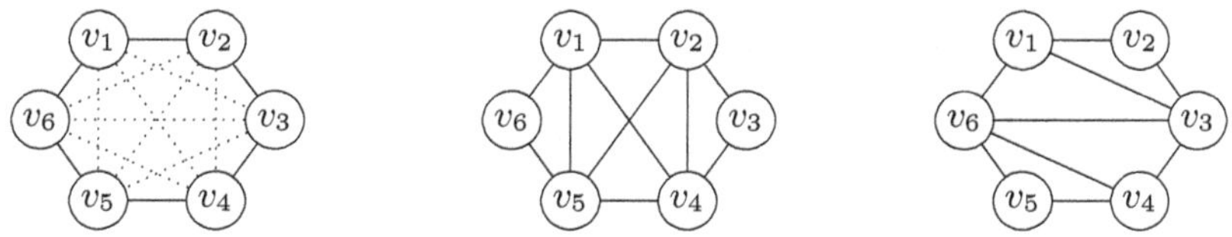


(a) Graph without chords (b) Chordal Graph (c) Minimal Chordal Graph

**Fig. 3.** Triangulation of Networks.

**Listing 1.3.** Search for minimal chordal graphs

```
chord((I;K),(K;J)) ← edge(I,J), edge(I,K), edge(K,J).
{edge(FROM,TO)} ← node(FROM), node(TO), FROM < TO.
⊥ ← edge(FROM,TO), not chord(FROM,TO).
#minimize{FROM,TO: edge(FROM,TO)}.
```

## 4 Robustness

Robustness is an inherently practical property that is of great relevance to many real-world application contexts. It can be seen as a proactive measure, and has been defined as "the ability of a system to resist change without adapting its initial stable configuration" [33]. In the context of constraint networks, a robust solution is said to be one that is more likely to remain valid after change [2]. A related concept is that of stability, where a solution is said to be stable if, after some perturbation, the now invalid solution can be repaired within a minimal number of steps [2]. A subtle yet notable distinction.

Robustness in QCNs has first been studied by Sioutis et al. [28]. In short, a robust scenario is defined as one that shares the highest similarity to all other scenarios. The motivation is that any change imposed on a robust scenario is more likely to result in one of the other scenarios, and thus the solution remains valid. However, as this is an exhaustive definition, all possible scenarios first have to be retrieved. Formally, a robust scenario is defined as follows.

**Definition 2 (Robust Scenario).** *Given a QCN $\mathcal{N}$ with the set of satisfiable scenarios $[\![\mathcal{N}]\!]$, a scenario $S$ of $\mathcal{N}$ is said to be robust if it shares the highest similarity with most other scenarios within $[\![\mathcal{N}]\!]$. The similarity between two scenarios is measured by the number of shared constraints.*

Our encoding, presented in Sect. 3, can be directly applied here, as it already finds all satisfiable scenarios $[[\mathcal{N}]]$. To facilitate a more practical definition of robustness [32], let us first introduce some relevant concepts.

**Definition 3 (Perturbation).** *Given a QCN $\mathcal{N} = (V, C)$, a perturbation of a scenario $S$ of $\mathcal{N}$ changes an assigned base relation $b \in C(v, v')$ into $b' \in C(v, v')$ where $b \neq b'$. We say that $\mathcal{N}$ has been* perturbed.

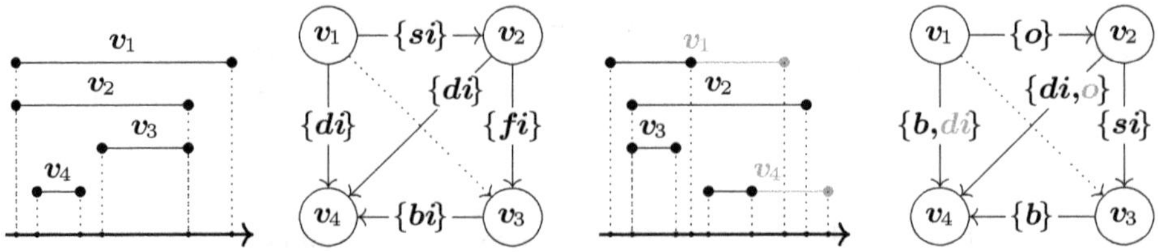


(a) Scenario without resistances (b) Scenario with resistances

**Fig. 4.** Fragile and robust solutions to QCN of Fig. 2 depicted as intervals.

**Listing 1.4.** Extention for QCN-robustness using resistance to perturbations

```
1 impossible(I,J,R) ← label(I,K,R1), label(K,J,R2),
     not compose(R1,R2,COMP), baserel(COMP).
2 resists(I,J,A) ← constraint(I,J,A), A!=R, label(I,J,R),
     not impossible(I,J,A).
3 #maximize {1,R: resists(I,J,R)}.
```

A scenario *survives* a perturbation if the perturbed QCN remains satisfiable [32]. We will refer to all perturbations a scenario can survive as its *resistance*, where the notion of resistance is a measure of robustness for a given scenario. Figure 4 illustrates scenarios and their resistances. Notice how scenario 4b has resistances concerning $v_1$ and $v_4$ highlighted in gray. The configuration can survive their expansion to the extent that the alternative base relations, depicted here gray in the QCN, are satisfied. The scenario depicted in Fig. 4a must exactly follow this configuration, as there are no alternatives.
We emulate perturbation resistance of a given scenario by checking all possible alternative base relations next to the one that was initially assigned. We include these potential perturbations in our encoding. In particular, `resists/3` represents all possible alternative assignments for an edge (Listing 1.4 line 2) that are satisfiable, and thus not `impossible` (line 1). Intuitively, one can think of these alternatives as glimpses into neighbouring scenarios that are reached in case this perturbation actually does occur. Maximizing these resistances results in robust scenarios.

## 5 Repair

The example in Fig. 5 shows an inconsistent QCN. Assume, for now, that the inconsistency is caused by introducing the constraint $v2$ *during* $v3$, because $v_2$ cannot be both *during* $v_3$ and *before* or *finished by* $v_4$ while $v_3$ is *during* $v_4$. Typically, the cause for inconsistency is not known beforehand and may be resolved in a number of ways. We can apply diagnostic techniques to identify a minimal set of *repairs*, i.e., edges that when repaired make the QCN consistent

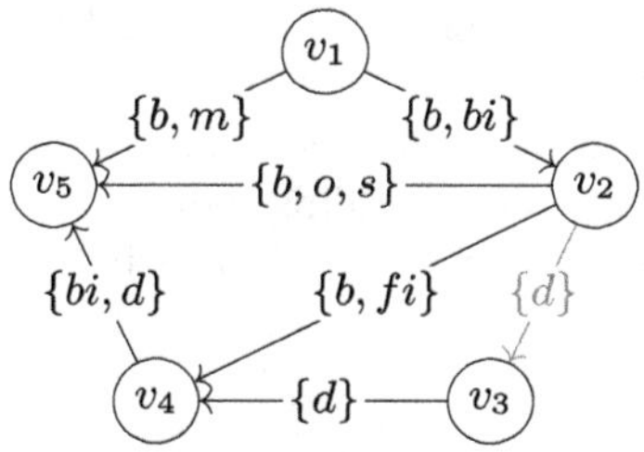


```
Answer: 1
ab(edge(3,4))
Answer: 2
ab(edge(2,3))
Answer: 3
ab(edge(2,4)) ab(edge(2,5))
Answer: 4
ab(edge(2,4)) ab(edge(4,5))
```

**Fig. 5.** Inconsistent QCN and repairs, example inconsistency at `edge(2,3)`.

again. Figure 5 depicts the output of such a diagnostic process, finding 4 options for repair.

Identifying the minimal number of repairs for an inconsistent QCNs has been formally defined as the MAX-QCN problem [9]. Here, we adopt a recently proposed reformulation of this definition [25].

**Definition 4 (QCN repair problem).** *Given a QCN $\mathcal{N}$ with constraints $C(i,j)$, $1 \leq i,j \leq |\mathcal{N}|$. The QCN repair problem consists of finding a minimal set of* repairs $C(i,j) \mapsto \mathsf{B}$ *that, when applied to $\mathcal{N}$, make $\mathcal{N}$ consistent.*

This formulation extends the MAX-QCN problem in that we are not only interested in the number of required repairs, but what these repairs actually are. We facilitate this by extending our encoding presented in Sect. 3 with the option to repair constraints whenever there is no consistent solution, adopting aspects of approaches for *Knowledge-Bases* [19] and *Model-Based Diagnosis* [34].

Specifically, we introduce a mechanism for *constraint suspension* to the encoding, borrowed from the field of diagnosis [10,34]. If no satisfying solution can be found for a QCN $\mathcal{N}$, then we assume that some constraints of $\mathcal{N}$ are wrong, i.e., *abnormal.* If a constraint is abnormal, then the specified constraints do not hold and are suspended, resulting in a diagnosis and a subsequent repair.

We encode this using predicate `ab/1` that identifies the edges as *abnormal*, forming a set $\Delta$ as diagnosis. Constraints are suspended by making the denial of Listing 1.2 line 2 conditional. The restriction on the choice of base relations applies only to edges not marked as abnormal. Note that any abnormal edge is labelled with a consistent base relation $C(i,j) \mapsto \{b\}$ previously not contained

**Listing 1.5.** Extention for QCN-Repair using Constraint Suspension

```
{ab(edge(FROM,TO))} ← edge(FROM,TO).
⊥ ← label(F,T,R), not ab(edge(F,T)),
    not constraint(F,T,R).
#minimize{1,E: ab(E)}.
```

in $C(i, j)$, concluding the repair. Therefore, the number of abnormal edges $|\Delta|$ corresponds to the number of necessary repairs.

Now, finding a solution to the QCN repair problem involves identifying the scenario that minimizes the number of necessary repairs. This can be accomplished using the `#minimize` directive (Listing 1.5, line 4). Alternatively, one could employ a multi-shot ASP solving approach, similar to that proposed by Wotawa and Kaufmann for Model-Based Diagnosis [34], where, starting at $i = 0$, the number of allowed repairs is succinctly increased until a solution is found. This approach also allows excluding super-sets when finding alternative minimal sets of repairs, as shown in Fig. 5. Preliminary results indicate that the iterative approach performs well when the number of necessary repairs is small.

## 6 Experimental Evaluation

In this section, we compare our declarative program to other known methods and, in particular, the state-of-the-art for solving the MAX-QCN problem. To this end, we perform an empirical evaluation on Allen's Interval Algebra (IA) [1] already introduced in the examples above and on the Region Connection Calculus (RCC-8) [22], both being prevalent qualitative calculi in practice and the literature. As the name suggests, RCC-8 contains only 8 base relations, compared to the 13 in IA. As a result, RCC-8 instances tend to span a smaller search space than their IA counterparts.

The dataset comprises random RCC-8 and IA networks that were created using the $\mathsf{A(n, d, l)}$ model [23]. Here, $n$ specifies the number of variables in the QCN, $d$ the average degree, i.e., constraints incident to a single variable, and $l$ the average label size, i.e., number of base relations per constraint, set to $l = |\mathsf{B}|/2$. To facilitate comparison, for *repairs*, we take the data set from Condotta et al. [9], consisting of 600 unsatisfiable QCNs of ascending degree. For *robustness*, we generated 100 IA QCNs following the same model from instantiated scenarios.

In the evaluation, we conducted experiments on following specifications: 32 x Intel(R) Xeon(R) Gold 6334 CPUs, each running at 3.60 GHz, distributed across 2 sockets. The server was equipped with 16 modules of 32 GB DDR-4 memory. The measurements were performed using Clingo version 5.5.2 in the default settings as a solver, running on a single core. As Partial MaxSAT solver we used RC2 [16] as part of the PySAT toolkit [15].

### 6.1 Robustness

Other robustness measures use tractable solutions of the QCN to calculate the resistances [32] or identify critical constraints using so-called minimum size prime scenario, analogous to the concept of prime implicates in classical logic [24].

Finding the most *robust* scenario, as in a stable model with the highest amount of resistance, becomes more challenging for QCN which are less constrained. Dense networks with many constraints, for example $\mathsf{A(20, 14, 4)}$, typically have a smaller number of solutions, allowing faster optimization. Figure 6a

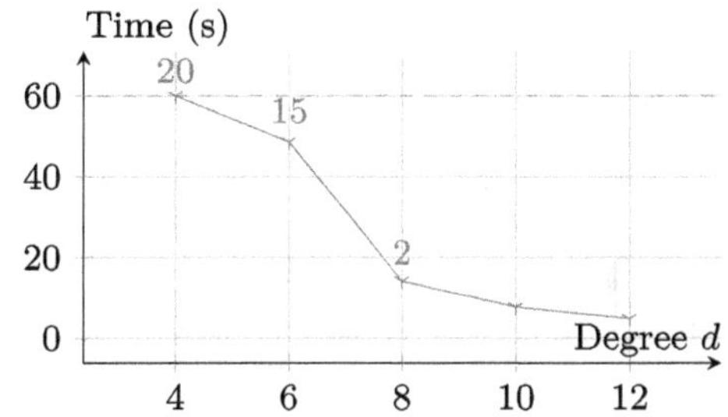


(a) Average runtime per IA QCN ($n = 20$) instance to most robust solution. Annotations of Timeout count included

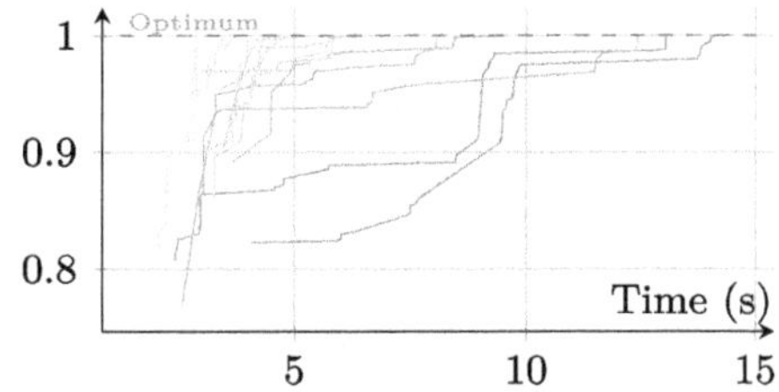


(b) Timeseries of IA $\mathsf{A}(20, 8, 4)$ instances suboptimal solutions found during optimization relative to optimum.

**Fig. 6.** Optimization towards **robust** scenarios of QCN instances.

shows how less constrained instances are harder to solve for robustness than denser constraint networks. Because the space of possible solutions increases, so does the time spent to find the optimal solution. In cases where the QCN yields too many scenarios, Answer Set Optimization can still navigate through the space of sub-optimal solutions. Figure 6b outlines how the robustness of the solutions found develops over time, where 1 is the global optimum, and timeouts (2) have been omitted. Hence, we can identify resistant, but possibly suboptimal, solutions during the optimization process in an anytime fashion.

### 6.2 Repair

The current state-of-the-art for finding the minimal number of repairs for an inconsistent QCN is based on a concise Partial MaxSAT encoding [27]. Although our ASP encoding is capable of producing intermediate cardinally suboptimal repairs, here we only focus on the optimal solution. Figure 7 displays the results of our experimental comparison. As expected, IA is more combinatorial challenging compared to RCC-8, which is reflected in overall longer run time. In summary, we can see that the Partial MaxSAT encoding generally yields faster solving times. Note, however, that the time required to generate the MaxSAT encoding was not taken into account for these results (cf. [27]) and, thus, the actual solving times of both approaches are likely closer than our results suggest.

### 6.3 Discussion

Repairing QCNs using our ASP encoding is generally slower than dedicated solvers but scales similarly to the Partial MaxSAT encoding on small calculi. For complex calculi like IA, repair nearly matched state-of-the-art performance. One advantage of our ASP program is that it is self-contained and does not rely on an external procedure for encoding, which makes it arguably more intuitive. Aggregation of solutions in more concise representations is a typical strategy to

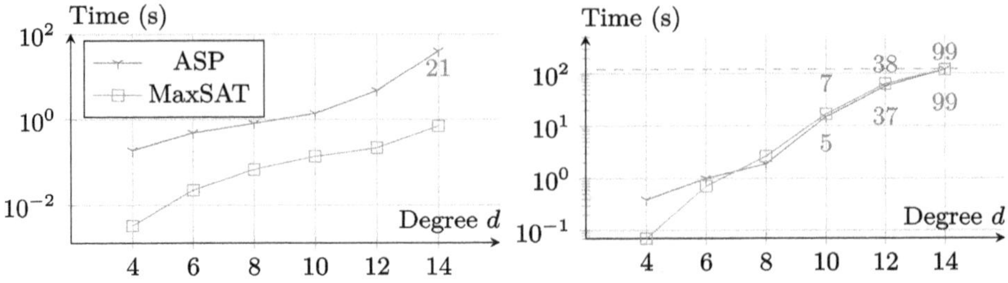


(a) RCC-8 dataset with size $n = 30$

(b) IA dataset with size $n = 20$

**Fig. 7.** Average runtime per QCN instance to first minimal cardinal **repair** in seconds required (timeout 120 s). Annotations of timeout count included.

improve efficiency when solving QCNs. Hence, in future work, we could consider representation of groups of scenarios using tractable sub-relations, in favor of atomic base relations in the answer sets, particularly to reduce the amount of possible solutions that need to be considered during optimization.

## 7 Conclusion

In this paper, we present an answer set programming (ASP) encoding of representing and reasoning about qualitative constraint networks (QCNs). ASP allows for concise programs that are general with respect to the constraint languages used. Our encoding further demonstrates flexibility to distinguish solutions based on their quality. Through optimization, we can identify the most *robust* solution of a consistent QCN. Moreover, our anytime variant also outputs intermediate, suboptimal solutions, enhancing practicality.

For inconsistent QCNs, our encoding enables optimization for minimal *repairs* using constraint suspension and diagnosis algorithms. Employing triangulation accelerates the search by reducing the number of stable models found.

Finally, our approach facilitates easy customization for various optimization problems. For instance, users can adjust the integrity constraints and optimization clauses to focus on specific subsets of edges in the graph, tailoring the solution to their specific requirements.

In summary, our ASP approach offers concise, flexible, and efficient solutions for representing and reasoning about qualitative constraint networks.

**Acknowledgments.** The work presented in this paper has been carried out in context of the VoLL-KI project (grant 16DHKBI091), funded by Bundesministeriums für Bildung und Forschung (BMBF).

## References

1. Allen, J.F.: Maintaining knowledge about temporal intervals. Commun. ACM **26**(11), 832–843 (1983)

2. Barber, F., Salido, M.A.: Robustness, stability, recoverability, and reliability in constraint satisfaction problems. Knowl. Inf. Syst. **44**(3), 719–734 (2014). https://doi.org/10.1007/s10115-014-0778-3
3. Baryannis, G., Tachmazidis, I., Batsakis, S., Antoniou, G., Alviano, M., Papadakis, E.: A generalised approach for encoding and reasoning with qualitative theories in answer set programming. Theory Pract. Log. Program. **20**(5), 687–702 (2020)
4. Berry, A., Blair, J.R., Heggernes, P., Peyton, B.W.: Maximum cardinality search for computing minimal triangulations of graphs. Algorithmica **39**, 287–298 (2004)
5. Brenton, C., Faber, W., Batsakis, S.: Answer set programming for qualitative spatio-temporal reasoning: methods and experiments. Dagstuhl (2016)
6. Brewka, G., Eiter, T., Truszczyński, M.: Answer set programming at a glance. Commun. ACM **54**(12), 92–103 (2011)
7. Chmeiss, A., Condotta, J.F.: Consistency of triangulated temporal qualitative constraint networks. In: Proceedings of the 23rd International Conference on Tools with Artificial Intelligence (ICTAI 2011), pp. 799–802 (2011). https://doi.org/10.1109/ICTAI.2011.125
8. Cohn, A.G., Renz, J.: Qualitative spatial representation and reasoning. Found. Artif. Intell. **3**, chap. 13, 551–596 (2008)
9. Condotta, J.F., Mensi, A., Nouaouri, I., Sioutis, M., Said, L.B.: A practical approach for maximizing satisfiability in qualitative spatial and temporal constraint networks. In: 2015 IEEE 27th International Conference on Tools with Artificial Intelligence (ICTAI), pp. 445–452. IEEE (2015)
10. Davis, R.: Diagnostic reasoning based on structure and behavior. Artif. Intell. **24**(1–3), 347–410 (1984)
11. Dylla, F., et al.: A survey of qualitative spatial and temporal calculi: algebraic and computational properties. ACM Comput. Surv. **50** (2017). article 7
12. Dylla, F., Mossakowski, T., Schneider, T., Wolter, D.: Algebraic properties of qualitative spatio-temporal calculi. In: Tenbrink, T., Stell, J., Galton, A., Wood, Z. (eds.) COSIT 2013. LNCS, vol. 8116, pp. 516–536. Springer, Cham (2013). https://doi.org/10.1007/978-3-319-01790-7_28
13. Freksa, C.: Qualitative spatial reasoning. Technical report (1990)
14. Gebser, M., Kaminski, R., Schaub, T.: Complex optimization in answer set programming. Theory Pract. Log. Program. **11**(4–5), 821–839 (2011)
15. Ignatiev, A., Morgado, A., Marques-Silva, J.: PySAT: a python toolkit for prototyping with SAT oracles. In: SAT, pp. 428–437 (2018). https://doi.org/10.1007/978-3-319-94144-8_26
16. Ignatiev, A., Morgado, A., Marques-Silva, J.: RC2: an efficient MaxSAT solver. J. Satisf. Boolean Model. Comput. **11**(1), 53–64 (2019). https://doi.org/10.3233/SAT190116
17. Janhunen, T., Nimelä, I.: The answer set programming paradigm. AI Mag. **37**(3), 13–24 (2016)
18. Janhunen, T., Sioutis, M.: Allen's interval algebra makes the difference. In: Hofstedt, P., Abreu, S., John, U., Kuchen, H., Seipel, D. (eds.) INAP/WLP/WFLP -2019. LNCS (LNAI), vol. 12057, pp. 89–98. Springer, Cham (2020). https://doi.org/10.1007/978-3-030-46714-2_6
19. Kuhlmann, I., Gessler, A., Laszlo, V., Thimm, M.: A comparison of ASP-based and SAT-based algorithms for the contension inconsistency measure. In: Dupin de Saint-Cyr, F., Öztürk-Escoffier, M., Potyka, N. (eds.) Scalable Uncertainty Management. SUM 2022. LNCS, vol. 13562, pp. 139–153. Springer, Cham (2022). https://doi.org/10.1007/978-3-031-18843-5_10

20. Li, J.J.: Qualitative spatial and temporal reasoning with answer set programming. In: 2012 IEEE 24th International Conference on Tools with Artificial Intelligence, vol. 1, pp. 603–609. IEEE (2012)
21. Ligozat, G., Renz, J.: What is a qualitative calculus? A general framework. In: Zhang, C., W. Guesgen, H., Yeap, W.-K. (eds.) PRICAI 2004. LNCS (LNAI), vol. 3157, pp. 53–64. Springer, Heidelberg (2004). https://doi.org/10.1007/978-3-540-28633-2_8
22. Renz, J.: Maximal tractable fragments of the region connection calculus: a complete analysis. In: Proceedings of the Sixteenth International Joint Conference on Artificial Intelligence (IJCAI 1999), pp. 448–455. Morgan Kaufmann (1999). http://ijcai.org/Proceedings/99-1/Papers/065..pdf
23. Renz, J., Nebel, B.: Efficient methods for qualitative spatial reasoning. J. Artif. Intell. Res. (JAIR) **15**, 289–318 (2001)
24. Salhi, Y., Sioutis, M.: Prime scenarios in qualitative spatial and temporal reasoning. In: 30th International Symposium on Temporal Representation and Reasoning (TIME 2023). Schloss Dagstuhl-Leibniz-Zentrum für Informatik (2023)
25. Schwartz, T., Wolter, D.: Computing most likely scenarios of qualitative constraint networks. In: Seipel, D., Steen, A. (eds.) KI 2023: Advances in Artificial Intelligence. KI 2023. LNCS, vol. 14236, pp. 184–197. Springer, Cham (2023). https://doi.org/10.1007/978-3-031-42608-7_15
26. Sioutis, M.: Triangulation versus graph partitioning for tackling large real-world qualitative spatial networks. In: 2014 IEEE 26th International Conference on Tools with Artificial Intelligence, pp. 194–201. IEEE (2014)
27. Sioutis, M.: Embarrassingly greedy inconsistency resolution of qualitative constraint networks. In: 30th International Symposium on Temporal Representation and Reasoning (TIME 2023). Schloss Dagstuhl-Leibniz-Zentrum für Informatik (2023)
28. Sioutis, M., Long, Z., Janhunen, T.: On robustness in qualitative constraint networks. In: Proceedings of the Twenty-Ninth International Conference on International Joint Conferences on Artificial Intelligence, pp. 1813–1819 (2021)
29. Sioutis, M., Wolter, D.: Dynamic branching in qualitative constraint networks via counting local models. In: 27th International Symposium on Temporal Representation and Reasoning (TIME 2020). Schloss Dagstuhl-Leibniz-Zentrum für Informatik (2020)
30. Sioutis, M., Wolter, D.: Qualitative spatial and temporal reasoning: current status and future challenges. In: Zhou, Z.H. (ed.) Proceedings of the Thirtieth International Joint Conference on Artificial Intelligence, IJCAI-21, pp. 4594–4601. International Joint Conferences on Artificial Intelligence Organization, August 2021. https://doi.org/10.24963/ijcai.2021/624, survey Track
31. Wallgrün, J.O., Frommberger, L., Wolter, D., Dylla, F., Freksa, C.: Qualitative spatial representation and reasoning in the sparq-toolbox. In: Barkowsky, T., Knauff, M., Ligozat, G., Montello, D.R. (eds.) Spatial Cognition V Reasoning, Action, Interaction. Spatial Cognition 2006. LNCS, vol. 4387, pp. 39–58. Springer, Berlin, Heidelberg (2007). https://doi.org/10.1007/978-3-540-75666-8_3
32. Wehner, J., Sioutis, M., Wolter, D.: On robust vs fast solving of qualitative constraints. J. Heuristics 1–25 (2023)
33. Wieland, A., Wallenburg, C.M.: Dealing with supply chain risks: linking risk management practices and strategies to performance. Int. J. Phys. Distrib. Logist. Manag. **42**(10), 887–905 (2012). https://doi.org/10.1108/09600031211281411
34. Wotawa, F., Kaufmann, D.: Model-based reasoning using answer set programming. Appl. Intell. 1–19 (2022)

# Efficiently Training Neural Networks for Imperfect Information Games by Sampling Information Sets

Timo Bertram[1(✉)], Johannes Fürnkranz[1,2], and Martin Müller[3]

[1] Johannes Kepler University, Linz, Austria
{tbertram,juffi}@faw.jku.at
[2] LIT AI Lab, Linz, Austria
[3] University of Alberta, Edmonton, Canada
mmueller@ualberta.ca

**Abstract.** In imperfect information games, the evaluation of a game state not only depends on the observable world but also relies on hidden parts of the environment. As accessing the obstructed information trivialises state evaluations, one approach to tackle such problems is to estimate the value of the imperfect state as a combination of all states in the information set, i.e., all possible states that are consistent with the current imperfect information. In this work, the goal is to learn a function that maps from the imperfect game information state to its expected value. However, constructing a perfect training set, i.e. an enumeration of the whole information set for numerous imperfect states, is often infeasible. To compute the expected values for an imperfect information game like *Reconnaissance Blind Chess*, one would need to evaluate thousands of chess positions just to obtain the training target for a single state. Still, the expected value of a state can already be approximated with appropriate accuracy from a much smaller set of evaluations. Thus, in this paper, we empirically investigate how a budget of perfect information game evaluations should be distributed among training samples to maximise the return. Our results show that sampling a small number of states, in our experiments roughly 3, for a larger number of separate positions is preferable over repeatedly sampling a smaller quantity of states. Thus, we find that in our case, the quantity of different samples seems to be more important than higher target quality.

**Keywords:** neural networks · imperfect information games

## 1 Introduction

Imperfect information games, which are characterised by unobservable aspects, are an important part of Game AI research. In recent years, they have received increased attention due to the inherent complexity of managing uncertainty. This category encompasses a wide array of games, spanning from classical card


A. Hotho and S. Rudolph (Eds.): KI 2024, LNAI 14992, pp. 17–29, 2024.
https://doi.org/10.1007/978-3-031-70893-0_2

games like Poker and Bridge to adaptions of traditional board games such as Dark Hex and Reconnaissance Blind Chess and real-time video games like Starcraft, Dota II and Counter-Strike. Thus, we see much interest - commercially and scientifically - in mastering this category of games. However, methods that conquered many classical perfect information games, e.g. AlphaZero [20], do not easily carry over to imperfect information games [16]. In imperfect information settings, decisions are typically based on a fusion of the public information and an implicitly learned or directly computed expected value of the hidden information. While there are several approaches to learning evaluations implicitly (see Sect. 3), we here focus on learning them explicitly in a supervised fashion.

At every decision point of an imperfect information game, the set of all possible states from one observer's perspective is called an *information set.* For sequential games such as *Poker* and *Reconnaissance Blind Chess,* it is possible to enumerate the states in the information set, thus allowing for approximation of the total evaluation of the imperfect information state as an aggregation of the single evaluations in the set. However, a perfect enumeration of the set can be vastly too costly. Thus, we aim to learn an evaluator which encapsulates the relationship between an imperfect information state and its aggregated target evaluation, enabling fast evaluation of states. In training, however, the same problem persists; generating perfect target values for states requires an unreasonable amount of evaluations. To reduce this load, we only use parts of the information set for target generation and investigate how many individual states are required for training.

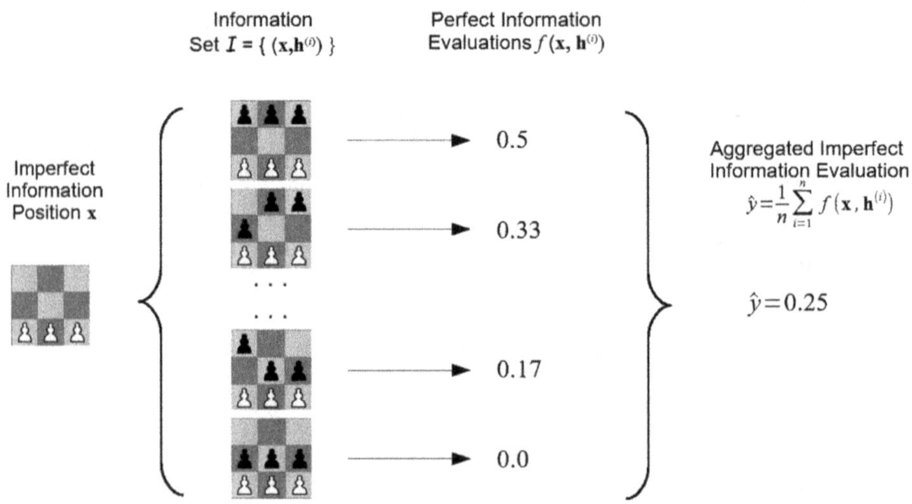


**Fig. 1.** Estimating the value of an imperfect information position (left) as the average of the perfect information evaluations of all positions in the information set.

This setting is motivated by work in *Reconnaissance Blind Chess* (RBC), a chess variant where parts of the board are not observable, and thus agents have to evaluate imperfect information game states. Many leading programs approximate these values by evaluating the states in the information set with a conventional perfect information chess engine and averaging the results. While these averages can serve as training signals for a model, evaluating every single board state in each information set is unreasonable. In this work, we question how we can best invest a fixed budget of total game state evaluations; should they be spread out to obtain inaccurate estimates for more training data or should they provide more reliable training signals for fewer data points? We begin this work by outlining the problem more formally (Sect. 2) and give a brief overview of related work and problems (Sect. 3). Subsequently, we empirically investigate the problem in two different imperfect information games:

- **Heads-Up Poker**: Here we evaluate a 2-card hand without knowledge about the community cards or the opponent's cards, sampling from information sets of possible completions to estimate the win rate of a hand.
- **Reconnaissance Blind Chess (RBC)** [7]: In this chess variant, the opponent's moves are often uncertain because of limited information. We receive an estimated evaluation of the imperfect information game state based on possible perfect game evaluations obtained from a conventional chess engine.

We summarise our results in Sect. 5 and give an outlook on potential future extensions in Sect. 6.

## 2 Problem Statement

We formalise the problem as follows: Given is a dataset of examples $\mathcal{D} = (\mathbf{x}_i, y_i) \subset X \times Y$, where each label $y_i = f(\mathbf{x}_i, \mathbf{h}_i)$ is determined by a function $f$, dependent not only on the observable information $\mathbf{x}_i$, but also on the hidden information $\mathbf{h}_i$. Our goal is to find a function $g(\mathbf{x})$ which approximates $f(\mathbf{x}, \mathbf{h})$, such that $\forall i \in \{1, .., |D|\} : g(\mathbf{x}_i) \approx f(\mathbf{x}_i, \mathbf{h}_i)$. This task is non-trivial, and such a function $g$ does not generally exist, as the same observable $\tilde{\mathbf{x}}$ can occur multiple times with different labels because, in general, $f(\tilde{\mathbf{x}}, \mathbf{h}^{(1)}) \neq f(\tilde{\mathbf{x}}, \mathbf{h}^{(2)})$ for $\mathbf{h}^{(1)} \neq \mathbf{h}^{(2)}$.

Our motivation for this problem originates from imperfect information games, where the information set represents all possible game states given one player's information. In several such games, remarkable performance has been achieved by basing the imperfect information gameplay, whether implicitly or explicitly, on perfect information evaluations of states in an information set [1,3,5]. For example, the value of a player's hand in *Poker* can be estimated as the expected value of the hand over all possible variations of the community and opponent's cards. Similarly, many strong *RBC* agents rely heavily on chess engines for evaluating conventional chess positions [7,8,15] and approximate the imperfect information state with the expected values of the states in the information set. Figure 1 illustrates this situation with a schematic $3 \times 3$ chess board where only the white pieces are public and the evaluation of this board is obtained by averaging the (hypothetical) evaluations of all possible configurations of black pieces.

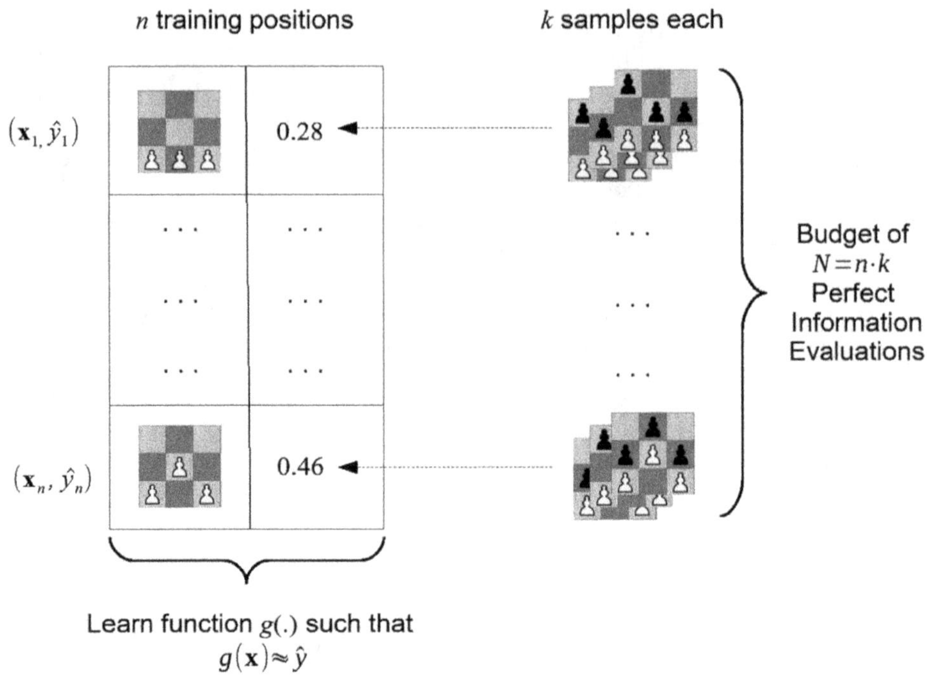


**Fig. 2.** Learning an imperfect information evaluation function from $n$ examples, for which the target evaluation is estimated from $k$ position, using a constant budget of $N = n \cdot k$ perfect information evaluations.

It is important to acknowledge the limitations of basing imperfect information policy fully on perfect information evaluations, as it is trivial to construct counterexamples where this fails. Despite its fundamental problems, the idea of estimating the evaluations by sampling from information sets has been successful in many algorithms. For example, using perfect information search on sampled states has been used in games such as Bridge [9], Skat [12] and Scrabble [19]. In particular, Blüml et al. [3] have noticed that perfect information evaluations for imperfect information games perform surprisingly strongly. On a smaller scale, one can also relate this concept to the central idea of Monte-Carlo Tree Search. There, nodes in the game tree are evaluated based on stochastic rollouts, which results in a noisy but still useful signal. Similarly to the problem at hand, one can use more sophisticated, but also slower, rollout strategies to receive more accurate signals or more frequent fast and noisy rollouts.

The central objective of this work is to learn the function $g$ which receives the public information of a state $\mathbf{x}$ and approximates the expected value of that state (Fig. 2), without the need for its (potentially expensive) explicit computation by iterating over the information set:

$$g(\mathbf{x}) \approx \hat{y} = \sum_{\mathbf{h}} P(\mathbf{h}|\mathbf{x}) \cdot f(\mathbf{x}, \mathbf{h}) \tag{1}$$

Here, $\mathbf{h} \in \mathcal{I}$ are all possible configurations of private information that are part of the information set $\mathcal{I}$, $f$ is an evaluator of a perfect information state and $P$ is a function which gives the probability of each hidden state for the given configuration $\mathbf{x}$. In practice, $P$ can be influenced by, among others, stochastic environments or adversary's (hidden) policies. In all cases, it is assumed to be an unobservable and unalterable part of the domain. In our experiments, as in Fig. 1, we assume that all possible determinations are equally likely, i.e. $P(\mathbf{h}|\mathbf{x}) = {}^{1}/_{|\mathcal{I}(\mathbf{x})|}$ but one could also directly learn meaningful weights for the positions in the information set from past behaviour or observations [2].

A simple strategy to learn $g$ is to collect samples of the form $(\mathbf{x}_i, \hat{y}_i)$, i.e., to compute the exact value $\hat{y}_i$ as in Eq. (1) for many training positions $\mathbf{x}_i$, and to use supervised learning to learn the function $\hat{y}_i = g(\mathbf{x}_i)$ from these samples. However, this approach generally is too costly due to the potentially large size of information sets, so obtaining a single $\hat{y}_i$ can require thousands of evaluations. Alternatively, $\hat{y}$ can be approximated by randomly sampling only a few of the possible $\mathbf{h}^{(j)}$, resulting in less accurate training signals $\dot{y}$ at a lower computational cost.

In our work, we aim to answer a fundamental question: Given a fixed budget of $N$ perfect information evaluations, how should we generate the training data for the learner? Options range from generating $N$ different training examples $\mathbf{x}_i$, each evaluated with one random sample, over using a fixed number of $k$ evaluations to generate targets for $n = {}^{N}/_{k}$ positions, up to exhausting the budget with exactly computing $\hat{y}$ for as many examples as possible. This trade-off between the training set size $n$ (the number of distinct $\mathbf{x}_i$) and label quality (the number of evaluations $k$ used to estimate the intended target values $\hat{y}_i$ for each $\mathbf{x}_i$) forms the core focus of this paper.

## 3 Related Work

The problem formulated in Sect. 2 is multifaceted and occurs in several learning paradigms, thus we can only give a brief overview of how it manifests in practice.

Several learning settings may be viewed as special cases of this formulation. Conventional supervised learning emerges when $\mathbf{h}_i = \emptyset, \forall i$, i.e. when no hidden information determines $y_i$. Similarly, learning from noisy labels can be formulated with a single hidden variable $\mathbf{h}_i$, which determines whether the original label remains intact or is corrupted. Knowledge of this hidden information makes the underlying function $f$ deterministic, but $g$ does not have access to the information about the corruption.

Our setting is also closely related to research in the area of noisy labels [11,21], which extends to crowdsourcing [10,18] and aggregating labels from different labellers. It also shares commonalities with active learning [17], which in our case is a problem of deciding whether to re-sample an existing example to improve label quality or to obtain a new sample to increase overall training data quantity. Importantly, most of this research aims to improve data distribution to the labellers or to reduce bias post-sampling, which differs significantly from choosing a sampling frequency a priori. In addition, noisy classification differs substantially from noisy regression.

In the context of imperfect information games, numerous approaches exist to, explicitly or implicitly, evaluate an information set. Techniques such as Perfect Information Monte Carlo [6, 13] combine evaluations of different perfect information searches into a policy for the imperfect information state and Information Set Monte Carlo Tree Search [22] operates on information sets. Counterfactual Regret Minimization [23], as well as its successors, and ReBeL [4] learn the utility of individual information sets through self-play. Recent work by Blüml et al. [3] samples individual world states and constructs imperfect information policies based on their evaluations. In essence, most techniques for solving imperfect information games involve estimating the value of information sets, further motivating the importance of the question which we aim to answer.

Finally, the general idea that sampling more states from information sets will lead to a more accurate estimate of its overall value is simply an instance of the law of large numbers and thus finds application in a variety of problems. This trade-off between quantity and quality of evaluations is analogous to the choice of rollout policy in classical Monte Carlo Tree Search [5], where one has to decide between random rollouts (fast, and thus allowing a larger quantity, but less informative) and more sophisticated rollout-policies (slower, thus limiting their number, but better at approximating true behaviour). However, one has to consider that MCTS differs slightly because UCB will over time re-visit promising states, thus allowing the resampling of important ones.

## 4 Experiments

Here, we present a series of experiments designed to investigate the trade-off between target quality and training data quantity. The learner has a limited budget of total evaluations $N$ and can decide how many evaluations $k$ should be spent on each training example. Each evaluation yields one value, which are aggregated into a training target by averaging.

### 4.1 Texas Hold'em Poker

The first experiment aims at a real-world setting where we have to balance the label accuracy with the number of total training examples seen. Here, the learner aims to estimate the win probability of a given 2-card hand of cards in two-player heads-up poker. This is a direct implementation of the problem outlined in Sect. 2. Pre-computed win-rate tables for any given hand exist, so one should regard this as a proof-of-concept rather than a useful application.

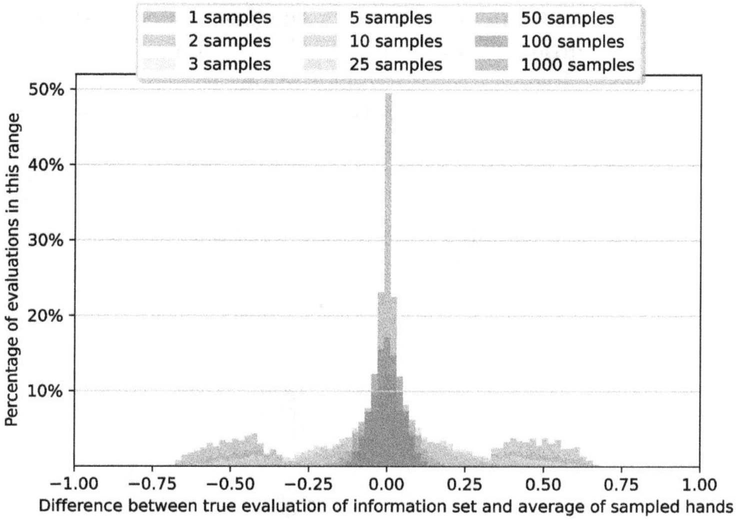


**Fig. 3.** Monte-Carlo estimate of error in evaluating the change of winning with a given hand pre-flop in heads-up poker. Estimations are computed by averaging over n samples of possible opponent hands and rivers.

**Setup.** In principle, for a given hand $\mathbf{x}$, $g(\mathbf{x})$ could be directly computed as the average over all possible hidden contexts $\mathbf{h_x}$, but doing so requires a large amount of computation. Without accounting for symmetry, a player can have $\binom{52}{2} = 1326$ unique Poker hands. One would need to compute all possible arrangements of the remaining cards into two opponent cards and five community cards, i.e. $\binom{52}{2} \cdot \binom{50}{2} \cdot \binom{48}{5} = 2{,}781{,}381{,}002{,}400$ total combinations. For each of these combinations, one needs to evaluate which player won the game and average this for all configurations that pertain to the same player's hand to estimate the overall winning probability of that hand. While public data for the win-chances of a hand exists, such data is only available for the most popular games and computing them is much costlier in other games with higher degrees of uncertainty or more expensive state evaluations. Thus, we aim to decrease the computational cost by only sampling parts of the information set instead of enumerating it entirely, and thus allow for the extension of the concept to a larger variety of applications.

When training the learner, we sample $k$ different configurations of cards for the given hand, evaluate the result of each configuration (0, 1 or 0.5), and train the network to predict the mean of all $k$ samples. Sampling more combinations leads to a smaller difference between the estimate and the ground truth, but results in a smaller amount of total hands seen when equating by the total number of evaluations.

**Results.** As a first estimate, Fig. 3 shows the discrepancy between an estimated hand strength through evaluations and the mathematical true win chance. Notable, with only a single sampled configuration, it is impossible to exactly receive the true win chance of most hands as the only possible results are 0, 0.5, and 1, thus resulting in three error clusters for one sample.

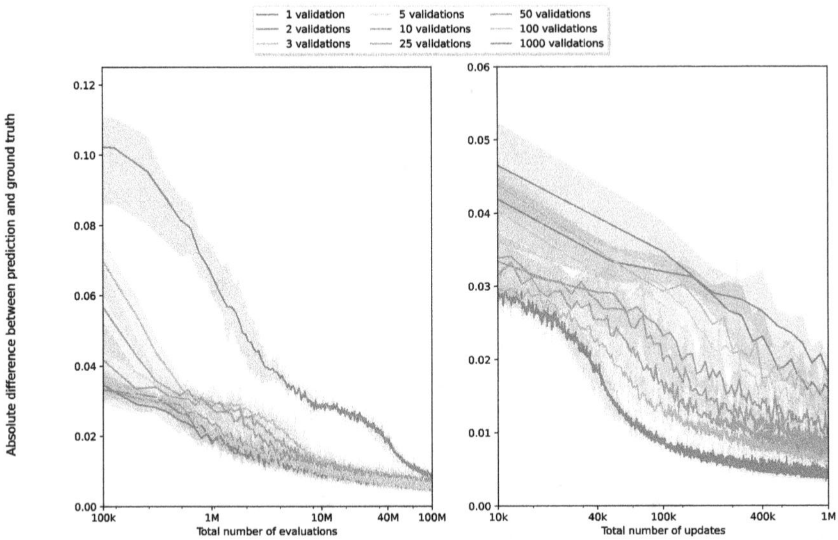


**Fig. 4.** Average training curves of learning to evaluate a poker hand with different numbers of evaluations per training example. The $x$-axis is logarithmically scaled either by the total number of hand evaluations (top) or by the total number of update steps made (bottom).

The training process (Fig. 4) shows that training with fewer evaluations per sample leads to much quicker progress when regarding the performance in relation to the total number of evaluations, but when the examples have higher-quality evaluations, each update is more meaningful. When comparing the best versions (Fig. 5), we see that even when equating for the total number of evaluations, using a single evaluation leads to worse peak results than using two, three, five, and ten sampled evaluations. As it should be, equating training updates makes more evaluations perform strictly better than less. All in all, these results suggest that multiple samples are useful for this setting, but naturally, spending too much computation on a single example degrades the overall performance as the total training size diminishes.

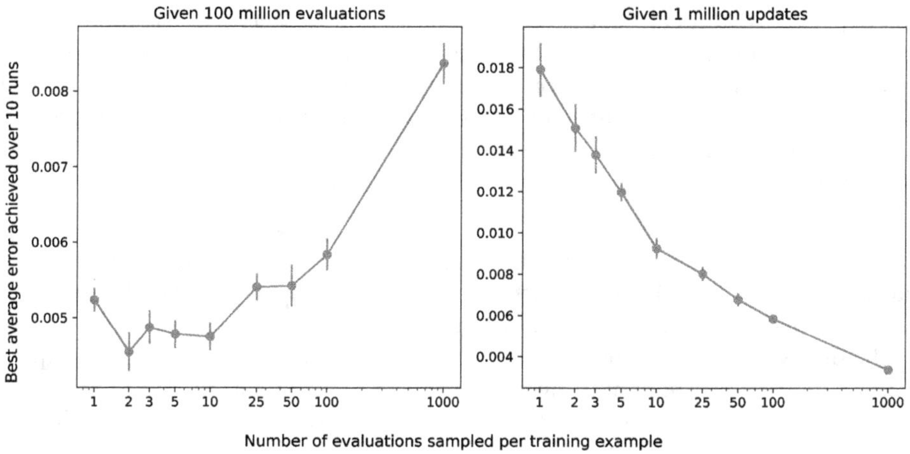


**Fig. 5.** Average lowest received error for the different options of hand validations, given a total budget of either 100M hand evaluations or 1M training updates.

We extend our experiments in the next section, where we regard a similar setting. For that, pre-computed evaluations do not exist and approximating through sampling is a realistic setting.

### 4.2 Reconnaissance Blind Chess

*Reconnaissance Blind Chess.*[1] (*RBC*) is an imperfect information adaption of chess, where players receive limited information about the opponent's moves. When training agents to play this game, it is highly useful to be able to evaluate a specific situation (i.e. the received observations at one point in time), and evaluation functions for regular chess are readily available (e.g., from open-source programs such as Stockfish[2]. Thus, computing the average evaluation of all states in an information set is an intuitive approach, but doing so is largely unfeasible due to the information set size.

**Setup.** For this experiment, training data is created offline for each $k$, thus generating a fixed training set for each setting. Each learner has a fixed budget of 1 million state evaluations that can be arbitrarily distributed among different information sets. Based on the previous results, sensible values of $k$ were chosen as {1, 2, 3, 5, 10, 25, 50, 100, 1000}, thus resulting in datasets of approximately {1M, 500k, 333k, 200k, 100k, 40k, 20k, 10k, 1k} examples respectively.[3] Importantly, the number of potential public-information states to get evaluated in

[1] https://rbc.jhuapl.edu/.

[2] https://stockfishchess.org/.

[3] The exact numbers vary slightly because the information set can consist of fewer states than $k$, thus exhausting it completely.

this experiment is much higher than in the previous Sect. 4.1. For *Poker*, there are only 1326 unique 2-card hands. However, in *RBC*, the number of potential observations which form one information set is estimated to be $10^{139}$ [14], much larger than our training datasets, thus minimising the probability of overlapping training example and increasing the importance of meaningful target value estimations.

**Results.** Again approximating the sampling error through Monte-Carlo estimates (Fig. 6), we receive the expected result: Sampling only a small number of states can lead to a large discrepancy between the approximation and the ground truth, but we see diminishing returns, such that sampling more than 50 positions only leads to slight improvements in the approximation. We thus expect efficiency increases with repeated sampling, but more than 50 states should lead to meaningfully degrading performance.

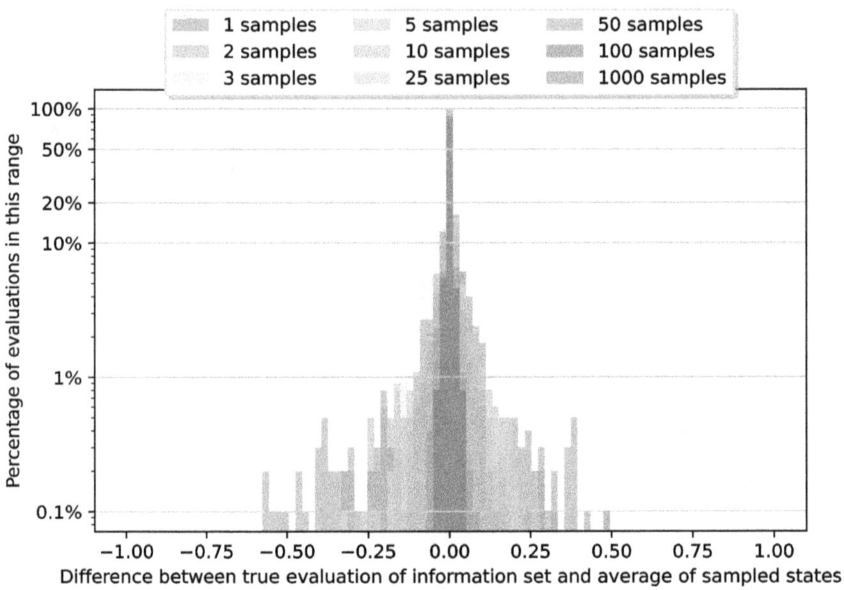


**Fig. 6.** Monte-Carlo estimate of error in evaluating the odds of winning for a given observation. Estimations are computed by averaging over $k$ samples of possible board states, true evaluation is defined as the average over the whole information set. Note that the y-axis is in a logarithmic scale to improve readability.

This observation is confirmed by the results in Figs. 7 and 8. The training curves of Fig. 8 show that using singular samples results in poor overall accuracy, but vast oversampling leads to too few total training examples. The extreme choices of $k$ (1 and 1000) perform poorly, but moderate sampling is useful.

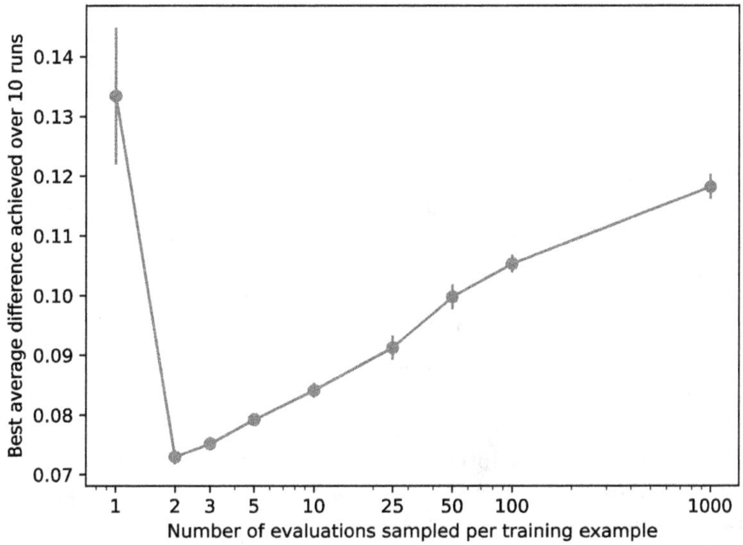


**Fig. 7.** Average lowest received error for training datasets created with a total budget of 1M sampled boards. For each training example, $k$ different boards are sampled from the information set, leading to $1M/k$ different training examples.

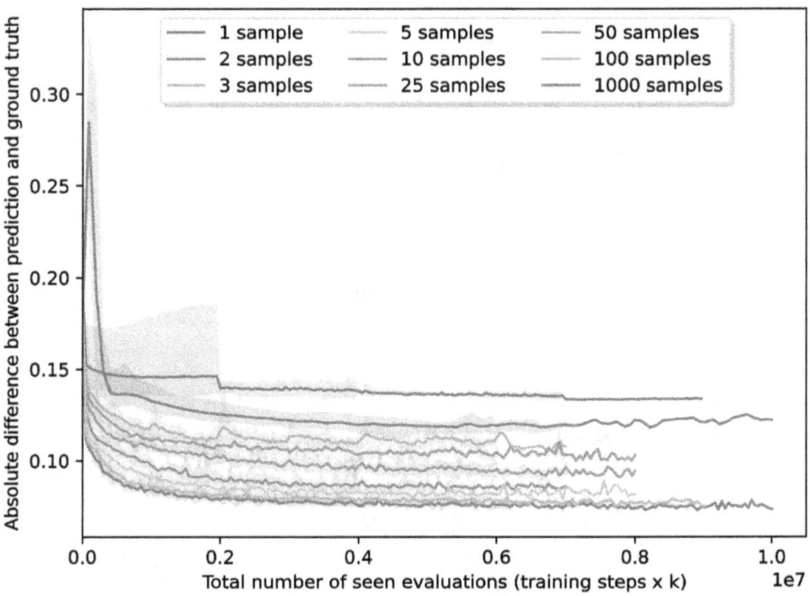


**Fig. 8.** Mean training progress of learning to evaluate imperfect information states with different numbers of boards sampled per training example. The x-axis is scaled by the total number of evaluations seen. Curves vary in length because the Neural Networks are individually trained until no further improvements are seen.

## 5 Summary and Conclusion

With this work, we provided experimental results on the influence of sampling different numbers of states from an information set to learn an evaluation of the whole set. For a given task, a total budget of $N$ evaluations is given, which can be distributed across information sets. Thus, we investigated the trade-off between the overall number of training samples generated and the accuracy of their associated training targets.

Firstly, the trade-off is influenced by the cost of generating evaluations and the cost of making an update to the learner, so the choice of $k$ needs to be related to their balance. We find that in *Heads Up Poker* and *Reconnaissance Blind Chess* (Sects. 4.1 and 4.2), multiple evaluations consistently improve the performance and efficiency of the learner. In both domains, sampling two evaluations per training example leads to the overall best results, and only using one sample did not perform well in comparison. As the results for both tasks were similar, we speculate that these findings will translate to more scenarios, but more work is required to validate this.

## 6 Future Work

We see multiple intriguing lines of further work based on these findings. First, we here assumed no agency over the process of sampling from the information sets and no online variation in sample numbers. Removing either of those assumptions will likely lead to better results, and some strategies have previously been outlined by Sheng [18] for categorical tasks. In addition, while our general formulation holds for other distributions of states, we used a uniform distribution of states for our experiments. This is a sensible assumption for *Poker*, but information sets do not follow a uniform distribution in *RBC*. Knowledge of this, or even access to a proxy of such a distribution, would lead to more accurate estimations in real-world tasks. Whether a non-uniform distribution changes the best choice of sampled evaluations will be investigated in the future.

## References

1. Bertram, T., Fürnkranz, J., Müller, M.: Supervised and reinforcement learning from observations in reconnaissance blind chess. In: Proceedings of the 2022 IEEE Conference on Games (CoG), pp. 608–611. IEEE (2022)
2. Bertram, T., Fürnkranz, J., Müller, M.: Weighting information sets with Siamese neural networks in reconnaissance blind chess. In: Proceedings of the 2023 IEEE Conference on Games (CoG). IEEE (2023)
3. Blüml, J., Czech, J., Kersting, K.: Alphaze**: alphazero-like baselines for imperfect information games are surprisingly strong. Front. Artif. Intell. **6**, 1014561 (2023)
4. Brown, N., Bakhtin, A., Lerer, A., Gong, Q.: Combining deep reinforcement learning and search for imperfect-information games. Adv. Neural Inf. Process. Syst. **33**, 17057–17069 (2020)

5. Browne, C.B., et al.: A survey of Monte Carlo tree search methods. IEEE Trans. Comput. Intell. AI Games **4**(1), 1–43 (2012)
6. Furtak, T., Buro, M.: Recursive Monte Carlo search for imperfect information games. In: 2013 IEEE Conference on Computational Inteligence in Games (CIG), pp. 1–8. IEEE (2013)
7. Gardner, R.W., et al.: The first international competition in machine reconnaissance blind chess. In: Escalante, H.J., Hadsell, R. (eds.) Proceedings of the NeurIPS 2019 Competition and Demonstration Track, vol. 123, pp. 121–130. PMLR, Vancouver, Canada (2019)
8. Gardner, R.W., et al.: The machine reconnaissance blind chess tournament of NeurIPS 2022. In: Proceedings of the NeurIPS 2022 Competitions and Demonstrations Track, pp. 119–132. PMLR (2023)
9. Ginsberg, M.L.: GIB: imperfect information in a computationally challenging game. J. Artif. Intell. Res. **14**, 303–358 (2001)
10. Karger, D.R., Oh, S., Shah, D.: Budget-optimal task allocation for reliable crowdsourcing systems. Oper. Res. **62**(1), 1–24 (2014)
11. Khetan, A., Lipton, Z.C., Anandkumar, A.: Learning from noisy singly-labeled data. In: Proceedings of the 6th International Conference on Learning Representations (ICLR). OpenReview.net, Vancouver, BC, Canada (2018)
12. Long, J., Sturtevant, N., Buro, M., Furtak, T.: Understanding the success of perfect information Monte Carlo sampling in game tree search. In: Burgard, W., Roth, D. (eds.) Proceedings of the 25th AAAI Conference on Artificial Intelligence, pp. 134–140 (2010)
13. Long, J., Sturtevant, N., Buro, M., Furtak, T.: Understanding the success of perfect information Monte Carlo sampling in game tree search. In: Proceedings of the AAAI Conference on Artificial Intelligence, vol. 24, pp. 134–140 (2010)
14. Markowitz, J., Gardner, R.W., Llorens, A.J.: On the complexity of reconnaissance blind chess. arXiv preprint arXiv:1811.03119 (2018)
15. Perrotta, G., et al.: The second NeurIPS tournament of reconnaissance blind chess. In: Kiela, D., Ciccone, M., Caputo, B. (eds.) Proceedings of the NeurIPS 2021 Competitions and Demonstrations Track, vol. 176, pp. 53–65. PMLR (2021)
16. Schmid, M., et al.: Player of games. arXiv preprint **2112.03178** (2021)
17. Settles, B.: Active Learning. Synthesis Lectures on Artificial Intelligence and Machine Learning. Morgan & Claypool Publishers, San Rafael (2012)
18. Sheng, V.S., Provost, F., Ipeirotis, P.G.: Get another label? Improving data quality and data mining using multiple, noisy labelers. In: Proceedings of the 14th ACM SIGKDD International Conference on Knowledge Discovery and Data Mining, pp. 614–622 (2008)
19. Sheppard, B.: Mastering scrabble. IEEE Intell. Syst. **14**(6), 15–16 (1999)
20. Silver, D., et al.: A general reinforcement learning algorithm that masters chess, shogi, and go through self-play. Science **362**(6419), 1140–1144 (2018)
21. Snow, R., OConnor, B., Jurafsky, D., Ng, A.Y.: Cheap and fast–but is it good? Evaluating non-expert annotations for natural language tasks. In: Proceedings of the 2008 Conference on Empirical Methods in Natural Language Processing, pp. 254–263 (2008)
22. Whitehouse, D., Powley, E.J., Cowling, P.I.: Determinization and information set Monte Carlo tree search for the card game Dou Di Zhu. In: 2011 IEEE Conference on Computational Intelligence and Games (CIG'11), pp. 87–94 (2011)
23. Zinkevich, M., Johanson, M., Bowling, M., Piccione, C.: Regret minimization in games with incomplete information. Adv. Neural Inf. Process. Syst. **20** (2007)

# A Note on Linear Time Series Prediction

Christopher Bonenberger[1,2(✉)], Markus Schneider[1], Wolfgang Ertel[1], and Friedhelm Schwenker[2]

[1] Ravensburg-Weingarten University of Applied Sciences (Institut für Künstliche Intelligenz), Weingarten, Germany
bonenbch@rwu.de

[2] Universität Ulm (Institut für Neuroinformatik), James-Franck-Ring, 89081 Ulm, Germany

**Abstract.** We consider the problem of univariate time series prediction from an elementary machine learning point of view. Beginning with the question of whether and how Principal Component Analysis (PCA) can be used for time series prediction, we describe a simple methodology and attempt to classify PCA-based prediction in terms of statistics, signal processing and dynamical systems theory. Moreover, we extend the unsupervised scenario to a self-supervised linear regression scenario and develop a unifying perspective. In this regard, we review several related techniques, namely autoregressive (AR) and moving-average (MA) models, Singular Spectrum Analysis (SSA), Wiener filtering, and the discrete Fourier transform (DFT). By presenting these methods in a unified way, we can show how PCA-based time series prediction can be categorized in different settings of stochastic and deterministic models. Finally, we show the distinct relation between PCA-based prediction and (finite-order) MA processes and propose a refined methodology.

**Keywords:** Time series · Linear Prediction · Linear Regression · PCA · ARMA · DFT · SSA · Wiener Filter

## 1 Introduction

Principal Component Analysis (PCA) is a more than 120 years old technique and it is still considered to be "the workhorse of exploratory data analysis and unsupervised machine learning" [23]. It forms the basis of many methods and ideas in machine learning (ML), especially with regard to representation learning and pattern recognition. However, it seems that the possibility of time series prediction based on PCA is not widely known. It does not seem to be apparent in how far PCA allows the prediction temporal data and—most importantly—whats the theoretical implications and limitations of such a procedure. Though, we believe that a coherent theory in this regard can be helpful, first and foremost, because it builds bridges between machine learning, time series analysis, classical signal processing and dynamical systems theory. Additionally, the recourse on comprehensible and well-understood techniques—like PCA—drives the development and understanding of more advanced and complex methodologies.


A. Hotho and S. Rudolph (Eds.): KI 2024, LNAI 14992, pp. 30–44, 2024.
https://doi.org/10.1007/978-3-031-70893-0_3

### 1.1 Contributions

Hereinafter, we consider several different methodologies for linear and univariate time series prediction based on linear recurrence relations (LRR). LRRs are primarily used in the context of Singular Spectrum Analysis (SSA, see [20]), yet, we consider it in the context of classical PCA. We contextualize the resulting methodology in terms of traditional time series analysis and show that PCA-based time series prediction is tightly related to moving-average mdoels (MA models, e.g. [24]). In short, we are able to show that PCA-based prediction is optimal with respect to a low rank representation of a moving-average process. In addition, we draw the connection to discrete Fourier analysis. To complete the picture we consider an analogous self-supervised linear regression scenario and recapture its connections to autoregressive (AR) models and Wiener filtering, which eventually do all culminate in LRR-based prediction.[1] Finally, exploiting the attained new insights, we outline a generalized methodology for unsupervised time series prediction that allows for targeted handling of statistical assumptions.

### 1.2 Related Works

In the realm of signal processing, where PCA is often referred to as Karhunen–Loève transform (KLT, e.g. [43], p. 567), plenty of questions regarding the analysis of temporal data by means of PCA, including PCA-based time series prediction, have been exhaustively discussed (e.g. in the context of Gaussian processes, [45], p. 122). Most of the other approaches to PCA-based prediction relate to *functional PCA* (e.g. [1], a brief overview is given in [28], p. 326) whereas the main subject of this paper is a methodology for PCA-based prediction of *discrete* time series. We thereby draw upon the estimation of LRRs as it is implemented in the context of SSA-based time series prediction [21]. Hence, the most important related methodology is SSA and SSA-based time series prediction (reviewed from a rather novel perspective in Sect. 2.3, drawing from [6]). A formalism to estimate LRR coefficients given a distinct signal subspace was proposed by [15] and forms the basis for SSA-based and PCA-based time series prediction. The resulting technique (see Sect. 3.1) may be considered as a subspace filtering method ( [15,21], more recently [19]). We dissolve this formalism from the context of SSA and delay representations, putting it to the more general context of classical PCA. Our main innovation in this regard is the provision of a clear mathematical connection between PCA-based prediction and MA models and a generalization of SSA and SSA-based time series prediction.

## 2 Related Methods

In the following, we review the basics of ARMA models, SSA and Wiener filters. However, since linear prediction is closely related to (adaptive) linear filtering

[1] For all these methodologies, a prediction of the next time step amounts to an inner product between the preceding samples and a data-adaptive coefficient vector.

circular convolution | linear convolution | valid convolution

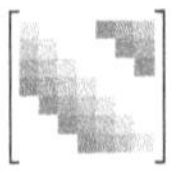 $= \mathbf{F}^{-1}\mathbf{\Lambda}_{\hat{\mathbf{g}}}\mathbf{F}$

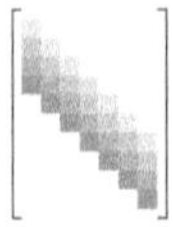 $= \mathbf{F}^{-1}\mathbf{\Lambda}_{\hat{\mathbf{g}}}\mathbf{F}\mathbf{M}^{\mathsf{T}}$

 $= \mathbf{M}\mathbf{F}^{-1}\mathbf{\Lambda}_{\hat{\mathbf{g}}}^{\mathsf{H}}\mathbf{F}$

**Fig. 1.** Examples on Circulant and Toeplitz matrices based on a kernel $\mathbf{g} \in \mathbb{R}^L$ defined as $g_i = i$. The matrix $\mathbf{M}$ formalizes truncation of a circulant matrix in order to realize linear and valid convolutions. Note that the latter is based on a transposed circulant

(see [29], p. 27), we begin with some basics regarding convolutions at the same time introducing notation. A *circular convolution* $\mathbf{x} \circledast \mathbf{y}$ between two signals $\mathbf{x}$ and $\mathbf{y}$ in $\mathbb{R}^D$ is expressed by means of a circulant matrix $\mathbf{C}_\mathbf{x}$ via $\mathbf{x} \circledast \mathbf{y} = \mathbf{C}_\mathbf{x}\mathbf{y}$, where $\mathbf{C}_\mathbf{x}$ holds the vector $\mathbf{x}$ in its first column.[2] A circulant matrix is diagonalizable via Fourier matrices $\mathbf{F} \in \mathbb{C}^{D\times D}$ with components $[\mathbf{F}]_{m,n} = \exp(-2\pi\imath(m-1)(n-1)/D)$, i.e., $\mathbf{C}_\mathbf{x} = \mathbf{F}^{-1}\mathbf{\Lambda}_{\hat{\mathbf{x}}}\mathbf{F}$, where $\mathbf{\Lambda}_{\hat{\mathbf{x}}}$ is a diagonal matrix with the discrete Fourier transform $\hat{\mathbf{x}} = \mathbf{F}\mathbf{x}$ of $\mathbf{x}$ on its diagonal (see [22], p. 32). Notably, the matrix $\mathbf{F} : \mathbb{R}^D \to \mathbb{R}^D$ implements the discrete Fourier transform (DFT). All together, we express the discrete convolution theorem as

$$\mathbf{x} \circledast \mathbf{y} = \mathbf{C}_\mathbf{x}\mathbf{y} = \mathbf{F}^{-1}\mathbf{\Lambda}_{\hat{\mathbf{x}}}\mathbf{F}\mathbf{y} = \mathbf{F}^{-1}\mathbf{\Lambda}_{\hat{\mathbf{x}}}\hat{\mathbf{y}} = \mathbf{F}^{-1}\left(\hat{\mathbf{x}} \odot \hat{\mathbf{y}}\right), \tag{1}$$

where $\odot$ denotes the Hadamard product. Correspondingly, a *linear convolution* $\mathbf{x} * \mathbf{y}$ can be expressed by appropriate zero padding of both sequences to length $2D-1$. We formalize zero padding as $\mathbf{x}_\mathsf{p} = \mathbf{M}^\mathsf{T}\mathbf{x} \in \mathbb{R}^{2D-1}$, where $\mathbf{M} \in \mathbb{R}^{D\times 2D-1}$ is a truncated identity matrix. Hence, according to the above equation we find

$$\mathbf{x} * \mathbf{y} = (\mathbf{M}^\mathsf{T}\mathbf{x}) \circledast (\mathbf{M}^\mathsf{T}\mathbf{y}) = \mathbf{F}^{-1}\mathbf{\Lambda}_{\mathbf{F}\mathbf{M}^\mathsf{T}\mathbf{x}}\mathbf{F}\mathbf{M}^\mathsf{T}\mathbf{y} = \mathbf{F}^{-1}\left(\hat{\mathbf{x}}_\mathsf{p} \odot \hat{\mathbf{y}}_\mathsf{p}\right). \tag{2}$$

The corresponding linear operator is $\mathbf{F}^{-1}\mathbf{\Lambda}_{\hat{\mathbf{x}}_\mathsf{p}}\mathbf{F}\mathbf{M}^\mathsf{T} \in \mathbb{R}^{2D-1\times D}$. Moreover, *correlation* is a convolution with one of the signals being reversed, e.g. for a signal $\mathbf{x}$ the circular sample autocorrelation is $\check{\mathbf{r}} = \mathbf{F}^{-1}\left(\hat{\mathbf{x}} \odot \overline{\hat{\mathbf{x}}}\right)$ (here and in the following a check indicates an estimate). For signals $\mathbf{x} \in \mathbb{R}^L$ and $\mathbf{y} \in \mathbb{R}^D$ with $L \leq D$ a valid convolution can be defined as $\mathbf{M}(\mathbf{x} \circledast \mathbf{y}) \in \mathbb{R}^{D-L+1}$, where $\mathbf{M} \in \mathbb{R}^{D-L+1\times D}$ is again a truncated identity matrix. Each of these operations can be expressed by means of an appropriate linear operator, formed by a truncated circulant. A visualization for each of the convolution operators is provided in Fig. 1.

## 2.1 AR and MA Models

A main characteristic of AR and MA models is the importance of noise. For both models the noise term may be regarded as the driving force to a otherwise fixed system. For every AR model it is possible to find an equivalent MA model and vice versa. However, following the principle of parsimony MA models are more

[2] In case that one sequence is shorter that the other, we presuppose zero padding, i.e., the signal in question is concatenated with a zero vector of appropriate length.

appropriate to describe processes with a power spectrum characterized by wide peaks, while lower order AR models are able to capture sharp spectral peaks.[3]

**Autoregressive Models:** With a $P$-th order autoregressive model the value of an observation $x_t$ at time $t$ is explained from a weighted sum of $P$ preceding observations plus a white noise term $\varepsilon_t$.[4] Denoting the weights (model parameters) as $\mathbf{a} = (a_P \ \cdots \ a_1)^\mathsf{T}$ we define an AR($P$) model as

$$x_t = a_1 x_{t-1} + \cdots + a_P x_{t-P} + \varepsilon_t = (x_{t-1} \ \cdots \ x_{t-P})^\mathsf{T}\mathbf{a} + \varepsilon_t. \tag{3}$$

Given the model and an observation $\mathbf{x} \in \mathbb{R}^P$ we may predict the next time step as $y = \mathbf{x}^\mathsf{T}\mathbf{a}$. If, however, the model parameters are unknown we may estimate these from an observation $\mathbf{x} \in \mathbb{R}^D$ by solving the least-squares problem (e.g. [17])

$$\min_{\mathbf{a}\in\mathbb{R}^P} \left\{ \|\mathbf{X}\mathbf{a} - \mathbf{y}\|_2^2 \right\} \text{ with } \mathbf{X} = \begin{bmatrix} x_P & x_{P-1} & \cdots & x_1 \\ x_{P+1} & x_P & \cdots & x_2 \\ \vdots & \vdots & & \vdots \\ x_{D-1} & x_{D-2} & \cdots & x_{D-P} \end{bmatrix}, \ \mathbf{y} = \begin{pmatrix} x_{P+1} \\ \vdots \\ x_D \end{pmatrix}. \tag{4}$$

This is easiest understood as a standard linear regression, where the dependent variable $x_{t+1}$ is supposed to be linearly dependent on $P$ preceding terms [16]. The solution can be found from the Yule–Walker equations (detailed in [17])

$$\begin{bmatrix} r_0 & r_1 & r_2 & \cdots & r_{P-1} \\ r_1 & r_0 & r_1 & \cdots & r_{P-2} \\ \vdots & \vdots & \vdots & & \vdots \\ r_{P-2} & r_{P-3} & r_{P-4} & \cdots & r_1 \\ r_{P-1} & r_{P-2} & r_{P-3} & \cdots & r_0 \end{bmatrix} \begin{pmatrix} a_1 \\ a_2 \\ \vdots \\ a_{P-1} \\ a_P \end{pmatrix} = \begin{pmatrix} r_1 \\ r_2 \\ \vdots \\ r_{P-1} \\ r_P \end{pmatrix} = \mathbf{R}\mathbf{a} = \mathbf{r}. \tag{5}$$

Here, $r_p$ is the autocorrelation at lag $p$, i.e., $\mathbf{R} \in \mathbb{R}^{P\times P}$ is the autocorrelation matrix. The model order $P$ is usually determined by consideration of the partial autocorrelation function.[5] Since $\mathbf{R}$ is a Toeplitz matrix the estimate $\check{\mathbf{a}} = \mathbf{R}^{-1}\mathbf{r}$ can be found in $\mathcal{O}(P^2)$ via the Levinson–Durbin algorithm (see [41], p. 96).

One step ahead prediction of a times series $\mathbf{x} \in \mathbb{R}^P$ based on an estimated AR polynomial $\check{\mathbf{a}} \in \mathbb{R}^P$ according to Eq. (3) with $\varepsilon_t = \mathbb{E}\{\varepsilon_t\} = 0$ amounts to

$$\check{x}_{P+1} = \check{a}_P x_1 + \cdots + \check{a}_1 x_P = (x_P \ \cdots \ x_1)^\mathsf{T}\check{\mathbf{a}}. \tag{6}$$

[3] Mostly, the two models are combined to an ARMA or rather a autoregressive *integrated* moving-average (ARIMA) model (see [10]).

[4] We presuppose unit variance Gaussian white noise, i.e., $\varepsilon_t$ is a discrete-time stochastic process whose autocorrelation sequence $\{r_t\}_{t\in\mathbb{Z}}$ is zero except for $r_0 = 1$.

[5] The $k$-th element $[\mathbf{p}]_k = p_k$ of the partial autocorrelation sequence $\mathbf{p}$ is estimated from the $k$-th coefficient of an AR($k$) model, i.e., $\check{p}_k = \left[\mathbf{R}^{-1}\mathbf{r}\right]_k$ with $\mathbf{R} \in \mathbb{R}^{k\times k}$.

**Moving-Average Models:** A $Q$-th order moving-average model with model parameters $\mathbf{b} = (b_1 \; \ldots, \; b_Q)^\mathsf{T}$ is a weighted sum of white noise terms $\varepsilon_t$, i.e.,

$$x_t = b_1\varepsilon_{t-Q+1} + b_2\varepsilon_{t-Q+2} + \cdots + b_Q\varepsilon_t. \tag{7}$$

In terms of signal processing a MA model hypothesizes time series data to be filtered (colored) white noise. Hence, for a time series $\mathbf{x} \in \mathbb{R}^D$ the above equation can be written via a valid convolution operator $\mathbf{B} \in \mathbb{R}^{D \times D+Q-1}$, i.e.,

$$\begin{pmatrix} x_1 \\ x_2 \\ \vdots \\ x_D \end{pmatrix} = \begin{bmatrix} b_1 & \cdots & b_Q & 0 & \cdots & 0 \\ 0 & b_1 & \cdots & b_Q & \cdots & 0 \\ & & \ddots & & \ddots & \\ 0 & \cdots & 0 & b_1 & \cdots & b_Q \end{bmatrix} \begin{pmatrix} \varepsilon_{-Q+2} \\ \varepsilon_{-Q+3} \\ \vdots \\ \varepsilon_D \end{pmatrix} = \mathbf{x} = \mathbf{B}\boldsymbol{\varepsilon}. \tag{8}$$

In the present context it is noteworthy that, given a signal $\mathbf{x} \in \mathbb{R}^D$, MA model estimation (or rather spectral density estimation) can be stated as

$$\max_{\mathbf{g} \in \mathbb{R}^L} \left\{ \|\mathbf{G}\mathbf{x}\|_2^2 \right\} \text{ s.t. } \|\mathbf{g}\|_2^2 = 1 \text{ with } \mathbf{G} = \mathbf{F}^{-1}\boldsymbol{\Lambda}_{\hat{\mathbf{g}}}\mathbf{F}\mathbf{M}^\mathsf{T} \in \mathbb{R}^{D+L-1 \times D}, \tag{9}$$

i.e., maximizing *linear correlation* between $\mathbf{g}$ and $\mathbf{x}$ yields a spectral estimate for MA processes and can be associated to a linear phase estimate $\check{\mathbf{b}}$ [7].

Knowing the true model parameters, theoretically, one might forecast signals $\mathbf{x}$ according to Eq. (7) based on a noise estimate $\check{\boldsymbol{\varepsilon}} = \mathbf{B}^\dagger\mathbf{x}$ ($^\dagger$ denoting the pseudoinverse), which should be updated with $\varepsilon_{t+1} = \mathbb{E}\left\{\varepsilon_{t+1}\right\} = 0$ (cf. [24], p. 77). However, estimates of $\mathbf{b}$ are ambiguous with respect to phase, i.e., it is impossible to estimate the phase response of the process, because the true driving noise terms are inaccessible.[6] Yet, estimates $\check{\mathbf{b}}$ can be found by Maximum Likelihood estimation ([24], p. 128 and [38]) or a Cholesky factorization of the autocovariance matrix (e.g. [33]).

### 2.2 The Wiener Filter for Prediction

We consider an adaptive discrete linear filter $\mathbf{G}$, which is optimized with respect to minimum squared prediction error, i.e., the coefficients $\mathbf{g} \in \mathbb{R}^L$ that define the valid[7] convolution operator $\mathbf{G}$ are found from

$$\min_{\mathbf{g} \in \mathbb{R}^L} \left\{ \|\mathbf{G}\mathbf{x} - \mathbf{y}\|_2^2 \right\} \text{ with } \mathbf{G} = \mathbf{M}\mathbf{F}^{-1}\boldsymbol{\Lambda}_{\hat{\mathbf{g}}}^\mathsf{H}\mathbf{F} \in \mathbb{R}^{D-L+1 \times D}. \tag{10}$$

In general, this leads to the discrete-time Wiener–Hopf equations (see [37], p. 125). However, in the case of prediction, i.e., when

$$\mathbf{x} = \left(x_1 \cdots x_{D-1}\right)^\mathsf{T} \text{ and } \mathbf{y} = \left(x_{L+1} \cdots x_D\right)^\mathsf{T},$$

[6] The standard methodology to MA model forecasting relies on the recursive estimation of past noise terms ([24], p. 82). This requires an invertible estimate $\check{\mathbf{b}}$.

[7] Often the Wiener filter is defined based on a linear convolution.

it is readily shown that the Wiener–Hopf equations and the Yule–Walker equations coincide (i.e., the optimization problem in Eq. (10) is equivalent to Eq. (4), [26], p. 245).[8] Hence, the optimal coefficients are found the least-squares solution $\breve{\mathbf{g}} = \mathbf{X}^{\dagger}\mathbf{y}$ (with $\mathbf{X}$ as in Eq. (4)) or via the Levinson–Durbin algorithm.

### 2.3 Singular Spectrum Analysis

At a glance, SSA allows for signal decomposition (e.g. into trend, oscillation and noise), signal smoothing and prediction. The canonical approach to SSA splits the algorithm into four consecutive steps (embedding, SVD, grouping and diagonal averaging, see [20]). Notably, given a time series $\mathbf{x} \in \mathbb{R}^D$ the first two steps basically formulate PCA applied to a delay representation

$$\mathbf{x} \mapsto \acute{\mathbf{X}} = \begin{bmatrix} x_1 & x_2 & \cdots x_{D-L-1} \\ x_2 & x_3 & \cdots & x_{D-L} \\ \vdots & \vdots & & \vdots \\ x_L & x_{L+1} & \cdots & x_D \end{bmatrix} = \begin{bmatrix} | & | & & | \\ \acute{\mathbf{x}}_1 & \acute{\mathbf{x}}_2 & \cdots & \acute{\mathbf{x}}_{D-L-1} \\ | & | & & | \end{bmatrix} \tag{11}$$

of the time series at hand (motivated by Taken's theorem, see [11]). If the conditions for Taken's theorem are fulfilled[9] the delay representation $\acute{\mathbf{X}}$ is an *embedding*, i.e., the observations $\acute{\mathbf{x}}_i$ in $\acute{\mathbf{X}}$ resemble the original dynamics up to a diffeomorphism. However, SSA may also be understood as a form of adaptive filtering (see [6,9,25,30]). Following [6], we focus on the latter approach as it allows to understand SSA as a member to the family of methods under consideration. Given a time series $\mathbf{x} \in \mathbb{R}^D$, SSA means to seek a linear transform $\mathbf{G} \in \mathbb{R}^{D-L+1\times D}$ that is optimal with respect to variance maximization, i.e., we maximize the norm of the valid convolution between $\mathbf{g}$ and $\mathbf{x}$ via

$$\max_{\mathbf{g}\in\mathbb{R}^L} \left\{ \|\mathbf{G}\mathbf{x}\|_2^2 \right\} \text{ s.t. } \|\mathbf{g}\|_2^2 = 1, \text{ where } \mathbf{G} = \mathbf{M}\mathbf{F}^{-1}\mathbf{\Lambda}_{\hat{\mathbf{g}}}\mathbf{F} \in \mathbb{R}^{D-L+1\times D}. \tag{12}$$

The solution to Eq. (12) yields an eigenvalue decomposition of a scattering matrix $\mathbf{Z} = \acute{\mathbf{X}}\acute{\mathbf{X}}^{\mathsf{T}} \in \mathbb{R}^{L\times L}$ (see [6]). As shown in [6] the eigenvectors $\mathbf{g}_1, \ldots, \mathbf{g}_L$ of $\mathbf{Z}$ constitute a *frame* (see [35]) with corresponding *synthesis operator*

$$\mathbf{\Phi} = \left[\mathbf{G}_1^{\mathsf{T}} \cdots \mathbf{G}_L^{\mathsf{T}}\right] \text{ with } \mathbf{G}_l = \mathbf{M}\mathbf{F}^{-1}\mathbf{\Lambda}_{\hat{\mathbf{g}}_l}\mathbf{F}$$

and with $K = \min\{L, D\}$ the corresponding frame operator is (cf. [6])

$$\mathbf{\Theta} = \mathbf{\Phi}\mathbf{\Phi}^{\mathsf{T}} = \operatorname{diag}\{1, 2, \ldots, K, \ldots, K, \ldots, 2, 1\} \in \mathbb{R}^{D\times D}, \tag{13}$$

that is, $\mathbf{\Theta}$ is a diagonal matrix with entries as defined above. An SSA-based signal reconstruction $\mathbf{w}$ with respect to a set of eigenvectors, represented by the

[8] This problem setting is also underlying *linear predictive coding* (see [13], p. 185).

[9] Most importantly, $L \geq 2D_{\mathcal{M}} + 1$ assuming that the system dynamics are bound to a $D_{\mathcal{M}}$-dimensional attractor.

index set $\mathcal{I}$, amounts to signal analysis and synthesis (projection) analogously to PCA, yet, with respect to a frame instead of an orthonormal basis, such that

$$\mathbf{w} = \sum_{k\in\mathcal{I}} \boldsymbol{\Theta}^{-1}\mathbf{G}_k^\mathsf{T}\mathbf{G}_k\mathbf{x}. \tag{14}$$

SSA-based time series prediction relies on the estimation of linear recurrence relations from the eigenvectors of $\mathbf{Z}$ (see [21]). More precisely, a time series forecast is computed as $\breve{w}_{t+1} = \mathbf{w}^\mathsf{T}\breve{\mathbf{q}}$ based on the reconstructed series $\mathbf{w} = \left(w_{t-D+1} \cdots w_t\right)^\mathsf{T}$. The linear recurrence relation $\breve{\mathbf{q}}$ is defined based on a subset of $r$ eigenvectors $\mathbf{Z}\mathbf{u}_i = \lambda_i\mathbf{u}_i$ (e.g. [32]). Details will be provided in Sect. 3.1.

## 3 Unsupervised and Self-supervised Linear Prediction

Now, we reconsider the LRR estimation for PCA, derive the connection between MA models and PCA-based prediction and propose a generalized formalism.

### 3.1 PCA and Linear Recurrence Relations

We follow classical ML literature (e.g. [3]) and understand PCA as the *analysis* of signals with respect to the *principal components*, where the latter are directions of maximum variance. Given a data set $\mathbf{X} \in \mathbb{R}^{D\times N}$ containing $N$ signal of length $D$, we solve the constrained optimization problem

$$\max_{\mathbf{u}\in\mathbb{R}^D} \left\{\left\|\mathbf{u}^\mathsf{T}\mathbf{X}\right\|_2^2\right\} \text{ s.t. } \|\mathbf{g}\|_2^2 = 1 \implies \mathbf{X}\mathbf{X}^\mathsf{T}\mathbf{u} = \lambda\mathbf{u}.$$

Presupposing zero mean signals the matrix $\mathbf{X}\mathbf{X}^\mathsf{T}$ is the covariance matrix, whose eigenvectors—the principal components—point into directions of maximum variance. This may also be expressed in terms of the Singular Value Decomposition (SVD) of the matrix $\mathbf{X}$ into $\mathbf{X} = \mathbf{U}\boldsymbol{\Sigma}\mathbf{V}^\mathsf{T}$ (see [39], pp. 64–66). We write the projection of an observation $\mathbf{x}$ onto a subset of $r$ principal components as $\mathbf{w} = \mathbf{U}_r\mathbf{U}_r^\mathsf{T}\mathbf{x}$, where $\mathbf{U}_r \in \mathbb{R}^{r\times D}$ holds a subset of columns from $\mathbf{U}$.

Following [15] we consider the eigenvectors of $\mathbf{X}\mathbf{X}^\mathsf{T}$ for time series prediction, however, $\mathbf{X}$ is not necessarily a delay representation. We assume that the time series or process under consideration is bound to a subspace $\mathcal{U}_r$ spanned by $r$ leading principal components.[10] In this regard, we consider a time series $\mathbf{x} \in \mathbb{R}^D$ whose last component $x_D$ is unknown. Building on the assumption that $\mathbf{x} \in \mathcal{U}_r = \operatorname{span}\{\mathbf{u}_1, \ldots, \mathbf{u}_r\}$ we seek the coefficients $\mathbf{c} \in \mathbb{R}^r$ such that $\mathbf{U}_r\mathbf{c} = \mathbf{x}$. With a matrix $\mathbf{E}$, holding the first $D-1$ rows of the identity matrix and the last row $\mathbf{e}_D^\mathsf{T}$ of the identity matrix $\mathbf{I}$, we write

$$\mathbf{U}_r\mathbf{c} = \mathbf{x} = \begin{bmatrix}\mathbf{E}\mathbf{U}_r\mathbf{c}\\ \mathbf{e}_D^\mathsf{T}\mathbf{U}_r\mathbf{c}\end{bmatrix} = \begin{bmatrix}\mathbf{E}\mathbf{x}\\ \mathbf{e}_D^\mathsf{T}\mathbf{x}\end{bmatrix}, \text{ where } \begin{bmatrix}\mathbf{E}\\ \mathbf{e}_D^\mathsf{T}\end{bmatrix} = \mathbf{I} \in \mathbb{R}^{D\times D}, \tag{15}$$

[10] We assume the eigenvalues to be sorted in descending order by size.

i.e., we split the problem into two parts. Now a least-squares coefficient estimate is found from the upper part of the system via pseudoinverse

$$\check{\mathbf{c}} = (\mathbf{E}\mathbf{U}_r)^\dagger\, \mathbf{E}\mathbf{x} \text{ so that } \check{x}_D = \mathbf{e}_D^\mathsf{T}\mathbf{U}_r\check{\mathbf{c}}.$$

This allows us to write PCA-driven time series prediction based on estimated coefficients $\check{\mathbf{q}} \in \mathbb{R}^{D-1}$ of the linear recurrence relation (see [21], p. 73) as

$$\check{x}_{t+1} = \check{\mathbf{q}}^\mathsf{T} \begin{pmatrix} x_{t-D+2} \\ \vdots \\ x_t \end{pmatrix} \quad \text{with } \check{\mathbf{q}} = \frac{1}{1 - \left\|\mathbf{e}_D^\mathsf{T}\mathbf{U}_r\right\|_2^2} \sum_{i=1}^{r} \mathbf{e}_D^\mathsf{T}\mathbf{u}_i\mathbf{E}\mathbf{u}_i. \tag{16}$$

Here, following [15], the pseudoinverse is written in terms of the *verticality coefficient* $v = \|\mathbf{e}_D^\mathsf{T}\mathbf{U}_r\|_2^2$, i.e., under the assumption that $\mathbf{e}_D \notin \mathcal{U}_r$ (that means $x_{t+1}$ depends on preceding observations) for $r < D$ a PCA-based prediction yields

$$\check{x}_{t+1} = \check{\mathbf{q}}^\mathsf{T}\underline{\mathbf{x}} = \mathbf{e}_D^\mathsf{T}\mathbf{U}_r\,(\mathbf{E}\mathbf{U}_r)^\dagger\,\underline{\mathbf{x}} = \frac{1}{1-v}\mathbf{e}_D^\mathsf{T}\mathbf{U}_r\mathbf{U}_r^\mathsf{T}\mathbf{E}^\mathsf{T}\underline{\mathbf{x}} \tag{17}$$

with the shorthand $\underline{\mathbf{x}} = \mathbf{E}\mathbf{x} \in \mathbb{R}^{D-1}$ (the component to be predicted is removed).

### 3.2 PCA and Moving-Average Models

In the following, we will study PCA-based time series prediction in the context of moving-average models, i.e., we assume that a given time series stems from a finite-order MA process. As for Wold's theorem (e.g. [14], p. 29) this is not a too restrictive assumption. In short, Wold's theorem states that any stationary random process can be modeled by means of an infinite-order MA model. Finite-order MA models, however, are not suited for processes with sharp (narrow-band) spectral peaks. According to Eq. (17) and Eq. (8), prediction of a time series in the context of moving-average models amounts to[11]

$$\check{x}_{t+1} = \mathbf{e}_D^\mathsf{T}\mathbf{U}_r\,(\mathbf{E}\mathbf{U}_r)^\dagger\,(\mathbf{E}\mathbf{x}) = \mathbf{e}_D^\mathsf{T}\mathbf{U}_r\,(\mathbf{E}\mathbf{U}_r)^\dagger\,\mathbf{E}\mathbf{B}\boldsymbol{\varepsilon}, \tag{18}$$

with $\mathbf{B}$ and $\boldsymbol{\varepsilon}$ as described in Sect. 2.1. Working with empirical data the columns of $\mathbf{U}_r$ are the eigenvectors of $\mathbf{X}\mathbf{X}^\mathsf{T}$. However, considering data that is drawn from an MA process we can show that in expectation $\mathbf{U}_r$ holds the left singular vectors of the associated MA operator $\mathbf{B}$ (a valid convolution operator), because[12]

$$\mathbb{E}\left\{\mathbf{X}\mathbf{X}^\mathsf{T}\right\} = \mathbb{E}\left\{\mathbf{B}\boldsymbol{\varepsilon}(\mathbf{B}\boldsymbol{\varepsilon})^\mathsf{T}\right\} = \mathbb{E}\left\{\mathbf{B}\boldsymbol{\varepsilon}\boldsymbol{\varepsilon}^\mathsf{T}\mathbf{B}^\mathsf{T}\right\} = \mathbf{B}\mathbf{B}^\mathsf{T}.$$

Considering the singular value decomposition of $\mathbf{B}$ into $\mathbf{B} = \mathbf{U}\boldsymbol{\Sigma}\mathbf{V}^\mathsf{T}$ we define the low rank approximation $\mathbf{B}_r = \mathbf{U}_r\boldsymbol{\Sigma}_r\mathbf{V}_r^\mathsf{T}$ analogously to Sect. 3.1 such that

$$\check{x}_{t+1} = \mathbf{q}^\mathsf{T}\mathbf{E}\mathbf{B}\boldsymbol{\varepsilon} \approx \mathbf{e}_D^\mathsf{T}\mathbf{U}_r\,(\mathbf{E}\mathbf{U}_r)^\dagger\,\mathbf{E}\mathbf{U}_r\boldsymbol{\Sigma}_r\mathbf{V}_r^\mathsf{T}\boldsymbol{\varepsilon}.$$

[11] Note that $\mathbf{E}$ as defined in Sect. 3.1 removes the last element of $\mathbf{x}$ when evaluating $\mathbf{E}\mathbf{x}$, i.e., $x_{t+1}$ is not involved in the prediction of $\check{x}_{t+1}$.

[12] Recall that we draw $\boldsymbol{\varepsilon}$ from a unit variance Gaussian white noise process.

Since we require that $\mathbf{e}_D \notin \mathcal{U}_r$ the columns of $\mathbf{EU}_r$ are independent for $r < L$ and $(\mathbf{EU}_r)^\dagger \mathbf{EU}_r = \mathbf{I}$ such that a one step ahead prediction amounts to

$$\check{x}_{t+1} = \mathbf{e}_D^\mathsf{T}\mathbf{U}_r\mathbf{\Sigma}_r\mathbf{V}_r^\mathsf{T}\boldsymbol{\varepsilon} = \mathbf{e}_D^\mathsf{T}\mathbf{B}_r\boldsymbol{\varepsilon}, \tag{19}$$

which establishes a direct connection between PCA-based prediction and MA models. Notably, for $r = L$ we find that $\mathbf{e}_D^\mathsf{T}\mathbf{U}_r(\mathbf{EU}_r)^\dagger = \mathbf{0}$, which means that in this case $\check{x}_{t+1} = 0$.[13] This is consistent, since consideration of the complete MA process in the setting of Eq. (15) necessarily yields $\check{x}_{t+1} = 0$. It is therefore reasonable to extrapolate dominant local dynamics of the signal at hand.

Now, according to Eq. (19) predictions based on Eq. (17) are made with respect to an optimal[14] rank-$r$ approximation of the MA process associated to $\mathbf{B}$. From this point of view the projection onto the axis of maximum variance is enriched by a signal processing perspective. In fact, we consider the most dominant *modes* (eigenvectors) of the process at hand and drive these forward by means of a least-squares approximation. These subspaces optimally capture the governing frequency domain characteristics of the given process under a time frequency resolution trade-off (cf. Eq. (9), related to [40]). Notably the covariance matrix $\mathbf{R} = \mathbf{BB}^\mathsf{T}$ is a symmetric Toeplitz matrix and its eigenvectors approximate harmonic oscillations when $D \gg Q$ [22]. For a stationary and periodic (or rather cyclostationary) process, however, the sample autocovariance matrix $\check{\mathbf{R}}$ is a circulant matrix and its eigenvectors are exactly equal to the discrete Fourier modes (the columns of $\mathbf{F}$, cf. Eq. (1)), i.e., for periodic processes the DFT yields an optimal representation with respect to variance maximization.

In summary, working with $L$-dimensional stationary time series data, the optimal representation found via PCA approximates the dynamics of an $L$-th order MA process. PCA-based prediction amounts to continuation of these dynamics via projections onto the subspace spanned by the principal components, which goes along with the well-known limitations of MA models.

### 3.3 A Unifying Perspective

In order to further categorize PCA-based prediction we put all the methodologies presented above into a general ML framework of self-supervised and unsupervised learning. In fact, all the above methods can be cast as one of the following optimization problems. Given a set of $N$ time series $\mathbf{x}_n \in \mathbb{R}^D$ we seek to

$$\max_{\mathbf{g}\in\mathbb{R}^L}\left\{\sum_{n=1}^{N}\|\mathbf{Gx}_n\|_2^2\right\} \text{ s.t. } \|\mathbf{g}\|_2^2 = 1 \text{ or } \min_{\mathbf{g}\in\mathbb{R}^L}\left\{\sum_{n=1}^{N}\left\|\mathbf{Gx}_n^{(t-1)} - \mathbf{x}_n^{(t)}\right\|_2^2\right\}, \tag{20}$$

where $\mathbf{G}$ is a convolution operator as shown in Fig. 1 and $\mathbf{x}^{(t)}$ denotes a time series $(\dots x_{t-1}\ x_t)$ with components up to time $t$. While the former optimization states unsupervised learning, the latter implements self-supervised learning.

[13] For $r = L$ Eq. (19), respectively $(\mathbf{EU}_r)^\dagger \mathbf{EU}_r = \mathbf{I}$, does not hold and $\mathbf{e}_D \in \mathcal{U}_r$.

[14] Optimality in the sense of SVD or rather the Eckart–Young–Mirsky theorem.

Notably, in case that $\mathbf{G} = \mathbf{g}^\mathsf{T}$ the unsupervised problem is equivalent to standard PCA, and the self-supervised problem amounts to linear regression. While it is known that the latter case is tied to AR models, we have now shown that the unsupervised scenario corresponds to MA modeling when dealing with stationary time series. This is conclusive, since the input to an MA process is not observable.

Before we transfer well-known findings of the regression setting to the unsupervised case, we reconsider the former. Usually AR model estimation is using all possible overlapping segments of length $P$ of one single time series. This way the information contained in a given signal is fully exploited. Instead of reshaping data, however, we may also define the model in terms of a corresponding structure $\mathbf{G}$, which represents the associated statistical assumptions (here, most of all, imposing shift-invariance). In case of AR models this is readily seen by comparison to the Wiener filter. As aforementioned both approaches can be associated to the Yule–Walker equations. A reference to self-supervised linear regression in a standard ML setting arises from the handling of data, i.e., given an appropriate data set (or restructuring given data appropriately) linear regression is equivalent to Eq. (4) respectively Eq. (12). Thus, a Wiener filter for prediction may be understood as a special case of linear regression and the equivalence to AR coefficient estimation adds a probabilistic perspective. This allows for a systematic deployment of this methodology, relying on the corresponding means of time series analysis. All together, this points up to the modeling aspect of $\mathbf{G}$, which is also reflected by the solutions to the above learning problems.

In general, both of the above optimization problems lead to a linear system[15] involving a scattering matrix $\mathbf{Z}$ (in general a covariance matrix), which incorporates the statistical assumptions going along with $\mathbf{G}$. More precisely,

$$\mathbf{Z}\mathbf{g} = \lambda\mathbf{g} \text{ (unsupervised) or } \mathbf{Z}\mathbf{g} = \mathbf{r} \text{ (self-supervised)},$$

where the structure of $\mathbf{Z}$ (and $\mathbf{r}$) is determined by the structure of $\mathbf{G}$. As an example a Toeplitz matrix $\mathbf{Z}$, which implies stationarity, results from a linear convolution $\mathbf{G}$ and a circulant matrix $\mathbf{Z}$ appears in case of a cyclic convolution $\mathbf{G}$. When $\mathbf{G} = \mathbf{g}^\mathsf{T}$, however, $\mathbf{Z}$ is the sample covariance matrix. Solutions for a more general class of structures can be found analogously to [5]).

Now, falling back to the theory presented in Sect. 3.2 we are able to extend this perspective to unsupervised learning. Starting from classical PCA it has been shown that LRRs estimated from the eigenvectors of the sample autocovariance matrix have a distinguished meaning in the context of MA processes. Going along with this we may seek to make more efficient use of available data by following the above considerations. Imposing shift-invariance in form of a valid convolution is a first step in this direction. This leads to basic SSA. In fact, considering the two above optimization problems as adaptive filtering problems, it becomes evident that SSA may simply be understood as an unsupervised

[15] Both problems in Eq. (20) are solved by setting zero the derivative of the goal function respectively the Lagrangian, where $\mathbf{G}$ is defined according to Fig. 1 (following [5] one may define convolutions as a matrix polynomial of cyclic permutation matrices).

Wiener filter (or rather a Wiener filter bank) implementing a *valid convolution*. However, a recourse on statistical modeling suggests to leverage the estimation of LRRs based on an estimate of the autocorrelation sequence as it is done for AR coefficient estimation (cf. Eq. (5)), i.e., setting $\mathbf{Z} = \check{\mathbf{R}}$.[16] Moreover, the statistical perspective allows for a comparably simple interpretation of the involved parameters. Regarding SSA the choice of a proper embedding dimension based on Taken's theorem is extremely difficult—if possible at all—compared to the interpretation of the same parameter as the hypothesized MA model order $Q$. Additionally, the perspective outlined here allows a simple design of the underlying model. Here, we are solely dealing with basic convolutions[17], however, more complex models are possible using appropriate structures of $\mathbf{G}$ (e.g. periodicity, dilation and stride, see [5,8]). We refer to the resulting methodology as *structured PCA* (sPCA). Beyond estimation of LRRs from eigenvectors of a structured covariance matrix $\mathbf{Z}$, the same structures may be applied analogously to SSA (cf. Eq. (14)), namely in order to filter the time series at hand. This procedure is analogous to basic SSA (Sect. 2.3), only that the frame structure varies with the choice of $\mathbf{G}$. For a periodic convolution the associated frame is composed from the eigenvectors $\mathbf{g}_l$ of a Toeplitz scattering matrix $\mathbf{Z}$ and

$$\mathbf{\Phi} = \left[\mathbf{G}_1^\top \cdots \mathbf{G}_L^\top\right] \text{ with } \mathbf{G}_l = \mathbf{F}^{-1}\mathbf{\Lambda}_{\hat{\mathbf{g}}_l}\mathbf{F} \text{ we find } \mathbf{\Theta} = L\mathbf{I} \in \mathbb{R}^{D\times D}$$

according to Eq. (13), while frame projections may be implemented according to Eq. (14) (using an FFT). Again, this formalism allows the implementation of a variety of statistical assumptions based on appropriate "filters" $\mathbf{G}$. Remarkably, for cyclic convolutions with $L = D$ the associated frame $\mathbf{\Phi}$ turns out to be a DFT filter bank (e.g. [34], p. 170), because in this case $\mathbf{Z}$ is circulant.

### 3.4 Practical and Numerical Aspects

Finally, we address some numerical issues and provide visualizations to the above theory. Figure 2 shows an example[18] on time series prediction based on the aforementioned methods. One apparent difference between self-supervised and unsupervised prediction is smoothness of predictions in the latter case.[19] Notably in this example the SSA-based LRR is almost equal to the sPCA-based, which is analogous to the equivalence of AR coefficient estimation and the Wiener filter. However, the respective predictions using SSA and sPCA vary due to the frame projections in the course of SSA. A visualization of the eigenvectors underlying the sPCA-based prediction is given in Fig. 3 alongside the eigenvectors of the autocovariance matrix that corresponds to basic PCA in a classical ML setting

[16] In this regard, there exist several different approaches to estimate the autocorrelation matrix $\mathbf{R}$ (in Sect. 3.4 we use a Bartlett-like estimator to compute $\mathbf{r}$ following [7]).

[17] This is similar to Toeplitz SSA (see [42]). Yet, our method goes beyond Toeplitz SSA, as it implements frame projections consistent with prescribed structures.

[18] Code is available under https://github.com/cmab92/linear_prediction_ki2024.

[19] The Wiener filter results from a least-squares solution, which coincides with AR coefficients based on the same ACF estimate that underlie sPCA-based prediction.

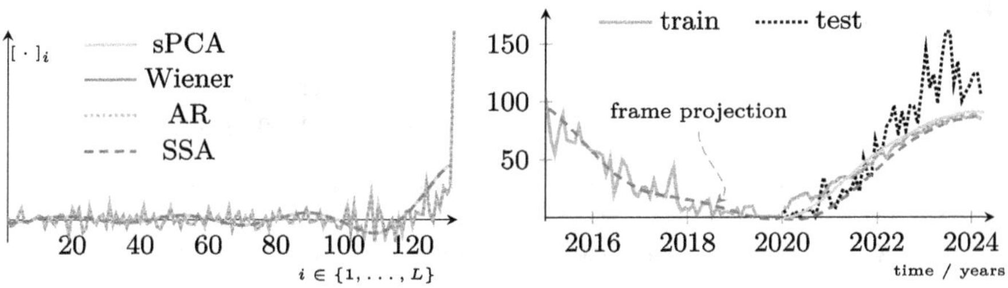


**Fig. 2.** An example on prediction using the monthly mean sunspot numbers [2] (right). The training set contains observations from the year 1749 up to the year 2020. Here, $L = 132$, which corresponds to 11 years (the solar cycle has a period time of about 11 years). The left graph shows the corresponding LRR coefficient estimates.

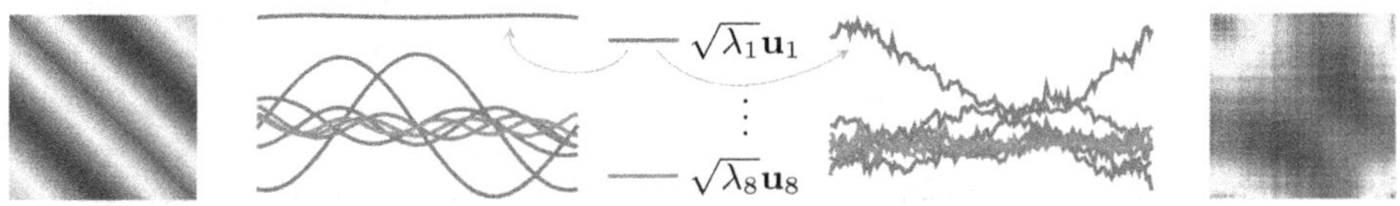


**Fig. 3.** The scattering matrices $\mathbf{Z} \in \mathbb{R}^{L \times L}$ for classical PCA (right) and sPCA (based on an autocorrelation matrix estimated according to [7]) and their first 8 eigenvectors. The latter underlies the LRR prediction in Fig. 2. Notably the eigenvectors of $\acute{\mathbf{X}}\acute{\mathbf{X}}^\top \in \mathbb{R}^{L \times L}$ (cf. Sect. 2.3) used for SSA-based prediction are similar to the sPCA eigenvectors.

(we split the time series into $\lfloor D/L \rfloor L$ signals of length $L$). In fact, classical PCA also yields an estimate of the autocorrelation matrix, yet, converge is slow and the resulting LRRs are unsuitable for prediction, which substantiates the proposed approach via structured PCA. Notably, estimated LRR polynomials may be unstable. This is a well-known problem in adaptive filtering and AR coefficient estimation (e.g. [29], p. 237) and there are many useful results available (e.g. [4,21,27,36]). Another issue is the number of eigenvectors used for LRR estimation (the parameter $r$). We recommend the optimal threshold according to [18], which has proven to be suitable in related applications (e.g. [12,31,44]).

## 4 Conclusion and Future Work

This work establishes a novel perspective on linear time series prediction with a focus on PCA-based estimation of linear recurrence relations. Most importantly, we established an exact connection between delay embeddings, MA processes and PCA-based forecasting. This also facilitates the choice of parameters and a deeper understanding of SSA. Moreover, we introduced a general framework that allows signal analysis and prediction subordinate to a prescribed correlation structure. A comprehensive numerical study of structured PCA for time series prediction with different models and a further exploration of the connection of PCA and MA processes are pending.

## References


1. Aguilera, A.M., Ocaña, F.A., Valderrama, M.J.: Forecasting time series by functional PCA. Discussion of several weighted approaches. Comput. Stat. **14**, 443–467 (1999)
2. Royal Observatory of Belgium, B.: Sunspot data from the world data center silso (2024). https://www.sidc.be/silso/datafiles
3. Bishop, C.M.: Pattern recognition and machine learning, vol. 2, pp. 645–678. Springer (2006)
4. Boche, H., Pohl, V.: Structural properties of the wiener filter-stability, smoothness properties, and fir approximation behavior. IEEE Trans. Inf. Theory **52**(9), 4272–4282 (2006)
5. Bonenberger, C., Ertel, W., Schneider, M.: $\kappa$-circulant maximum variance bases. In: Edelkamp, S., Möller, R., Rueckert, E. (eds.) KI 2021. LNCS (LNAI), vol. 12873, pp. 17–29. Springer, Cham (2021). https://doi.org/10.1007/978-3-030-87626-5_2
6. Bonenberger, C., Ertel, W., Schwenker, F., Schneider, M.: Singular spectrum analysis and circulant maximum variance frames. Adv. Data Sci. Adapt. Anal. **14**(03n04), 2250008 (2022)
7. Bonenberger, C., Schneider, M., Schwenker, F., Ertel, W.: A novel approach to spectral estimation and moving average model parameter estimation. IEEE Signal Processing Letters (2023)
8. Bonenberger, C.M., Schwenker, F., Ertel, W., Schneider, M.: Cyclic nonlinear correlation analysis for time series. IEEE Access **10**, 114223–114231 (2022)
9. Bozzo, E., Carniel, R., Fasino, D.: Relationship between singular spectrum analysis and fourier analysis: theory and application to the monitoring of volcanic activity. Comput. Math. Appl. **60**(3), 812–820 (2010)
10. Brockwell, P.J., Davis, R.A.: Time Series: Theory and Methods. Springer, New York (1991). https://doi.org/10.1007/978-1-4419-0320-4
11. Broomhead, D.S., King, G.P., et al.: On the qualitative analysis of experimental dynamical systems. Nonlinear Phenomena Chaos **113**, 114 (1986)
12. Brunton, S.L., Brunton, B.W., Proctor, J.L., Kaiser, E., Kutz, J.N.: Chaos as an intermittently forced linear system. Nat. Commun. **8**(1), 19 (2017)
13. Byrne, C.L.: Signal Processing A Mathematical Approach. CRC Press, Boca Raton (2015)
14. Clarkson, P.M.: Optimal and Adaptive Signal Processing. Routledge, London (2017)
15. Danilov, D.: Principal components in time series forecast. J. Comput. Graph. Stat. **6**(1), 112–121 (1997)
16. Durbin, J.: Estimation of parameters in time-series regression models. J. R. Stat. Soc. Ser. B Stat. Methodol. **22**(1), 139–153 (1960)
17. Eshel, G.: The yule walker equations for the AR coefficients (2003). http://www-stat.wharton.upenn.edu/~steele/Courses/956/Resource/YWSourceFiles/YW-Eshel.pdf
18. Gavish, M., Donoho, D.L.: The optimal hard threshold for singular values is $4/\sqrt{3}$. IEEE Trans. Inf. Theory **60**(8), 5040–5053 (2014)
19. Ghorbani, M., Chong, E.K.: Stock price prediction using principal components. PLoS ONE **15**(3), e0230124 (2020)

20. Golyandina, N.: Particularities and commonalities of singular spectrum analysis as a method of time series analysis and signal processing. Wiley Interdiscip. Rev. Comput. Stat. **12**(4), e1487 (2020)
21. Golyandina, N., Zhigljavsky, A.: Singular Spectrum Analysis for Time Series. Springer, Berlin, Heidelberg (2013). https://doi.org/10.1007/978-3-642-34913-3
22. Gray, R.M., et al.: Toeplitz and circulant matrices: a review. Found. Trends® Commun. Inf. Theory **2**(3), 155–239 (2006)
23. Greenacre, M., Groenen, P.J., Hastie, T., d'Enza, A.I., Markos, A., Tuzhilina, E.: Principal component analysis. Nat. Rev. Methods Primers **2**(1), 100 (2022)
24. Hamilton, J.D.: Time Series Analysis. Princeton University Press, Princeton (2020)
25. Harris, T., Yuan, H.: Filtering and frequency interpretations of singular spectrum analysis. Physica D **239**(20–22), 1958–1967 (2010)
26. Haykin, S.S.: Adaptive Filter Theory. Pearson Education India, London (2002)
27. de Hoon, M.J., van der Hagen, T.H., Schoonewelle, H., van Dam, H.: Why yule-walker should not be used for autoregressive modelling. Ann. Nucl. Energy **23**(15), 1219–1228 (1996)
28. Jolliffe, I.: Principal Component Analysis, vol. 2. Springer, Berlin, Heidelberg (2002). https://doi.org/10.1007/0-387-22440-8_13
29. Kantz, H., Schreiber, T.: Nonlinear Time Series Analysis, vol. 7. Cambridge University Press, Cambridge (2004)
30. Kume, K.: Interpretation of singular spectrum analysis as complete eigenfilter decomposition. Adv. Adapt. Data Anal. **4**(04), 1250023 (2012)
31. Kutz, J.N., Brunton, S.L., Brunton, B.W., Proctor, J.L.: Dynamic mode decomposition: data-driven modeling of complex systems. SIAM (2016)
32. Leles, M.C., Moreira, M.G., Vale-Cardoso, A.S., Nascimento, C.L., Sbruzzi, E.F., Guimaraes, H.N.: Comparison between basic and Toeplitiz SSA applied to non-stationary time-series. Stat. Interface **12**(4), 527–536 (2019)
33. McMurry, T.L., Politis, D.N.: Estimating ma parameters through factorization of the autocovariance matrix and an ma-sieve bootstrap. J. Time Ser. Anal. **39**(3), 433–446 (2018)
34. Mertins, A., Mertins, D.A.: Signal Analysis: Wavelets, Filter Banks, Time-Frequency Transforms and Applications. John Wiley & Sons, Inc., Hoboken (1999)
35. Morgenshtern, V.I., Bölcskei, H.: A short course on frame theory. In: Mathematical Foundations for Signal Processing, Communications, and Networking, pp. 737–789. CRC Press (2017)
36. Moses, R.L., Liu, D.: Determining the closest stable polynomial to an unstable one. IEEE Trans. Signal Process. **39**(4), 901–906 (1991)
37. Orfanidis, S.J.: Introduction to Signal Processing. Prentice-Hall, Inc., Hoboken (1995)
38. Osborn, D.: Maximum likelihood estimation of moving average processes (1976). http://www.nber.org/chapters/c12707
39. Serpedin, E., Chen, T., Rajan, D.: Mathematical Foundations for Signal Processing, Communications, and Networking. CRC Press, Boca Raton (2011)
40. Slepian, D., Pollak, H.O.: Prolate spheroidal wave functions, fourier analysis and uncertainty-i. Bell Syst. Tech. J. **40**(1), 43–63 (1961)
41. Stoica, P., Moses, R.L., et al.: Spectral Analysis of Signals. Pearson Prentice Hall, Upper Saddle River, NJ (2005)
42. Vautard, R., Yiou, P., Ghil, M.: Singular-spectrum analysis: a toolkit for short, noisy chaotic signals. Physica D **58**(1–4), 95–126 (1992)

43. Vetterli, M., Kovačević, J., Goyal, V.K.: Foundations of Signal Processing. Cambridge University Press, Cambridge (2014)
44. Wang, C., Zhu, Z., Gu, H., Wu, X., Liu, S.: Hankel low-rank approximation for seismic noise attenuation. IEEE Trans. Geosci. Remote Sens. **57**(1), 561–573 (2018)
45. Wang, L.: Karhunen-Loeve expansions and their applications. London School of Economics and Political Science (United Kingdom) (2008)

# Data Augmentation in Latent Space with Variational Autoencoder and Pretrained Image Model for Visual Reinforcement Learning

Xuzhe Dang(✉) and Stefan Edelkamp

Czech Technical University in Prague, Prague, Czechia
dangxuzh@fel.cvut.cz

**Abstract.** In this paper we investigate alternative data augmentation strategies for Visual Reinforcement Learning and explore the potential benefits of fine-tuning a pretrained image encoder to enhance the learning process. We propose an innovative approach that applies data augmentation in the latent space, rather than directly manipulating pixel values. This method utilizes a Variational Autoen- coder, integrated with a pretrained image model, to facilitate the data augmentation process in a more abstract and feature-rich latent space. We use the DeepMind Control suite as a benchmark to evaluate the impact of our approach.

**Keywords:** Visual Reinforcement Learning · Deep Learning · Representation Learning

## 1 Introduction

Recent years have witnessed significant advancements in Reinforcement Learning (RL) research [29,39,41]. Yet, a predominant focus of these studies relies on the assumption of true state features, wherein it is presumed that the agent possesses complete information about itself and its environment. This assumption, however, often falters in real-world scenarios where access to such comprehensive information is not always feasible. Cameras, as cost-effective and convenient sensors, are frequently employed to collect information in real-world settings. Consequently, the development of an efficient, visually-based RL algorithm becomes crucial for addressing real-world challenges.

Learning from images, in contrast to utilizing true state features, presents significant challenges. This difficulty stems not only from the high-dimensional nature of image data but also from the complexity of the information they contain, including both pertinent and extraneous background details. Recent research emphasizes the critical role of data augmentation [38] in markedly enhancing the sample efficiency of Visual Reinforcement Learning (VRL) [21,44]. Techniques such as random shifts and background modifications of raw images


A. Hotho and S. Rudolph (Eds.): KI 2024, LNAI 14992, pp. 45–59, 2024.
https://doi.org/10.1007/978-3-031-70893-0_4

are commonly employed as methods of data augmentation. Moreover, the use of pretrained models from datasets like ImageNet as image encoders has been identified as beneficial in facilitating the learning process [37,42,49].

In our research, we delve into the exploration of alternative data augmentation strategies for Visual Reinforcement Learning (VRL), with a particular focus on the potential advantages of fine-tuning a pretrained image encoder to bolster the learning dynamics. Our proposed methodology introduces a novel approach to data augmentation that operates within the latent space, diverging from the traditional method of direct pixel manipulation. By employing a Variational Autoencoder (VAE) [20], in conjunction with a pretrained image model, we navigate the augmentation process through a more abstract, feature-enriched latent space. This innovative strategy is visually summarized in Fig. 1, offering an illustrative overview of our approach.

Adhering to the architectural framework and strategies of the DrQv2 algorithm[1], we implement pivotal modifications to better accommodate the complexities inherent in visual observations. At the heart of our innovation is the decision to fix the observation encoders, thereby ensuring a stable and consistent extraction of features from visual inputs. This marks a departure from the standard DrQv2 methodology, which primarily utilizes random shift augmentation to foster variability and robustness within the learning environment. Instead, we introduce our VAE-trained model as a replacement.

This strategic substitution is predicated on the VAE's capacity to produce a broad spectrum of augmented latent features from the observation encoder. Such capability facilitates a more intricate and comprehensive representation of the environment, which we hypothesize will lead to enhanced efficiency and performance in reinforcement learning tasks.

The incorporation of the VAE into the DrQv2 framework represents a crucial element of our methodology, leveraging the generative power of the VAE to supply the training process with a rich diversity of high-quality augmentations in latent space. This integration is designed to refine the DrQv2 algorithm's effectiveness in navigating and learning from complex visual scenarios, thereby advancing the state of the art in VRL.

## 2 Related Work

There is a substantial body of precursor work.

### 2.1 Visual-Base RL

Visual-based Reinforcement Learning has garnered considerable interest in the research community, driven by the proliferation and advantages of image sensors. The capability of agents to autonomously learn from visual observations represents a significant advancement in the field. Various simulators have been

[1] https://github.com/facebookresearch/drqv2.

developed specifically for visual-based RL or to provide image-based observations, catering to this growing area of study [10,34,40,47].

Traditionally, a number of established RL algorithms, such as Proximal Policy Optimization (PPO) [35], Asynchronous Advantage Actor-Critic (A3C) [28], and Soft Actor-Critic (SAC) [14], have been adapted for visual-based tasks. These adaptations often involve replacing fully connected layers with Convolutional Neural Networks (CNNs) to encode the image observations. A critical issue with these extensions is their lack of sample efficiency. The adaptation of these algorithms to visual tasks has not inherently solved the challenge of efficiently learning from limited data samples.

To address the issue of sample efficiency, recent research has shifted focus towards enhancing loss functions and introducing auxiliary tasks [2,23,36,44,50]. These strategies aim to provide more informative gradients and learning signals, thereby improving the learning efficiency from each data sample. Additionally, the application of data augmentation has gained traction as a method to artificially expand the training dataset, enabling the model to learn more robust and generalizable features from a given set of images [15,21,24,45].

### 2.2 Pretrained Image Encoder for RL

The successful application of pretrained models in domains such as computer vision and natural language processing (NLP) has sparked interest in similar approaches for Reinforcement Learning (RL). Specifically, the concept of using pretrained vision models, like ResNet trained on ImageNet, to encode visual observations into low-dimensional features, has gained traction among researchers [11,19,32,43,46].

Utilizing pretrained models like ResNet allows for the transformation of high-dimensional visual inputs into more manageable, lower-dimensional feature representations. This process is anticipated to facilitate more efficient learning in RL tasks by providing a rich, condensed summary of the visual data.

In an effort to develop robust visual representations, researchers have explored fine-tuning the models with expert or related video datasets, to providing more comprehensive foundation for downstream policy learning in RL [31,33,37]. Beyond feature encoding, pretrained models have also been proposed for goal specification in RL [9].

### 2.3 Data Augment

Data augmentation, pivotal for enhancing generalization in convolutional neural networks, has been a foundational element in training methodologies from their inception [5,7,8,22]. This process typically involves image transformations, including translations, scaling, and color shifts, each carefully selected to preserve the semantic essence of the object categories. Such transformations, integral for object classification tasks, ensure the consistency of the category's meaning while diversifying the dataset.

Recently, the advancement of generative models such as Variational Autoencoders (VAE) [20], Generative Adversarial Networks (GAN) [13], and diffusion models [18] has sparked interest in exploring synchronized data as a method for data augmentation [1,3,27,30,48]. By leveraging the generative capabilities of these models, researchers can create realistic, synthetic data that aligns closely with the original dataset in terms of distribution and characteristics, yet provides novel variations.

In addition to applying data augmentation directly to the original dataset, there's a growing interest in the investigation of data augmentation within the latent space [4,6,26]. This innovative approach involves manipulating the latent representations of data, which are abstract, compressed forms derived from the original input through the embedding layers of neural networks. Augmenting data in the latent space offers a unique advantage: it allows for modifications at a more abstract level, potentially uncovering novel and more diverse variations than those achievable through traditional pixel-level augmentations.

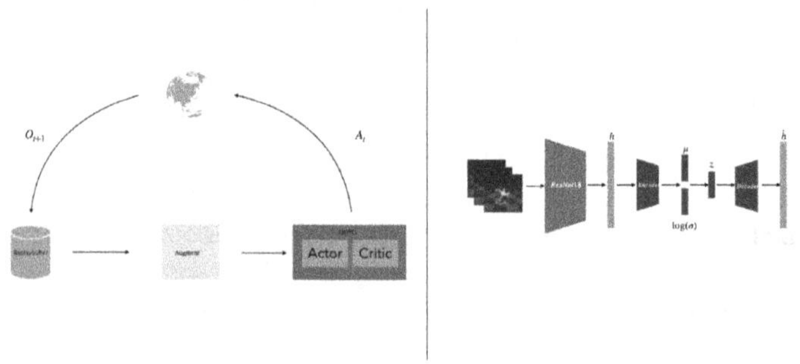


**Fig. 1.** (Left) DrQ-v2 is an off-policy actor-critic algorithm designed for image-based reinforcement learning. One of its key features is its ability to mitigate encoder overfitting. In the original algorithm, enhanced performance in handling visual inputs is achieved through the application of random shift augmentation to pixel observations drawn from the replay buffer. This augmentation technique significantly bolsters the algorithm's robustness. In our study, we employ reconstructed features as a means of augmentation. (Right) In our approach, rather than reconstructing the observations directly, we utilize a Variational Autoencoder (VAE) to reconstruct features that are extracted by a pretrained ResNet18

## 3 Preliminaries

We first look at the foundations for image-based control and self-supervised learning.

### 3.1 POMDP

Due to the inherent limitation of images in encapsulating all necessary information, we conceptualize image-based control problems within the framework of a Partially Observable Markov Decision Process (POMDP). Here, $S$ represents the state space, $O$ denotes the set of observations, and $A$ is the action space. Additionally, $r : S \times A \rightarrow \mathbb{R}$ is defined as the reward function, $P(S_{t+1} \mid S_t, a_t)$ as the state transition function, and $\gamma$ as the discount factor. Thus, the POMDP can be formally represented as $M = \langle S, O, A, r, P, \gamma \rangle$. The primary objective in this context is to devise an optimal policy $\pi_\theta^*$ that maximizes the expected cumulative returns with parameter $\theta$.

### 3.2 VAE

Variational Autoencoders (VAEs), as a form of self-supervised learning, have demonstrated their effectiveness in efficiently learning latent representations from high-dimensional data.

The architecture of a VAE comprises two key components: an encoder, which maps the input states into latent distributions, and a decoder, which reconstructs the states from these latent distributions. During the training process, the parameters of both the encoder and decoder are jointly optimized to maximize the loss function.

In our study, we assume that the prior distribution $p(z)$ is Gaussian, and the Kullback-Leibler (KL) divergence, denoted as $D_{kl}$, is utilized in the loss function. By introducing a hyperparameter $\beta$ to scale $D_{kl}$, we transform the standard VAE into a $\beta$-VAE [17], offering a modified approach to balance latent space regularization with reconstruction fidelity.

## 4 Method

In this section, we elucidate our methodology, emphasizing our approach to training a Variational Autoencoder (VAE) leveraging pretrained image models, and its subsequent integration with DrQv2, an advanced reinforcement learning (RL) framework.

### 4.1 Latent Space Representation and Augment

Visual Reinforcement Learning (VRL) faces unique challenges in data scarcity, distinguishing it from fields like computer vision or NLP. We counter this by integrating the first two layers of the pretrained ResNet18s [16] as our observation encoder, achieving significant advancements:

1. Efficient Training via High-Level Feature Utilization: The incorporation of ResNet18's initial layers, rich with high-level features from diverse datasets, markedly enhances control task training in VRL. These layers are pivotal in capturing essential visual cues early in the learning process [49].
2. Computational Efficiency: By adopting only the first two layers of ResNet18, we substantially reduce computational demands. This streamlined approach is crucial for the VRL domain, where swift processing is imperative for navigating dynamic environments and executing real-time decisions.

This methodology not only addresses the pressing issue of sample efficiency in VRL but also sets a precedent for employing deep learning techniques more judiciously.

In the realm of Visual Reinforcement Learning (VRL), the intrinsic limitation of single frames lacking explicit motion details, such as directional movement or velocity, compels the adoption of consecutive frame stacking as a technique for enriching observations.

This method introduces a significant challenge due to the resultant observation channels deviating from the standard three-channel RGB format anticipated by conventional pretrained image models. To bridge this gap, we implemented a technique known as convolutional channel expansion [42]. This approach modifies the input layer of the pretrained model to align with the augmented number of observation channels, ensuring the model's continued proficiency in processing both temporal and spatial dimensions of input data.

The convolutional channel expansion method entails the replication of the ResNet model's weight matrix $m$ times, followed by the scaling of these weights to $1/m$ of their initial values. Such a strategy permits the model to concurrently interpret temporal and spatial attributes of the observed data, significantly improving its capability to comprehend dynamic environments. A pivotal advantage of this method is its ability to enhance model performance without necessitating the retraining of the model's weights, thus maintaining the pre-trained network's integrity and operational efficiency.

Our research aims to explore the potential benefits of augmenting the latent space representation on the learning process in Visual Reinforcement Learning (VRL). This investigation seeks to determine whether enhancing the latent space through specific reconstruction techniques can lead to improved VRL performance.

**Latent Space Reconstruction via VAE:** Contrary to traditional approaches that focus on reconstructing the input frame itself, our study employs Variational Autoencoders (VAE) to reconstruct the high-level features extracted by the ResNet18 model. This methodological pivot emphasizes the refinement of the latent space representation, hypothesizing that a better understanding and reconstruction of these features can significantly contribute to the learning efficiency in VRL tasks.

**Data Collection and Training.** In our approach for each task within the DeepMind Control (DMC) suite, we employed a strategic mix of replays derived from three distinct policy types. This mix consisted of 33% from a well-trained policy, another 33% from a policy of similar training but with noise injected into its actions, and the final 34% from random policies. Such a diverse composition of replay sources is crucial for ensuring that our dataset encompasses a comprehensive range of scenarios, which is instrumental in capturing a wide array of potential features that might arise in different situations.

Furthermore, we introduced a modification to the data collection process. Instead of accumulating replays throughout the entire duration of each scenario, we limited the collection to only the first 100 steps. This constraint was implemented with the intention of gathering data from a greater variety of scenarios, rather than extensively sampling from a smaller set of longer episodes. By doing so, we aimed to enhance the diversity and richness of our dataset, promoting a more robust and well-rounded training environment for our models.

This methodological tweak is expected to contribute significantly to the development of models that are better equipped to handle a wide array of situations, thereby improving their overall adaptability and performance. A considerable volume of frames is collected through this process, forming the core dataset for the subsequent training of the Variational Autoencoder (VAE) models.

The primary goal is to employ these frames not just for feeding the VAE, but also to guarantee that the model is trained to accurately reconstruct the features derived from ResNet18, as opposed to the original frames. During the training phase, we keep the weights of ResNet18 fixed (frozen). The VAE model is iteratively updated by

$$\begin{aligned}\mathcal{L}(\theta, \phi; h) = -\mathbb{E}_{q_\phi(z|h)}[\log p_\theta(h|z)] + \\ \beta \times \mathrm{KL}(q_\phi(z|h)||p(z)),\end{aligned}$$

where $h = f_\xi(s)$, $f_\xi$ is the ResNet18, and $s$ is the stacked frames. In our study, we implemented the $\beta$-VAE approach to achieve a balance between the reconstruction loss and the Kullback-Leibler (KL) divergence.

## 4.2 Reinforcement Learning Backbone

In our experimental setup, DrQv2 was chosen as the primary algorithm for Visual Reinforcement Learning tasks.

DrQv2 stands as an evolved variant of the Deep Deterministic Policy Gradient (DDPG) algorithm, as initially described in [25]. It incorporates elements of clipped Double Q-learning, a technique presented by [12], which is instrumental in addressing the overestimation bias that is commonly observed in traditional DDPG frameworks during target Q-value estimations.

The core methodology of DrQv2 involves the simultaneous training of two separate Q-value functions, along with their respective target networks. These are collectively referred to as $Q$. This dual-Q structure is a key feature of DrQv2, enabling it to effectively reduce overestimation bias.

By adopting this approach, DrQv2 enhances the stability and accuracy of the learning process in visual reinforcement learning environments, which is particularly crucial when dealing with high-dimensional input spaces such as images.

The Q-value functions update is orchestrated via an $n$-step Temporal Difference (TD) target, formulated as

$$y = \sum_{i=0}^{n-1} \gamma^i r_{t+i} + \gamma^n \min_{k=1,2} Q_{\theta_k}(h_{t+n}, a_{t+n}). \tag{1}$$

This extension of conventional TD learning to an $n$-step sequence enables a more accurate and stable estimation of the Q-values.

The Q functions are updated by sampling a batch of steps from the replay buffer, following the loss function

$$L(\theta_k) = \mathbb{E}[(Q_{\theta_k}(h_t, a_t) - y)^2], \qquad \forall k \in \{1, 2\}, \tag{2}$$

and the policy $\pi$ is updated through

$$L(\pi) = -\mathbb{E}[\min_{k=1,2} Q_{\theta_k}(h_t, \pi(s_t))]. \tag{3}$$

In the formulation, $h = f_\xi(\text{aug}(s))$ represents the process where $f_\xi$, denoting the ResNet18, is applied to the augmented state $s$. Here, aug() refers to the random shift augmentation in the original method, and in our study, it also encompasses the application of the pretrained Variational Autoencoder (VAE) model.

## 5 Experiments

In this section, we evaluate our method on an extensive set of visual continuous control tasks from the DeepMind Control (DMC) suite [40]. Our objective is to assess the effectiveness of our data augmentation approach compared to the random shift augmentation method employed in DrQv2.

### 5.1 Setup

In accordance with the guidelines established by DrQv2, our research leveraged the DeepMind Control (DMC) suite as the benchmark to evaluate the efficacy of our model. The details of our experimental framework are as follows.

**Observation Framework.** Our method involved processing observations by stacking three consecutive frames, each frame sized at $84 \times 84$ pixels. This technique was chosen to capture a detailed perspective of the temporal dynamics occurring within the environment, providing a richer context for the model to understand and predict future states.

**Action Repetition.** To ensure a consistent action space and foster a more stable learning environment, each action decided by the policy was repeated for two consecutive timesteps. This repetition strategy not only simplifies the decision-making process for the agent but also aids in smoothing the trajectory of the agent's learning curve by reducing the variability of outcomes associated with each action.

**Cross-Comparison.** To enable an equitable and direct comparison with the standard DrQv2 setup, we integrated the ResNet18 encoder, initially utilized in training our VAE, into the DrQv2 framework. Importantly, we kept the weights of this encoder fixed throughout the training phase.

This adjustment enabled us to isolate the influence of our encoder selection on the agent's overall performance, ensuring that any performance disparities observed could be solely attributed to the encoder's inherent characteristics, rather than variations in learning dynamics.

In our research, we executed a series of experiments to assess the impact of data augmentation within the latent space. Our experimental setup involved training two distinct DrQv2 agents, each utilizing a fixed ResNet18 as the encoder for observations.

**Dual-Agent Design.** One of these agents was configured according to the traditional DrQv2 framework, which incorporates augmentation of observations through random shifts. This served as our control setup, adhering to the standard practices of DrQv2.

Conversely, the second agent was aligned with our innovative approach, wherein the training observations were reconstructed using a pretrained Variational Autoencoder (VAE) model. This method represented our experimental condition, designed to investigate the efficacy of latent space augmentation.

By adopting this dual-agent experimental design, we were able to directly compare the conventional random shift augmentation with our novel latent space augmentation approach. This comparison was pivotal in analyzing and understanding the respective influences of these two strategies on the performance and learning dynamics of the DrQv2 agents.

## 5.2 Training Detail

We conducted an evaluation of our method on six tasks of medium difficulty from the DeepMind Control Suite (DMC). In this evaluation, agents were trained using one million frames per task. Every 10,000 frames, an evaluation phase was initiated where the agents were tested over 10 episodes. The average returns from these episodes were calculated to gauge the agents' performance. To ensure the reliability of our results, each task was executed with three different random seeds, aiming to reduce the impact of randomness and variability on the training outcomes.

During the training process, we made a strategic alteration to the activation function in the output from the second layer of the ResNet18, switching from ReLU to Tanh. This modification was driven by the suitability of Tanh for the training of Variational Autoencoders (VAEs) with Mean Squared Error (MSE) as the loss function. Utilizing a pretrained VAE for data augmentation necessitated the freezing of the Batch Normalization (BN) layers to maintain the assumption of independence and identical distribution, despite the potential benefits BN layers could offer to the learning process.

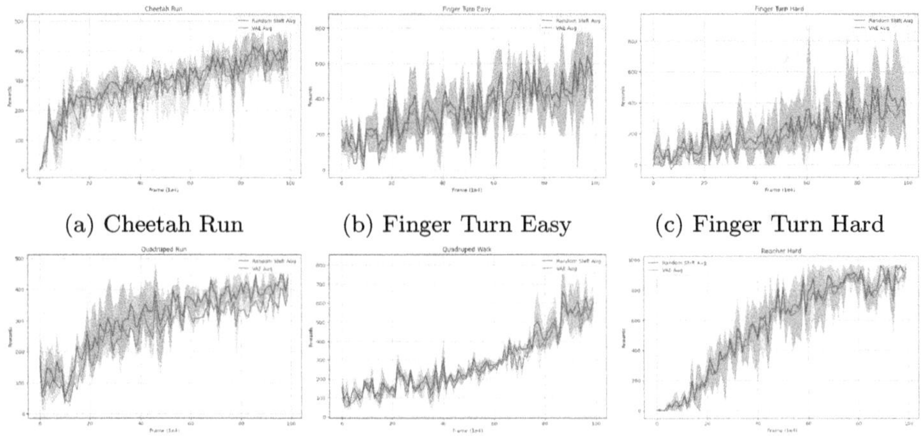

(a) Cheetah Run (b) Finger Turn Easy (c) Finger Turn Hard

(d) Quadruped Run (e) Quadruped Walk (f) Reacher Hard

**Fig. 2.** We compare two distinct data augmentation methods across six tasks in the DeepMind Control (DMC) suite. Both agents utilize a frozen ResNet18 architecture as their observation encoder. Each task is conducted with three different random seeds to ensure the robustness of our results

Moreover, our approach includes the implementation of random shift augmentation, where there is a deliberate probability that the agents are updated using the original, unmodified data. It's crucial to recognize that a VAE's reconstruction of frames is never perfectly accurate. With this in mind, we incorporated a mechanism during the VAE augmentation training that allows for a 1% chance of using the original frames directly. This method acknowledges the inherent limitations of frame reconstruction accuracy while enhancing the robustness of the training regimen.

### 5.3 Evaluation

In this section, we detail the results from our experiments on six medium-difficulty tasks within the DeepMind Control Suite (DMC).

These outcomes are depicted in Fig. 2. Our analysis demonstrates that both data augmentation strategies substantially improve the agent's proficiency in learning from and acting on visual observations. A notable observation from our analysis is the slightly enhanced performance of the agent trained with random shift augmentation compared to its counterpart trained with VAE (Variational Autoencoder) data augmentation.

**Hypothesis.** We hypothesize that the superior performance of the random shift augmentation method can be attributed to its sampling technique. In the VAE approach, a Gaussian distribution is utilized during the reparameterization of latent features. Consequently, this causes the augmented latent features to

**Table 1.** Cosine similarity between original and augmented latent features in Reacher Hard Task

| Data Augment Method | Mean | Std |
|---|---|---|
| Randomshift | 0.8741 | 0.0612 |
| VAE | 0.9744 | 0.0031 |

adhere to a Gaussian distribution, clustering most samples around the location of the mean ($\mu$).

In contrast, the random shift augmentation employs a uniform sampling method to crop the images, resulting in a more diverse set of features. This diversity likely provides a broader range of visual scenarios for the agent to learn from, thereby enhancing its ability to generalize and perform across different tasks.

To test our hypothesis regarding the differences in performance between random shift augmentation and VAE-based augmentation, we conducted a comparative analysis focused on the cosine similarity between latent features derived from original observations and those augmented by both methods. This analysis was carried out on two specific tasks: Quadruped Run and Reacher Hard. For each task, we collected replays of 5,000 steps and processed the data as follows:

**Encoding Original Frames.** We encoded the original frames to obtain the latent features, denoted as $h_o$. Encoding Random Shift Augmented Observations: We applied random shift augmentation to the original observations and encoded these augmented observations to obtain the latent features, denoted as $h_r$.

**Reconstruction Through VAE.** We used a VAE to reconstruct the original latent features $h_i$ to produce $h_v$, which represents the latent features derived from VAE-augmented observations.

For both sets of augmented features ($h_r$ from random shift and $h_v$ from VAE augmentation), we calculated the cosine similarity with the original features ($h_o$). We then computed the mean and standard deviation of these cosine similarity values to assess the degree of similarity between the original and augmented features.

**Table 2.** Cosine similarity between original and augmented latent features in Quadruped Run Task

| Data Augment Method | Mean | Std |
|---|---|---|
| Randomshift | 0.8856 | 0.0575 |
| VAE | 0.9694 | 0.0060 |

The empirical findings are presented in Tables 1 and 2. From the analyses' results, we can observe that the cosine similarity between the features of the original observations and those augmented by random shift has a larger standard deviation (std) value. This outcome suggests that the features resulting from random shift augmentation exhibit a higher level of diversity.

## 6 Conclusion and Outlook

Our study introduces a novel and innovative approach to data augmentation in the latent space by leveraging a Variational Autoencoder (VAE) in conjunction with a pretrained image model. This method represents a significant departure from traditional data augmentation techniques, which typically manipulate pixel values directly in the input space.

By operating in the latent space, our approach aims to produce more diverse and potentially more meaningful augmentations, thereby enriching the training data without altering the original input distribution dramatically.

We compared our innovative approach, integrating a Variational Autoencoder (VAE) with a ResNet18 encoder for latent space data augmentation, against the standard DrQv2 setup known for its simple but effective random shift augmentations in visual reinforcement learning. The results of our experiments show that the agent is able to learn with the augmented latent features and our method rivals traditional ones in terms of learning effectiveness.

Our general research agenda is to derive motion models through simulation for instructing a robot in longer-term task-motion planning, for which an improved and efficient visual reinforcement learning scheme is essential. We will integrate our refined inference algorithm as a component in more complex robotic planning frameworks. In fact, for a better cross-comparison, we extracted the learning to run in the DMC.

**Acknowledgements.** The presented work has been supported by the Czech Science Foundation (GAČR) under the research project number 22-30043S.

## References

1. Bowles, C., et al.: Gan augmentation: augmenting training data using generative adversarial networks (2018)
2. Cetin, E., Ball, P.J., Roberts, S., Celiktutan, O.: Stabilizing off-policy deep reinforcement learning from pixels. arXiv preprint arXiv:2207.00986 (2022)
3. Chadebec, C., Allassonnière, S.: Data augmentation with variational autoencoders and manifold sampling (2021)
4. Chen, C., Hammernik, K., Ouyang, C., Qin, C., Bai, W., Rueckert, D.: Cooperative training and latent space data augmentation for robust medical image segmentation (2021)
5. Chen, T., Kornblith, S., Norouzi, M., Hinton, G.: A simple framework for contrastive learning of visual representations. In: International Conference on Machine Learning, pp. 1597–1607. PMLR (2020)

6. Cheung, T.H., Yeung, D.Y.: MODALS: modality-agnostic automated data augmentation in the latent space. In: International Conference on Learning Representations (2021). https://openreview.net/forum?id=XjYgR6gbCEc
7. Ciregan, D., Meier, U., Schmidhuber, J.: Multi-column deep neural networks for image classification. In: 2012 IEEE Conference on Computer Vision and Pattern Recognition, pp. 3642–3649. IEEE (2012)
8. Cireşan, D.C., Meier, U., Masci, J., Gambardella, L.M., Schmidhuber, J.: High-performance neural networks for visual object classification. arXiv preprint arXiv:1102.0183 (2011)
9. Cui, Y., Niekum, S., Gupta, A., Kumar, V., Rajeswaran, A.: Can foundation models perform zero-shot task specification for robot manipulation? In: Learning for Dynamics and Control Conference, pp. 893–905. PMLR (2022)
10. Dosovitskiy, A., Ros, G., Codevilla, F., Lopez, A., Koltun, V.: Carla: an open urban driving simulator. In: Conference on Robot Learning, pp. 1–16. PMLR (2017)
11. Ebert, F., et al.: Bridge data: boosting generalization of robotic skills with cross-domain datasets. arXiv preprint arXiv:2109.13396 (2021)
12. Fujimoto, S., van Hoof, H., Meger, D.: Addressing function approximation error in actor-critic methods (2018)
13. Goodfellow, I.J., et al.: Generative adversarial networks (2014)
14. Haarnoja, T., Zhou, A., Abbeel, P., Levine, S.: Soft actor-critic: off-policy maximum entropy deep reinforcement learning with a stochastic actor. In: International Conference on Machine Learning, pp. 1861–1870. PMLR (2018)
15. Hansen, N., Su, H., Wang, X.: Stabilizing deep q-learning with convnets and vision transformers under data augmentation (2021)
16. He, K., Zhang, X., Ren, S., Sun, J.: Deep residual learning for image recognition. In: Proceedings of the IEEE Conference on Computer Vision and Pattern Recognition, pp. 770–778 (2016)
17. Higgins, I., et al.: beta-VAE: learning basic visual concepts with a constrained variational framework. In: International Conference on Learning Representations (2016)
18. Ho, J., Jain, A., Abbeel, P.: Denoising diffusion probabilistic models (2020)
19. Khandelwal, A., Weihs, L., Mottaghi, R., Kembhavi, A.: Simple but effective: clip embeddings for embodied AI. In: Proceedings of the IEEE/CVF Conference on Computer Vision and Pattern Recognition, pp. 14829–14838 (2022)
20. Kingma, D.P., Welling, M.: Auto-encoding variational bayes. arXiv preprint arXiv:1312.6114 (2013)
21. Kostrikov, I., Yarats, D., Fergus, R.: Image augmentation is all you need: regularizing deep reinforcement learning from pixels. arXiv preprint arXiv:2004.13649 (2020)
22. Krizhevsky, A., Sutskever, I., Hinton, G.E.: Imagenet classification with deep convolutional neural networks. Adv. Neural Inf. Process. Syst. **25** (2012)
23. Laskin, M., Srinivas, A., Abbeel, P.: CURL: contrastive unsupervised representations for reinforcement learning. In: International Conference on Machine Learning, pp. 5639–5650. PMLR (2020)
24. Laskin, M., Lee, K., Stooke, A., Pinto, L., Abbeel, P., Srinivas, A.: Reinforcement learning with augmented data. Adv. Neural Inf. Process. Syst. **33**, 19884–19895 (2020)
25. Lillicrap, T.P., et al.: Continuous control with deep reinforcement learning. arXiv preprint arXiv:1509.02971 (2015)

26. Liu, X., et al.: Data augmentation via latent space interpolation for image classification. In: 2018 24th International Conference on Pattern Recognition (ICPR), pp. 728–733 (2018). https://doi.org/10.1109/ICPR.2018.8545506
27. Mansourifar, H., Chen, L., Shi, W.: Virtual big data for GAN based data augmentation. In: 2019 IEEE International Conference on Big Data (Big Data), pp. 1478–1487 (2019). https://doi.org/10.1109/BigData47090.2019.9006268
28. Mnih, V., et al.: Asynchronous methods for deep reinforcement learning. In: International Conference on Machine Learning, pp. 1928–1937. PMLR (2016)
29. Mnih, V., et al.: Playing Atari with deep reinforcement learning. arXiv preprint arXiv:1312.5602 (2013)
30. Moreno-Barea, F.J., Jerez, J.M., Franco, L.: Improving classification accuracy using data augmentation on small data sets. Expert Syst. Appl. **161**, 113696 (2020). https://doi.org/10.1016/j.eswa.2020.113696, https://www.sciencedirect.com/science/article/pii/S0957417420305200
31. Nair, S., Rajeswaran, A., Kumar, V., Finn, C., Gupta, A.: R3M: a universal visual representation for robot manipulation. arXiv preprint arXiv:2203.12601 (2022)
32. Pari, J., Shafiullah, N.M., Arunachalam, S.P., Pinto, L.: The surprising effectiveness of representation learning for visual imitation. arXiv preprint arXiv:2112.01511 (2021)
33. Parisi, S., Rajeswaran, A., Purushwalkam, S., Gupta, A.: The unsurprising effectiveness of pre-trained vision models for control. In: International Conference on Machine Learning, pp. 17359–17371. PMLR (2022)
34. Savva, M., et al.: Habitat: a platform for embodied AI research. In: Proceedings of the IEEE/CVF International Conference on Computer Vision, pp. 9339–9347 (2019)
35. Schulman, J., Wolski, F., Dhariwal, P., Radford, A., Klimov, O.: Proximal policy optimization algorithms. arXiv preprint arXiv:1707.06347 (2017)
36. Schwarzer, M., Anand, A., Goel, R., Hjelm, R.D., Courville, A., Bachman, P.: Data-efficient reinforcement learning with momentum predictive representations. arXiv preprint arXiv:2007.05929 **2**(3) (2020)
37. Shah, R., Kumar, V.: RRL: resnet as representation for reinforcement learning. arXiv preprint arXiv:2107.03380 (2021)
38. Shorten, C., Khoshgoftaar, T.M.: A survey on image data augmentation for deep learning. J. Big Data **6**(1), 1–48 (2019)
39. Silver, D., et al.: Mastering the game of go without human knowledge. Nature **550**(7676), 354–359 (2017)
40. Tassa, Y., et al.: Deepmind control suite (2018)
41. Vinyals, O., et al.: Grandmaster level in StarCraft II using multi-agent reinforcement learning. Nature **575**(7782), 350–354 (2019)
42. Wang, C., Luo, X., Ross, K., Li, D.: VRL3: a data-driven framework for visual deep reinforcement learning. Adv. Neural Inf. Process. Syst. **35**, 32974–32988 (2022)
43. Xiao, T., Radosavovic, I., Darrell, T., Malik, J.: Masked visual pre-training for motor control. arXiv preprint arXiv:2203.06173 (2022)
44. Yarats, D., Fergus, R., Lazaric, A., Pinto, L.: Improving sample efficiency in model-free reinforcement learning from images. arXiv preprint arXiv:2102.06281 **4**(5), 1–12 (2021)
45. Yarats, D., Fergus, R., Lazaric, A., Pinto, L.: Mastering visual continuous control: improved data-augmented reinforcement learning. arXiv preprint arXiv:2107.09645 (2021)

46. Yen-Chen, L., Zeng, A., Song, S., Isola, P., Lin, T.Y.: Learning to see before learning to act: visual pre-training for manipulation. In: 2020 IEEE International Conference on Robotics and Automation (ICRA), pp. 7286–7293. IEEE (2020)
47. Yu, T., et al.: Meta-world: a benchmark and evaluation for multi-task and meta reinforcement learning. In: Conference on Robot Learning, pp. 1094–1100. PMLR (2020)
48. Yu, X., et al.: Diffusion-based data augmentation for nuclei image segmentation (2024)
49. Yuan, Z., et al.: Pre-trained image encoder for generalizable visual reinforcement learning. Adv. Neural Inf. Process. Syst. **35**, 13022–13037 (2022)
50. Zheng, R., et al.: TACO: temporal latent action-driven contrastive loss for visual reinforcement learning. arXiv preprint arXiv:2306.13229 (2023)

# Could the Declarer Have Discarded It? Refined Anticipation of Cards in Skat

Stefan Edelkamp[1,2(✉)]

[1] Department of Theoretical Computer Science and Mathematical Logic, Faculty of Mathematics and Physics, Charles University, Prague, Czechia
edelkamp@ktiml.mff.cuni.cz

[2] Computer Science Department, Artificial Intelligence Center, Faculty of Electrical Engineering, Czech Technical University, Prague, Czechia
edelkste@fel.cvut.cz

**Abstract.** In this paper we refine the concept of anticipation within a card game, taking the Nullspiel in Skat as a running example. We generate the belief space of all distributions of cards according to the assumption on plausible play of the declarer. Using a selection of open-card searches in a two-stage knowledge filtering we improve the play of the two opponents to find forced wins. In the voting scheme to combine the open-card analyses of the possible worlds, we additionally use the search tree size and depth to prefer short proofs. In hundreds of thousands human ouvert games replayed by our AIs over 99% matched the predictions of the open-card solver, with only 0.21% of games known lost for the declarer were not won by the AIs, both trademarks outperforming human play.

**Keywords:** Game Play · Uncertainty · Search

## 1 Introduction

Game playing research in AI has shown some of its biggest success stories in beating top-playing humans in table-top games like Backgammon, Checkers, Chess, Connect-Four, Go, Scrabble, Shogi, or Stratego [1,16–20,22].

Despite visible progress the situation in table-top card games, however, is less clear. While Heads-Up Limit Texas Hold'Em Poker is weakly solved for two players [2] for Full-Scale Poker this question is still open [21]. Multiplayer Poker bots [3] and Mahjong bots [14] often use machine learning-based approaches that, while not optimal, proved to be highly effective. Bots aimed at solving multiway games stray away from theoretic, symbolic models in favor of exploitative, neural models.

In complex multiplayer trick-taking card games like Bridge [9,15] only particular games have been played at superhuman performance. In Skat there are Full-Scale players occasionally beating top humans [7,8].


A. Hotho and S. Rudolph (Eds.): KI 2024, LNAI 14992, pp. 60–72, 2024.
https://doi.org/10.1007/978-3-031-70893-0_5

The card game Skat [10–12][1] is the national card game of Germany and over 200 years old. This trick-taking game of strategy is played in leagues across the world. Three players take turns placing bids on how many points they expect to be able to win in a round's tricks and then facing off against the round's declarer. The Nullspiel is a misere variant, where the declarer must not get any trick. In total, 32 cards are used for playing Skat. These cards include the Ace, King, Queen, and Jack (Ass, König, Dame and Bube in German) of each of the four suits as well as numbers 10 through 7 in those suits. In card decks that operate with the French suits, these four suits are clubs ♣, spades ♠, hearts ♡, and diamonds ♢. Each player is dealt 10 cards to begin (totaling 30 cards) with the 2 remaining cards being placed face down to be used for the skat. Players in the game have names for their roles in gameplay: the forehand (player to the dealer's left), middlehand (player to the forehand's left), and rearhand (player to the middlehand's left). The dealer in the game is first chosen at random and then the responsibility for dealing rotates clockwise.

In this paper we consider improved anticipatory reasoning in the game of Skat for constraining the space of possible beliefs. We also refine the voting mechanism among the recommended cards of the open-card solving. While also applicable to Trump games, we consider the case of the Nullspiel, especially the Ouvert variant. We start with an example, review known concepts, then, going into refinements. Experimental results show that the improved player plays close to optimal. According to the outcome of the open-card solver on the true deal in the Nullspiel we obtain 0.77% off the prediction for all games and 0.21% off for all lost games.

## 2 Motivation

While game play of humans in Skat is rather competitive, the general observation is that many players are not aware of winning probabilities. Even professional players are sometimes mislead. As an example, Emanuel Lasker, the one and only German Chess champion, published some strategies about Skat [12,13]. In addition to the suit and grand games he also dealt extensively with Null games. He first pointed out that in the middle- or rearhand suits in which where the lower card values are played with only a single void, e.g. 7, 7 + 9, 7 + 9 + J or 7 + 9 + J + K, do not represent a risk. The same applies if individual values of such a sequence are even lower. Weak points arise with larger gaps between the value, such as 7–10, for example. If a player holds such a gap, his Nullspiel can be brought down as soon as one of the two opponents in the concerned suit is blank, either already from the start of the game or after discarding during play.

Lasker assessed the risks of weak points in a Nullspiel and arrived at the following gradation from minor to greater: 1. An ace in group of 4 like 7 + 8 + 10 + A or 7 + 9 + J + A; 2. A blank 8; 3. Cards 7 + 10; 4a) Cards 8 + 9; 4b) A Q(ueen) in group of 3; 5. Cards 8 + 10; 6. A blank 9; 7. A K(ing) in group of 3. This risk ranking plays an important role in deciding which cards the declarer puts into the

[1] For the rules we refer the reader to https://www.pagat.com/schafkopf/skat.html.

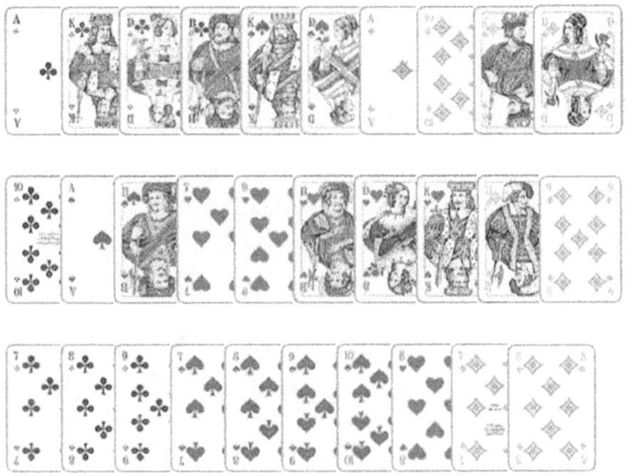

**Fig. 1.** Example case. Bottom player is playing Null Ouvert. Top player to issue

Skat. In doing so he should obviously prioritize the comparatively highest risks with priority, i.e. from bottom to top. In his book, Gößl [10] subjected Lasker's sequence to an exact verification on the basis of almost three million Null games that exhibit one of the eight weak points. This confirmed Lasker's ranking except with the two categories 4a and 6. Both have a higher risk than would result from Lasker's ranking would result from Lasker's order, so that overall the following risk ranking from low to high: $1 \to 2 \to 3 \to 4b \to 4a \to 5 \to 7 \to 6$.

Consider a small example to understand the concept of individual knowledge of the opponent players due to anticipation of the declarer's skat putting. In Fig. 1 a Null Ouvert is played, with forehand sitting at the top, and rearhand sitting at the bottom position. The result of the reasoning of the two opponents given that both cannot see $10+2$ cards, is illustrated in Fig. 2. It shows the cards that, when residing in her/his hand at the time of skat putting, would yield a 100% winrate for the declarer. Hence, in view of the forehand player the following six cards are not in the skat: $\clubsuit 10$; $\spadesuit J \spadesuit A$; $\heartsuit 7$; $\diamondsuit 9$; and $\diamondsuit J$. In view of middlehand player the following cards are not in the skat: $\clubsuit J, \clubsuit Q, \clubsuit K$; $\spadesuit D, \spadesuit K$; and $\diamondsuit 10$. As they are not in the Skat, in turn, these cards must be located in the hand of the partner player, reducing the uncertainty of the cards.

Therefore, our skat algorithm maintains a partitioning of the remaining cards hidden to a player into known and unknown cards, that is updated with each card that is played. Note that the epistemic initialization and progression of knowledge for each player on the location of the cards of the other player is central not only to Null but also to all games.

This plausibility analysis of good cards the player cannot have put considers the dropping of only one skat card at a time. There are, however, refined skat putting strategies of the declarer that include both cards in combination.

**Fig. 2.** Cards not in the skat for top and middle player

## 3 Basic Functionality

For the Nullspiel, Emanuel Lasker [12] suggested to combine statistics on winning probabilities of a hand $h$ in each suit as follows

$$P_w(h) = \prod_{suit \in \{\clubsuit,\spadesuit,\heartsuit,\diamondsuit\}} P_w(h, suit),$$

with suit probabilities $P_w(h, suit)$ for different variants of the Nullspiel. For other Skat variants, different ways to elicit winning probabilities from expert games based on feature selection and larger lookup tables apply [5].

In the Nullspiel the card ordering is A, K, Q, J, T, 9, 8, and 7. The declarer looses if s/he makes. In contrast, s/he can win by force if the hand is *safe.*

**Definition 1 (Safe Card).** *A declarer's card c is* safe, *if all gaps g in its suit with value lower than c are supported by at least the matching number of cards below g, with a special case for the declarer's turn in the first trick, where an extra support card is needed. The declarer's hand h is* safe *if all cards c in h are safe.*

Refined definitions of a safe hand accommodate the cards that have being played. Once the skat has been taken by the declarer, there are $\binom{12}{2} = 66$ options for putting it. Of course, not all are plausible.

A strategy to select a skat optimizes the *dropping gain.*

**Definition 2 (Dropping Gain).** *The* dropping gain *is the change in the winning probability when removing a card from the hand. If h is the hand before the drop and h′ is reduced hand after the card drop in suit suit, then we have*

$$drop(h, h') = P_w(h) - P_w(h') = P_w(h, suit) - P_w(h', suit).$$

In other words, for the choice of skat cards, we prefer the ones that improve the winning probability the most.

In imperfect information games, the belief state space is the set of possible states the game can be in. In the worst case for $|S|$ as the number of states in $S$, the belief state space can grow exponentially in $|S|$. In Ouvert games, however, the belief state space (in view of the opponents) is much smaller. A state includes all the hands of the player, the skat, the issuer, the table cards, and even the history of the played cards. It is a game of almost full information, so that exhaustive search is more effective. To reduce the uncertainty for the

```
decide_beliefs(card hidden, known, unknown, myhand, declarer, played)
  skat = partner = 0;
  for (i=0;i<32;i++) vote[i] = 0;
  maxvote = 0;
  while (unknown)
    card1 = choose(unknown);
    skat |= (1 << card1);
    unknown &= ~(1 << card1);
    rest_cards = unknown;
    while (rest)
      card2 = choose(rest);
      skat |= (1 << card2);
      rest_cards &= ~(1 << card2);
      partner = hidden & ~skat;
      if (issuer == player1)
        if (!opencardsolver(declarer,myhand,partner))
          vote[proposed_card]++;
      if (issuer == player2)
        if (!opencardsolver(declarer,partner,myhand))
          vote[proposed_card]++;
      skat &= ~(1 << card2);
    skat &= ~(1 << card1);
  for (i=0;i<32;i++)
    if (maxvote < vote[i]) maxvote = vote[i];
```

**Fig. 3.** Selecting an opponents' card in Ouvert game. The c-type implementation assumes a bitset representation for each selection of cards

opponents in the number of unknown hands down from $\binom{12}{10} = 66$ possible cards, we determine *good* cards that –under reasonable play– the declarer should not have put.

**Definition 3 (Good Card).** *A card c in hand h is* good *if the dropping gain is smaller than the highest dropping gain of any other card, i.e.,*

$$good(c) \text{ iff } drop(h, h\backslash\{c\}) < \max_{c' \in h} drop(h, h\backslash\{c'\}).$$

Once a card has being played, it is no longer unknown. For the remaining choices in the unknown cards, we call the open card solver.

In Fig. 3 we revisit the belief-space voting scheme from [4] that serves as a recommender to decide the next card to be played.

In case of an unambiguous majority vote, we suggest the card to play. In the unlikely case that the vote remains ambiguous, we use the rules from Nullspiel as a tie-breaker.

## 4 Algorithmic Improvements

We implemented a number of improvements to the above scheme.

### 4.1 Thresholds

As the definition of good cards does not apply any thresholds for the derivation of the dropping gain, the condition of skat cards that could have been put is too restrictive.

Therefore, we decided to include a hyperparameter on how far the dropping gain can be away from the best one. If the dropping gain between the best and the worst card is larger than $x = 15\%$, then we assume that this card may not have been put the declarer, and, therefore, is located in the partners hand.

**Definition 4 (Plausible Card).** *Let $x \in [0..1]$ be a probability threshold parameter. A card c in hand h is x-*plausible *if the dropping gain is not deviating more than x wrt. the highest dropping gain of any card s/he has shown, i.e.,*

$$plausible(c) \text{ iff } drop(h, h\backslash\{c\}) - \max_{c' \in h} drop(h, h\backslash\{c'\}) < x.$$

### 4.2 All-Card Analysis

In the approach of [4] for each open-card server run in each world of the belief space, there is only a single best card that is being proposed and subsequently processed. While this can be faster in cases where only one winning card is needed, it limits the fusion process across the different worlds of the voting, given that there might be several other cards that are equally winning and, thus, should also get votes. For a more reliable estimation on the number of worlds won by each card, we extended the search to all possible cards to be played. In other words, we compute the full statistics $win(c, s)$ for each card $c$ denoting all the belief states $s$ it is winning in.

**Definition 5 (Full Solution Stats).** *Let* ocs *be an open-card solver. For each card c, the full solution statistics $win(c, s)$ for a belief s denotes in how many worlds the card c in $(h_0, h_1, h_2, skat, played) \in s$ is* winning, *i.e.,*

$$win(c, s) = \sum_{(h_0,h_1,h_2,skat,played) \in s} [\texttt{ocs}(c, h_0, h_1, h_2, skat, played) = 1]$$

### 4.3 Game Tree Voting

Instead of giving all cards that win one vote, we decided to include the size of the proof tree to prioritize short solutions. This analysis of the search tree is generated for each card individually. We tried two criteria: a) the number of nodes in the search tree and b) the maximum search tree depth. The best results were obtained by combining a) and b) by adding two to the power of the search tree depth for each search tree node. Of course the knowledge of cards of played cards and prior knowledge can be used to constrain the belief.

**Definition 6 (Game-Tree Stats).** *Let* ocs *be any open-card solver. For each card c, the game-tree statistics $size(c, s)$ $depth(c, s)$ and $mix(c, s)$ for a belief s denotes the combined proof tree efforts of the card c in $(h_0, h_1, h_2, skat, played) \in$*

**Table 1.** Refined winning probability table for one suit in the Nullouvert Game

| ID | Group | mean value | with 0 in skat | with 1 in skat | with 2 in skat |
|---|---|---|---|---|---|
| 1 | ........ | 100.00% | 100.00% | 100.00% | 100.00% |
| 2 | .......7 | 100.00% | 100.00% | 100.00% | 100.00% |
| 3 | ......8. | 93.39% | 96.97% | 93.49% | 87.88% |
| 4 | ......87 | 100.00% | 100.00% | 100.00% | 100.00% |
| 5 | .....9.. | 52.42% | 48.42% | 54.76% | 51.67% |
| 6 | .....9.7 | 100.00% | 100.00% | 100.00% | 100.00% |
| 7 | .....98. | 59.78% | 69.87% | 56.30% | 45.41% |
| 8 | .....987 | 100.00% | 100.00% | 100.00% | 100.00% |
| 9 | ....T... | 21.00% | 27.85% | 24.31% | 31.28% |
| 10 | ....T..7 | 72.07% | 79.81% | 70.16% | 58.93% |
| ⋮ | ⋮ | ⋮ | ⋮ | ⋮ | ⋮ |
| 248 | AKQJ.987 | 100.00% | 100.00% | 100.00% | – |
| 249 | AKQJT... | 0.00% | 0.00% | 0.00% | 0.00% |
| 250 | AKQJT..7 | 50.84% | 49.84% | 50.84% | 100.00% |
| 251 | AKQJT.8. | 16.00% | 16.00% | 16.00% | 100.00% |
| 252 | AKQJT.87 | 100.00% | 100.00% | 100.00% | – |
| 253 | AKQJT9.. | 0.00% | 0.00% | 0.00% | 0.00% |
| 254 | AKQJT9.7 | 100.00% | 100.00% | 100.00% | – |
| 255 | AKQJT98. | 0.00% | 0.00% | 0.00% | – |
| 256 | AKQJT987 | 100.00% | – | – | – |

*s is* winning, *i.e.*,

$$depth(c,s) = \sum_{(h_0,h_1,h_2,skat,played)\in s} [maxdepth(\texttt{ocs}(c,h_0,h_1,h_2,skat,played))]$$

$$size(c,s) = \sum_{(h_0,h_1,h_2,skat,played)\in s} [size(\texttt{ocs}(c,h_0,h_1,h_2,skat,played))]$$

$$mix(c,s) = \sum_{(h_0,h_1,h_2,skat,played)\in s} [mix(\texttt{ocs}(c,h_0,h_1,h_2,skat,played))]$$

*where maxdepth, size, and mix are the different criteria applied to the proof tree of the open-card solver for one belief of card distributions for the current hands.*

### 4.4 Integrating the Skat Cards

The approach in [6] determines the winning probability for each suit, and pays no attention the cards already put. In the following we aim, however, at a refined table including the Skat being put (Table 1). These tables have been automatically generated by counting the ratio of games being won, in a set of millions of games [10].

This way we arrive a refined winning probability $P_w(h, suit, n)$, where $n$ denotes the number of cards in that suit that are put into the skat.

**Definition 7 (Refined Probability).** *For hand $h$ and skat cards $skat(\clubsuit)$, $skat(\spadesuit)$, $skat(\heartsuit)$, and $skat(\diamondsuit)$ and Null game type $t$ we compute the refined probability of the skat being put as*

$$P_{t,w}(h) = \prod_{suit \in \{\clubsuit, \spadesuit, \heartsuit, \diamondsuit\}} P_{t,w}(h, suit, |skat(suit)|),$$

*where $P_{t,w}(h, s, n)$ is the estimated winning probability of a pattern in a given suit with $n$ skat cards being put.*

Note that we denote only the number of cards in a suit put into the skat, and do not denote the exact cards being put, as we assume that the a plausible putting strategy of the declarer will use the highest ones. This table is used for the declarer in both bidding and skat putting.

In the evaluation of the belief state (shown in Fig. 4, we apply this information as a filter for the opponent player, anticipating and filtering the cards that were unlikely to be in the skat. A $32 \times 32$ matrix $F$ was derived with $F(i, j)$ denoting the probability of the combined skat $(i, j)$ being put by the declarer. We also use a Boolean Matrix $K(i, j)$ to filter implausible skats. The matrix $K$ was used as a filter before the open-card analysis was run. In each analysis of the belief space only the probable skats were analyzed, and again we applied a threshold of 10% deviation allowed wrt. the best possible value in the matrix.

## 5 Experiments

We implemented the improvements to the top-level Skat player [6] that was reported to have won a match against a top player in a match of 10 series of 36 games. We evaluated the player running three independent AIs for selections of hundred of thousands games. The running time for the full rerun of 100,000 games (on 10 cores of an AMD EPYC 7702P 64-Core Processor) was less than a day for Trump games (Grand and Suit) and less than 20 min for Null and Nullouvert games.

### 5.1 General Performance

We first look the effects of refined Skat selection. The playing strength of human and AI players in the Tables 2 and 3 for 100K grand games are comparable.

In 100K Null games for enforced human skat shown in Table 4 we obtain a slightly inferior play of the AI opponents compare to humans.

In Table 5 we also measured the performance of the player in 100K Suit games for enforced human skats. This data is mainly to show the current playing strength of the Full-Scale player.

```
decide_beliefs(card unknown, declarer, myhand, all, played)
 skat = 0, partner = won = maxvote = 0;
 u = unknown;
 while (unknown)
   card1 = select(unknown);
   skat |= (1 << card1);
   unknown &= ~(1 << card1);
   rest = unknown;
   while (rest)
     card2 = select(rest);
     rest &= ~(1 << card2);
     best = best_skat(declarer | (1 << card1) | (1 << card2));
     F[card1][card2] = prob_skat(declarer,card1,card2);
     if (best_skat - F[card1][card2] > 0.10)
       K[card1][card2] = true;
       continue;
     skat |= (1 << card2);
     partner = all & ~(skat);
     if (issuer == 1)
       won = fullrun(declarer,myhand,partner,played);
       if (!won && proposed)
         set = proposed_set;
         while(set) c = select(set); vote[c]++; set &= ~(1<<c);
         short[proposed]++;
     if (issuer == 2)
       won = fullrun(declarer,partner,myhand,played);
       if (!won && proposed)
         set = proposed_set;
         while(set) c = select(set); vote[c]++; set &= ~(1<<c);
         short[proposed]++;
     skat &= ~(1 << card2);
   skat &= ~(1 << card1);
 potcards = potworlds = 0;
 for (int j=0;j<32;j++)
   if (short[j] > 0)  potcards |= (1<<j); potworlds += short[j]
 for(int i=0;i<32;i++)
   if (maxvote < vote[i]) maxvote = vote[i]
 for(int j=0;j<32;j++)
   if (issuer == 1 || issuer == 2)
     if (vote[j] == maxvote && maxvote > 0)
       if (potworlds == vote[j])
         cost = 0; unknown = u
         while (unknown)
           card1 =  select(unknown)
           skat |= (1 << card1)
           unknown &= ~(1 << card1)
           rest = unknown;
           while (rest)
             card2 =  select(rest)
             rest &= ~(1 << card2)
             if (K[card1][card2]) continue
             skat |= (1 << card2)
             new_hand = all & ~(skat)
             if (issuer == 1) cost += proofcost(j,declarer,myhand,partner,played)
             else cost += proofcost(j,declarer,partner,myhand,played)
             skat &= ~(1 << card2)
         proofcost[j] = cost
 maxproofcost = infinity
 for(int j=0;j<32;j++)
   if (issuer == 1 || issuer == 2)
     if (vote[j] == maxvote && maxvote > 0)
       if (potworlds == vote[j])
         if (proofcost[j] > 0 && proofcost[j] < maxproofcost)
           maxvote = vote[j]
           maxproofcost = proofcost[j]
```

**Fig. 4.** Refined selection of an opponents' card in Null Ouvert game

**Table 2.** Replay of 100K human Grand games

| Pos | win | lost | %won | score lost | score player | score tot | per game |
|---|---|---|---|---|---|---|---|
| 1 | 38,024 | 2,928 | 92.9% | −519,456 | 4,945,420 | 4,425,976 | |
| 2 | 26,767 | 1,992 | 93.1% | −359,856 | 3,561,686 | 3,201,830 | |
| 3 | 28,478 | 1,809 | 94.0% | −325,266 | 3,800,164 | 3,474,898 | |
| tot | 93,269 | 6,729 | 93.3% | −1,204,578 | 12,307,282 | 11,102,704 | 111.03 |

**Table 3.** Replay of 100K human Null games with three AIs, enforced human skat

| Pos | win | lost | %won | score lost | score player | score tot | per game |
|---|---|---|---|---|---|---|---|
| 1 | 37,419 | 3,533 | 91.4% | −611,530 | 4,850,502 | 4,238,972 | |
| 2 | 26,779 | 1,980 | 93.1% | −355,128 | 3,559,070 | 3,203,942 | |
| 3 | 28,484 | 1,803 | 94.0% | −323,814 | 3,804,856 | 3,481,042 | |
| tot | 92,682 | 7,316 | 92.7% | −1,290,472 | 12,214,428 | 10,923,956 | 109.24 |

**Table 4.** Comparing Human and AI performance in 100K Nullgame for an enforced human skat

| Won | Human | %Won | AI | %Won |
|---|---|---|---|---|
| 1 | 61,853 | 61.9% | 63,726 | 63.7% |
| 0 | 38,147 | 28.1% | 36,274 | 26.3% |

Putting the skat with the AI increased the winning ratios significantly, outperforming humans. In Grand games the winning probability increased from 92.7% to 97.6% and for Nullgames it increased from 63.7% to 67.4% in Table 6. In 10% of the games the AI puts a completely different skat; in 54% of the game the skat was the same, otherwise there was an overlap of one card.

### 5.2 Case Study: Null Ouvert

The Null Ouvert is subject to the improvements by this paper's contributions. In 100K games we measured the number of wins of a human, an open-card solver that knows the skat cards.

In Table 7 we review the results of [6], where more than 10% of the games indicated by open card solver to be lost were lost. The result matched the open card prediction in about 75% of the cases.

In Table 8 we see that with $126 + 77 = 213$ of $100,000$ games only 0.21% were played different to the predicted loss of the open card solver that due to the uncertainty of hidden skat cards, the AI opponent players could not find. The total number of games that were played different to the prediction was 775 games or 0.775%.

**Table 5.** Replaying 100K human Suit games, enforced human skat

| WonHuman | OpenCardSolver | WonAI | Number | % |
|---|---|---|---|---|
| 0 | 0 | 0 | 11,214 | 11.2% |
| 0 | 0 | 1 | 5,369 | 5.3% |
| 0 | 1 | 0 | 520 | 0.5% |
| 0 | 1 | 1 | 2,573 | 2.5% |
| 1 | 0 | 0 | 7,157 | 7.1% |
| 1 | 0 | 1 | 11,035 | 11% |
| 1 | 1 | 0 | 1,718 | 1.7% |
| 1 | 1 | 1 | 60,406 | 60.4% |

**Table 6.** Replay of 100K mixed games with three AIs and AI skat selection

| Game | Number | Percentage | %Won |
|---|---|---|---|
| Suit | 23,796 | 24.0% | 87.4% |
| Grand | 71,577 | 72.1% | 97.6% |
| Null | 1,756 | 1.8% | 67.4% |
| Null Ouvert | 2,152 | 2.2% | 83.2% |
| | 99281 | 100.0% | 94,3% |

**Table 7.** Results in Null Ouvert Games of [6]

| OpenCardSolver | WonHuman | WonAI | Number | % |
|---|---|---|---|---|
| 0 | 0 | 0 | 186,656 | 64.3% |
| 0 | 0 | 1 | 56,709 | 19.5% |
| 0 | 1 | 0 | 30,537 | 10.5% |
| 0 | 1 | 1 | 16,182 | 5.6% |

**Table 8.** Replaying 100K human Nullouvert games

| WonHuman | OpenCardSolver | WonAI | Number | % |
|---|---|---|---|---|
| 0 | 0 | 0 | 9,427 | 9.4% |
| 0 | 0 | 1 | 136 | 0.13% |
| 0 | 1 | 0 | 272 | 0.27% |
| 0 | 1 | 1 | 759 | 0.75% |
| 1 | 0 | 0 | 1,915 | 1.9% |
| 1 | 0 | 1 | 77 | 0.07% |
| 1 | 1 | 0 | 290 | 0.29% |
| 1 | 1 | 1 | 87,124 | 87.12% |

## 6 Conclusion

Skat AIs are currently operating at top human playing strength, and professional players are occasionally be beaten [7]. To consistently beat top players, however, further improvements are needed.

What distinguished the strong AI from human top play is the estimation of accurate winning probabilities that is used for bidding, game selection and skat putting. For the case of the Nullspiel in this paper we have shown that winning probabilities and refined skat putting can be used for improved opponent play.

To this end, in this paper we have introduced improved reasoning based on the anticipation of the skat discarding strategy of the declarer. We extended the concept of card dropping gain to a knowledge filter. The Nullouvert variant was taken as our case study, given that it has a smaller Belief space, but the concept of plausible declarer behavior together with knowledge filtering and progression is applicable to all other Skat variants aswell. The voting was extended to prefer short proof trees. The experiments showed that there is little room left to perfect play of an open-card solver.

For future work to improve the player beyond human playing strength, we see different avenues. One is opponent model and to apply Nash theory of mixed strategies for critical actions like raising with a 10, or playing a low opening card. The other is to apply expert knowledge-enriched deep neural networks using this player for providing large-scale databases of training examples for bootstrapping. Both aspects might also be used to increase characteristics of the AI player, playing more or less aggressive, e.g., in the bidding and game selection stages.

**Acknowledgements.** I want to thank the expert player Rainer Gößl for the deep insights in Skat, who I frequently consulted during my conceptualization and implementation. I also thank Daniel Schäfer, Stefan Meinel, and Helmut Auer for their collaboration. The presented work has been supported by the Czech Science Foundation (GAČR) under the research project number 24-12046S.

## References

1. Allis, L.V.: A knowledge-based approach to connect-four. The game is solved: white wins. Master's thesis, Vrije Universiteit, The Netherlands (1998)
2. Bowling, M., Burch, N., Johanson, M., Tammelin, O.: Heads-up limit hold'em poker is solved. Commun. ACM **60**(11), 81–88 (2017)
3. Brown, N., Sandholm, T.: Superhuman AI for multiplayer poker. Science **365**(6456), 885–890 (2019)
4. Edelkamp, S.: Challenging human supremacy in Skat. In: Symposium on Combinatorial Search (SOCS), pp. 52–60 (2019)
5. Edelkamp, S.: Knowledge-based paranoia search. In: Conference on Games (CoG), pp. 1–8. IEEE (2021)
6. Edelkamp, S.: On the power of refined Skat selection. CoRR, abs/2104.02997 (2021)
7. Edelkamp, S.: Improving computer play in Skat with hope cards. In: Browne, C., Kishimoto, A., Schaeffer, J. (eds.) CG 2022. LNCS, vol. 13865, pp. 133–145. Springer, Cham (2022). https://doi.org/10.1007/978-3-031-34017-8_12

8. Furtak, T.M.: Symmetries and search in trick-taking card games. Ph.D. thesis, University of Alberta (2013)
9. Ginsberg, M.: Step toward an expert-level bridge-playing program. In: IJCAI, pp. 584–589 (1999)
10. Gößl, R.: Der Skatfuchs - Gewinnen im Skatspiel mit Mathematische Methoden. Dämmig, Chemnitz, Available from DSKV Altenburg (2019)
11. Grandmontagne, S.: Meisterhaft Skat spielen. Krüger Druck+Verlag (2005)
12. Lasker, E.: Das verständige Kartenspiel. August Scherl Verlag, Berlin (1929)
13. Lasker, E.: Strategie der Spiele - Skat. August Scherl Verlag, Berlin (1938)
14. Li, J., Wu, S., Fu, H., Fu, Q., Zhao, E., Xing, J.: Speedup training artificial intelligence for mahjong via reward variance reduction. In: IEEE Conference on Games (CoG), pp. 345–352 (2022)
15. Li, J., Zanuttini, B., Cazenave, T., Ventos, V.: Generalisation of alpha-beta search for AND-OR graphs with partially ordered values. In: IJCAI, pp. 4769–4775 (2022)
16. Perolat, J., et al.: Mastering the game of Stratego with model-free multiagent reinforcement learning. Science **378**(6623), 990–996 (2022)
17. Schaeffer, J., et al.: Solving checkers. In: IJCAI, pp. 292–297 (2005)
18. Sheppard, B.: World-championship-caliber Scrabble. Artif. Intell. **12**, 241–275 (2002)
19. Silver, D., et al.: Mastering the game of Go with deep neural networks and tree search. Nature **529**, 484 (2016)
20. Silver, D., et al.: Mastering Chess and Shogi by self-play with a general reinforcement learning algorithm. Technical report arXiv:1712.01815 (2017)
21. Sonawane, P., Chheda, A.: A survey on game theory optimal poker (2024)
22. Tesauro, G.: Temporal difference learning and TD-Gammon. Commun. ACM **38**(3), 58–68 (1995)

# A Framework for General Trick-Taking Card Games

Stefan Edelkamp[1,2(✉)]

[1] Department of Theoretical Computer Science and Mathematical Logic, Faculty of Mathematics and Physics, Charles University in Prague, Staré Město, Czech Republic
edelkamp@ktiml.mff.cuni.cz

[2] Computer Science Department, Artificial Intelligence Center, Faculty of Electrical Engineering, Czech Technical University in Prague, Prague, Czech Republic
edelkste@fel.cvut.cz

**Abstract.** Inspired by recent advances in Computer Skat and Bridge, this paper investigates automated play for several other trick-taking card games like Belote, Tarot, Doppelkopf, Spades, Hearts, Euchre, and Schafkopf. We present a general framework that instantiates all these card games with deals from their respective decks based on a programming interface to play them interactively at a human-adequate speed and level. We have included bidding, team building, and game selection, as well as general and specialized card recommenders applicable for the different stages of trick-taking. We study the impact of expert rules for enhanced play. The AIs are evaluated in different variants and against a general card player that lacks expert rules.

## 1 Introduction

While in most fully-observable, complete-information games modern AIs play better than humans [1,3,16,24–26], in incomplete-information games engines long trailed human playing strength. Recent research, however, has shown significant breakthroughs in playing the games of Poker [5] and Stratego [23], outperforming even top players.

We are interested in full-scale general multiplayer card games [4] like Belote, Tarot, Doppelkopf, Spades, Hearts, Euchre, or Schafkopf. Besides general-purpose exploration, additional to the mere rule of the games, we additionally allow the programmer to include of a limited form of expert knowledge to improve the play.

While general game playing competitions showed considerable progress [14] in AI, for card game playing, the input language yielded node generators that were too slow. General game playing alternatives like Google DeepMind's OpenSpiel are usally faster but for multiplayer trick-taking card games, these efficiencies might still not be enough, so that we propose a framework based on the bitvector set encodings for the cards.

For two-player incomplete-information games, Kubicek et al. [19] contributed refinements of DeepNash [23], and applied them in OpenSpiel to the card game Goofspiel (alias Game of Pure Strategy), where the complexity with $13! = 6,22 \cdot 10^9$ different games is, however, limited.


A. Hotho and S. Rudolph (Eds.): KI 2024, LNAI 14992, pp. 73–85, 2024.
https://doi.org/10.1007/978-3-031-70893-0_6

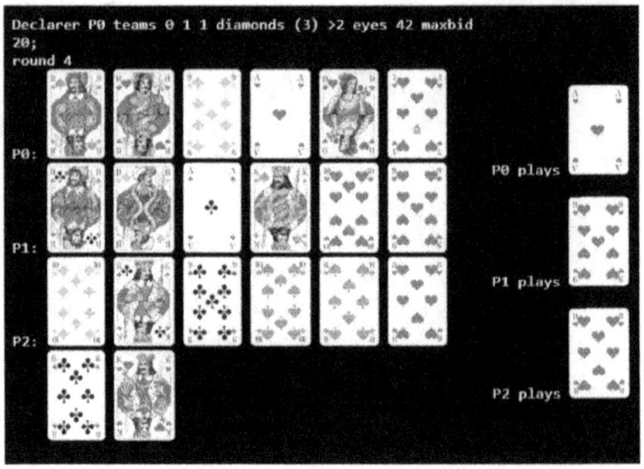


**Fig. 1.** Snapshot in a trick-taking card game. Our code support interactive text console with colored Unicode characters that we postprocess and substitute with card images.

This work is inspired by recent advances in playing a) Bridge, where in a restricted setting an AI haver shown superhuman performance [6], and b) Skat, where an AI was able to beat a top player in a match of $10 \times 36 = 360$ games [12]. To generalize and transfer these findings in the games of Skat and Bridge, this work additionally looks at multiplayer trick-taking games like Belote, Euchre, Bridge, Tarot, Schafkopf, Doppelkopf (two variants), and Spades. Figure 1 provides an example snapshot of such a trick-taking game.

Given that based on the Skat engine [12] we are developing a mobile application, the motivation for this work is to provide multi- and single-player AI prototypes for other games, too. This AI that we have selected as a starting point for our research, refers to a domain-dependent implementation with several statistical tables elicited from millions of top human games together with domain expert rules [8].

Nonetheless, the core of the AI includes several novel domain-independent algorithmic components, such as winning probabilities for extracted winning features [8]; camouflage cards to bluff the opponents, and playing conventions to transfer knowledge between cooperating opponents [11]; move-ordering techniques for sorting the search-tree traversal [12]; pruning equivalent cards, also known as partition search [17] knowledge vectors for belief-set enumeration [10]; bitvector set encoding for representing and exploring sets of cards [20], efficient open-card solvers with transposition tables [21]; voting scheme, for efficient endgame play [10] paranoia search variants, for finding forced wins in incomplete information games [11]; bidding strategies [18], and hope cards, which help to win in winning worlds [12]. Even Poker-style opponent models [2] are under consideration that try to exploit weaknesses and playing habits.

We implemented many of these advances for playing further card games. Due to the lack of a larger body of high-quality online games, tables for estimating winning probabilities for game selection are often not available. While we have summarized these initial stages into general building blocks, they are not the subject to this research, which is primarily concerned with evaluating general strategies for the trick-taking stage.

The main contribution of this paper is an efficient general card playing framework that leads to an advanced machine play for several trick-taking games with limited implementation overhead, and an ensemble of card recommenders. From our work

**Table 1.** Estimated number of deals in different card games; neglecting the position of the issuer.

| Game | Accurate Size | Approximate Size |
|---|---|---|
| Euchre | $\binom{24}{5} \cdot \binom{19}{5} \cdot \binom{14}{5} \cdot \binom{9}{5}$ | $\approx 1.24 \cdot 10^{14}$ |
| Skat | $\binom{32}{10} \cdot \binom{22}{10} \cdot \binom{12}{10}$ | $\approx 2.75 \cdot 10^{15}$ |
| Belote/Schafkopf | $\binom{32}{8} \cdot \binom{24}{8} \cdot \binom{16}{8} \cdot \binom{8}{8}$ | $\approx 9.95 \cdot 10^{16}$ |
| Bridge/Spades | $\binom{52}{13} \cdot \binom{39}{13} \cdot \binom{26}{13} \cdot \binom{13}{13}$ | $\approx 5.36 \cdot 10^{28}$ |
| Doppelkopf without 9 | $\binom{40}{10} \cdot \binom{30}{10} \cdot \binom{20}{10} \cdot \binom{10}{10} / 2^{20}$ | $\approx 4.48 \cdot 10^{15}$ |
| Doppelkopf with 9 | $\binom{48}{12} \cdot \binom{36}{12} \cdot \binom{24}{12} \cdot \binom{12}{12} / 2^{24}$ | $\approx 1.08 \cdot 10^{18}$ |
| Tarot | $\binom{78}{18} \cdot \binom{60}{18} \cdot \binom{42}{18} \cdot \binom{24}{18}$ | $\approx 3.84 \cdot 10^{80}$ |

with professional players, we know there are often subtle aspects of games that are not well covered by fully automated players. To allow coding of elicited knowledge of the domain experts we provide one additional programming interface, so that the play for the very first few tricks can be enhanced.

The work is structured as follows. First, we offer preliminaries on trick-taking in card games. Next, we describe the efficiency of data structures, including the card set representations, team building, and knowledge refinements. Afterwards present the general card game driver and search tree exploration in more detail. Then, in an ensemble of experts, we discuss the interplay of card recommendation modules and the degree of domain-independence achieved. The experiments show the efficiency of the framework across all of the games in self- and interactive play.

## 2 Preliminaries

For every deal, after bidding and game selection in the Skat and Bridge AIs there are three active players, divided into two cooperating opponents and one declarer (who may also play the *dummy*). The goal of this research is to extend the focus to a wider range of games with more players. Therefore, we generalized the implementation to a number of different games, including Belote (French national game), Schafkopf (German national game), Euchre (often played in England) Tarot (four-player game variant), Doppelkopf (several variants, including ones with double decks of $2 \times 20$ cards and $2 \times 24$ cards), and Spades (often played in the US). We also have implemented Hearts (also often played in the US), with each player being solist against three opponents. Note that the framework does not yet support non-trick-taking games with a Talon like Rommee, Canasta or Mahjong.

Let $p$ be the number of players, $n$ be the size of the deck, then $\lfloor n/p \rfloor$ is the number of cards per hand. Neglecting the players' turn, the number of deals is $\prod_{i=1}^{p} \binom{n-i\cdot\lfloor n/p \rfloor}{\lfloor n/p \rfloor}$. Table 1 displays these numbers for the various card games we consider. The number of deals is, of course, only a very small fraction of the number of possible game plays.

Introducing the rules for all these games would go beyond the scope of this paper[1].

**Definition 1 (Trick-Taking Game).** *A trick-taking card game* $\mathcal{G} = \langle \mathcal{C}, p, c, r \rangle$ *of* $p$ *players played with a deck of* $n = |\mathcal{C}|$ *cards in (multi)set* $\mathcal{C}$ *is played in several rounds,*

[1] See https://www.pagat.com and https://gathertogethergames.com.

*called tricks* $t \in T \subseteq \mathcal{C}^p$*, which are each evaluated to determine a* winner: $T \rightarrow \{0,1\}$ *of that trick. In the view of each players, there are two teams of players, so that membership to a team can be realized as mapping* $team : \{1,\ldots,p\} \rightarrow \{0,1\}$*. Each trick induces a cost* $c(t)$ *defined by the rules* $r$ *of the games, Trick costs are assigned to the winner (team).*

Hearts is for four individual player rather than two teams, and Euchre has left-over cards. In Doppelkopf the two teams are hinted by the bidding but usually uncovered during play. Sometimes the cost of a trick and the game value of the game depend on the entire history of the tricks. Announcements made by the player may change the scoring function. In addition, different tournament scoring systems apply. In games such as Bridge or Spades the objective of such games is tied to maximizing the number of tricks taken [15], while in games such as Belote, Skat, Doppelkopf, Tarot, or Hearts the main objective is to optimize the value of the cards contained in the tricks taken.

**Definition 2 (Suit/Trumps).** *On their respective turn players are free to play any card into a trick, but the card must follow suit* $S \in \mathcal{S} \subset \mathcal{C}$ *or trump* $\mathcal{T} \subseteq \mathcal{C}$*. A trump is a card that is elevated above its usual rank in trick-taking games. Typically, an entire suit is nominated as a trump suit, and these cards outrank all cards of non-trump suits.*

Trick-avoiding games, where one must not take tricks, like the Nullspiel in Skat, or trick-and-draw games, like Durak, in which the players refill their hands after each trick, are not considered in this paper. Such special rules can be added, as indicated in Hearts, where the objective is to *minimize* the number of hearts in the tricks.

## 3 Data Structures

**Card Encoding.** For the succinct representation and ordering of the cards, one of the first obstacles was the card bitset encoding, which provides fast constant-time operations in the search algorithms.

**Card Ordering.** We also derive the permutation $\pi = (\pi_1, \ldots, \pi_n)$ of the cards as a total order for the card's rank, so that $a_{\pi_1} > a_{\pi_2} \ldots > a_{\pi_n}$. We noticed that cards in different suits might not be directly comparable, but the key advantage of such an ordering is that with a single arithmetic operation, we can determihe the winner of two cards and, in turn, to determine the winner of the entire trick. In many games the total order is determined by the bidding process, and trough the choice of the trump color. In addition to fixing the contract to fulfil, trumps were often determined via choosing the suits (usually $\heartsuit$, ♠, ♣, $\diamondsuit$). However, trumps can also be a predefined selection of cards introducing a fifth group. In some games such as Belote, even the scoring function can be affected by the choice of trump. In the game Hearts, there is no trump at all. We call two adjacent cards $a_{\pi_i}$ and $a_{\pi_{i+1}}$ in the total order *equivalent*, if they are of the same suit and indistinguishable in the scoring system, which often goes down to accumulating individual card values. This order may change during the bidding, so that we start with a default order, and change it dynamically. The ordering usually respects the suits and trumps, so that cards of the same suit are adjacent and ordered by their rank. Partition Search [17] prunes the search tree. In Doppelkopf there are twin cards with different card indexes, but no dominance between them in trick-play, so that the first card always beats the second no matter what card index it has. This is solved by distinguishing the card index from the card face.

**Knowledge Vectors.** Card games have randomness in the deal and partial information due to the cards being usually hidden. There is knowledge that each player has to represent to indicate what he already knows about the remaining distribution of cards, induced by the tricks being played and sometimes also by the bidding. We use knowledge vectors $k_{i,j}$ to encode the cards that a player $P_j$ knows that the other player $P_i$ does **not** have, $1 \leq i, j \leq p$. Each knowledge vector $k_{i,j}$ represents a set of cards. There are no restrictions other than efficiency to use more advanced knowledge representation formalism and inference algorithm in terms of propositions or probabilities. We use vectors of players not having a card, but one can also record cards that a player has to have. The reason for using negation is that it is often faster: if a player does not obey suit, then all cards of the suit can be moved into the not-having vector. A concise knowledge encoding and reasoning is needed as knowledge vectors are also progressed in the incomplete-information search to provide earlier pruning options. The update is implemented as follows.

```
void Player::updateBelief(Game* game, card c)
  for (int p = 0; p < N_PLAYERS; p++) have_not[p] |= c;
  for (card d : set_cards(~have_not[game.turn]))
    if ((c & playable(c | d, game)) == 0) have_not[game.turn] |= d;
```

**Team Building.** Teams are usually formed during bidding. For example, in Bridge, we have declarer and opponents, and one player being the *dummy*, who belongs to the declarer's team and who puts the cards on the table heads-up. We handle such *ouvert* games by adapting the knowledge vectors, given that all the cards on the table are known not to be in the players' hand from the point of view of the remaining players. One additional implementation challenge in Doppelkopf is that after bidding even the teams are not necessarily known in advance. For this case, we selected each player to first play selfish (one-against-three in the search tree) like in the one-against-all setting in Hearts, maximizing own tricks, unless the partner is known, when parties join. Of course, improvements to this scheme can be applied.

## 4 General Card Game Play

The AI player interface is a set of (virtual abstract) functions and looks like follows. All the players operate fully independent of each other, and have no other information than provided by the server alias driver loop, who also receives all the player's answers.

**startGame** Called when a new game is started—parameters: position at table and hand cards array with card codes;
**placeBid** Called when AI has to confirm a bid—parameter: bidding value to increase;
**bidPlaced** Announcing bid to other players—parameter: bidding value;
**setTeams** called when AI has lost the bidding—parameters: team: Boolean array of players, array of highest bid of players;

**game_name** text for variant of game played;
**bid_name** text for bidding value;
**contract_name** text for point contract negotiated;
**set_team** setting a value in the team vector;
**winner_trick** determines the winner of a trick;
**initial_points** offset score given to the players;
**points_card** the score value of a card in a game;
**points_trick** the score value of a trick in a game;
**trump_mask** the subset of trump cards for the game;
**next_bid_turn** next player to bid;
**end_bidding** check if bidding has concluded;
**playable** given a hand and table cards, determine the reduced set of valid cards;
**end_trickgame** check if game is already over;
**score** determine the current score of a given player.

**Fig. 2.** Card game interface used for the general card game player.

**declareGame** Called when AI has requested has to declare a game—parameters: game declaration record;
**isHandGame** Asking AI if it requests a handgame—parameters: game declaration record;
**setGame** Information sent to every player after game is declared—parameters: game declaration record;
**playCard** called when AI has to play a card—no parameters;
**cardPlayed** called always when a card is played by a player—parameters: position of player who played the card, card code of played card.

The general driver loop is implemented in Fig. 3 and is aimed to serve all games, so that games may or may not have extra *dog* cards to be taken and discarded (as in Tarot and Skat). The interface for the domain-specific rules takes the (uni)code corresponding to a bid, the (uni)code corresponding to the cards, vectors of cards for the suits, and arrays of integers for game variant and contract chosen (usually used for bidding). Figure 2 shows the functions called by move generators and bidding modules. We also have a function to update the state of the game, the data structures like the table cards, the bids, the tricks, the points, the turn. We used bitmasks to distinguish between the different games that can be bid for, and one for the game card partitioning function mainly into suits. To avoid a recompilation we outsourced these parameter to a file. The card encoding and rules for the driver may be different from the player implementation with its fast move generation modules.

**Initial Card Distribution.** We have noticed that the distributions were consistent with the predictions of the hyper-geometric distribution, so we used a Mersenne twister to generate random deals for the players. Object inheritance and virtual functions are used to foster a vector of different AI or human players. The cards not dealt to the players are placed in the dog.

```
void GeneralCardGame(AI ai[N_PLAYERS], Game game)
  list<card> hands = deal(deck, distrib);
  int i = 0; card all = 0;
  for (auto h : hands)
    all |= ai[i].hand;
    ai[i].startGame(i,h);
    i++;
  int bid[N_PLAYERS], turn = 0;
  while (!end_bidding(game))
      int last_bid = 0;
      if (!game.bids.empty())
        last_bid = game.bids.front();
      max_bid = max(last_bid,max_bid);
      int b = ai[game.turn].placeBid(max_bid);
      for (int i = 0; i < N_PLAYERS; i++)
        ai[i].bidPlaced(b);
      if (b >= max_bid)
        bid[game.turn] = b;
        turn = game.turn;
      update_bid(game, b);
  card dog = allbits & ~allhands;
  for (int i = 0; i < N_PLAYERS; i++)
    ai[i].setTeams(game.team, bid);
  if (dog)
      if (!ai[game.declarer].isHandGame(game))
        ai[game.declarer].declareGame(dog,game);
  else ai[game.declarer].declareGame(0,game);
  for (int i = 0; i < N_PLAYERS; i++)
    ai[i].setGame(game);
  while (!end_trickgame(game))
      card c = ai[game.turn].playCard();
      ai[game.turn].hand &= ~c;
      for (int p = 0; p < N_PLAYERS; p++)
        ai[p].cardPlayed(game.turn, c);
      updateCard(game,c);
```

**Fig. 3.** General card game playing loop.

**Bidding.** The (simplified) bidding process continues unless the end of bidding has been determined, players are asked to posed a bit and are informed about the bid chosen. Also the currently maximum bid and the next turn in the bdding process is determined. Internally, all bids are kept in a stack, which allows to determine the game after a set of passes, as well as maximum bid of each player.

**Declaring a Game.** After bidding the teams are set and and depending on the game type chosen the dog can be used by the declarer to strengthen the hand cards. If a dog/skat/talon is put it overwrites the existing one. Also there might be point offsets given at the start of the game.

**Searching a Game.** In Fig. 4 we provide the implementation of a search node for a recursive AND/OR search for a given point limit and search-tree depth threshold. The game value will be found in binary search.We apply transposition table pruning at each end of a trick Search nodes for the other players is similar. The code is for four players, but it is simple to extend to an arbitrary number.

```
int Node(card playable, int player)
  int decl = 1 - team[player], opp = team[player];
  int best = table_lookup(hands, decl, decl_points);
  if (best != -1) return best;
  if ((playable) && count(played) - depth > threshold)
    playable = 1 << any(playable);
  while (playable)
    int index = select(playable,player);
    card bitindex = (a << index);
    hands[player] &= ~bitindex;
    played |= bitindex;
    table[player] = index;
    int rval = -1;
    int np = next_turn(player)
    if (count(played) % PLAYER == 0)
      int turn = winner(table);
      int score = Cost(table);
      if (team[turn] == 1) opp_points += score; else decl_points += score;
      int temp = table;
      table = emptytable;
      rval = (count(played) == PLAYSIZE)
          ? (decl_points > limit)
          : Node(playable_cards(hands[np], played),turn)
      table = temp;
      if (team[turn] == 1) opp_points -= score; else decl_points -= score;
    else
      rval = Node(g.playable_cards(hands[np], issuer(np), played, i),np);
    table[player] = -1;
    played &= ~bitindex;
    hands[player] |= bitindex;
    playable &= ~bitindex;
    if (rval == decl) {
      if (count(played) == depth)
        proposed_card = index;
        proposed_set = index;
        if (count(played) % N_PLAYERS == 0) table_add(hands,decl);
        return decl;
  if (count(played) % PLAYER == 0) table_add(hands,decl);
  return opp;
```

**Fig. 4.** Search node for a general card game.

# 5 Ensemble of Card Recommender Algorithms

The main aspect for trick-taking are the card recommenders for player $i \in \{1, \ldots, p\}$, the position in the trick $j \in \{1, \ldots, p\} \setminus \{i\}$, the current hand $h$, the knowledge about the other hands $k_{i,j}$, $1 \leq i, j \leq p$, and the score vector $c = (c_1, \ldots, c_p)$, condensed to two teams 0. We also have the history of the game available in the current state $s$.

We implemented different card recommenders, so that the card-game playing algorithm can be viewed as a *mixture* or *ensemble of experts* [13] $\mathcal{E}_1^{l_1}(i,j,s,h,k,c,t)$, $\mathcal{E}_2^{l_2}(i,j,s,h,k,c,t)$,..., $\mathcal{E}_k^{l_k}(i,j,s,h,k,c,t)$. Most of our recommender modules use an efficient open-card solver $\mathcal{O}(i,j,s,w,c,t)$ as ingredient, which computes the game value of an open-card game with world $w$. The solver can decide on a game or optimize it; it can also be used for solving suit factors [9] as in combinatorial game theory [22].

1. $\mathcal{E}^l_{\text{api}}$: For this recommender we allow users to interface with the system, which provides a default card recommendation depending on the trick position and game being played. This not absolutely necessary but helpful in the early stages of the game. So

we might take the smallest card that wins the trick, or the lowest if we lose the trick, play low trumps to take away high trumps from the opponents, or high trumps if there are still low trumps left, and so on.

2. $\mathcal{E}^l_{\text{pimc}}$: Perfect Information Monte-Carlo (PIMC) sampling. It uses the size $b$ of the belief-space as a parameter and evaluates each of the $b$ worlds with the open-card solver, in some cases limiting the search to a certain depth $d$, where the Minimax search is truncated. Depending on $d$, even the first card $l$ can be used to start the analysis, so that PIMC can be used as an automated default recommeder. We prefer to use hard constraints to eliminate worlds, but if one has obtained soft knowledge about the distribution of cards, in weighted sampling [7] one can change the distribution accordingly.
3. $\mathcal{E}^l_{\alpha\mu}$. This new algorithm [6] assumes full knowledge for one team and partial knowledge for the other team. It applies belief-space sampling with $b$ as a parameter matching the functionality in PIMC, but instead updates the partial ordering of winning vectors progressed upwards in an incomplete-information Minimax game tree. Together with the nodes we no longer have singleton scores but score vectors that need to be merged (either in a fusion or a max/min operation). We have extended the update in $\alpha\mu$ from winning to score vectors, maintaining a partial order. If hyperparameters are adapted, it can be applied to the beginning of a game, while it usually will be applied later on.
4. $\mathcal{E}^l_{\text{paranoia}}$. This algorithm finds forced wins and advances knowledge in the search tree, but branches on cards and card distribution. If a win is forced, the recommended card will be played with priority. To find sharp cards, if time allows, we compute the card with the best score. Since the computational requirements for this search option are substantial, its starting card $l$ can be larger and applies to middle and end game. There are different paranoia algorithms for one team and the other, allow approximation so a forced win if we neglect extreme distributions.
5. $\mathcal{E}^l_{\text{voting}}$. Here the entire belief space is spanned, so that all states consistent with the knowledge are generated. This allows to guarantee that no case is missed, so that the analysis is exact (assuming the rational players). The vote is a predefined mixture of worlds won, points gained, and hope cards found. Voting is applied in later stages of the game. Based on the chosen mix, it is actually an ensemble of sub-recommendation modules $\mathcal{E}^l_{\text{score}}$, $\mathcal{E}^l_{\text{won}}$, and $\mathcal{E}^l_{\text{hope}}$.

In our implementation we mainly use the ensemble $\mathcal{E}_{\text{hybrid}}$ of the expert $\mathcal{E}^l_{\text{api}}$, $\mathcal{E}^l_{\text{paranoia}}$ and $\mathcal{E}^l_{\text{voting}}$. above for a trade-off of exploration efficiency and playing strength. The mixture of experts is dependent on the card being played. We include the notion of confidence, that we want a recommender to be certain to a degree of confidence $\alpha$ (based on some criteria, often $> 90\%$). We do not apply a learned mix but a prioritization scheme, in which forced wins are of course prioritized and it is that the API module is used by default for the early stages of the games in case the analysis takes too long. Being called in advance allows to order the search trees in some cases.

## 6 Experiments

The individual games are called with parameters

- `<seed>` a random number generation initializer;
- `<repetitons>` the number of games in the experiment;
- `<humanselect>`; a bitvector for selecting Human (0) or AI (1) for all players;
- `<aiselect>` is a bitvector for selecting $AI_1$ (0) or $AI_2$ (1) for all players;
- `<strategy1>` base10 vector for selecting strategy in $AI_1$ for all players;
- `<strategy2>` base10 vector for selecting strategy in $AI_2$ for all players.

We record the game being declared the position of the player, and the obtained outcome of the game. We also print the parameter being fixed for the search. More values can be recorded. The games won by a team is distributed to the players of the team, and the according player strategy is awarded to compare different strategies.

The framework is written in GNU `g++` (version 9.3), optimized with parameter `-O2`. The players share one core of the Intel(R) Core(TM) i7-8565U CPU @ 1.80 GHz in an Ubuntu Linux system. We use a Makefile (and symbolic links) to select the game to play, and a JSON file for the parameters needed for the otherwise generic search algorithm. As command-line options we use: *seed* pseudo-random number generator initialization, with seed = 0 hitting the clock; *repetitions*: number of times to repeat the experiment; *hselect*: bitvector for selecting human players; *aiselect*: bitvector for the selection of the chosen framework; and $s_i$: base-10 vectors for strategy chosen for $AI_i$, $i \in \{0, 1\}$. In the first framework we include expert rules, in the second we do not.

One core aspect we were interested in was exploration efficiency. For example, we investigated how fast the open-card solver is for the initial set of games, and at which trick number paranoia search or endgame play can be started. Especially in card games with 54 and 72 cards, the number of tricks is substantial. We are mostly interested in finding the optimal value of the game, since this often leads to the sharpest card to be played, in some cases we restrict to deciding the game, i.e., whether it was won or lost for the player. We carefully implemented new algorithms for incomplete information card games like $\alpha\mu$ but, surprisingly, they did not turn out to be universally stronger wrt. carefully parameterized algorithms with voting schemes. These algorithms also consumed a lot of computational power, making them usable only in endgames.

### 6.1 Selfplay

Table 2 shows an evaluation of 100 self-playing games in ensemble of experts $\mathcal{E}_{\text{hybrid}}$. Even though our bidding systems were simple, in the 100 games in Bridge and Belote we played all the possible suits with different contracts; in Skat all suits, grand and null; in Tarot we played garde, petit and garde sans and garde contree; and in Doppelkopf we played normal games as well as solos in queens and jack. Hearts and Spades dont have different games to play. In Skat, 18% games were folded, which is due to the simplified round-and-robin bidding, as in the corresponding Skat AI [8] at most 5% are folded.

In Table 3 we study the efficiencies of the open-card solver module to illustrate the exploration speed. In all games the solver generates more than a million nodes per

**Table 2.** Results for 100 self-playing games in ensemble of experts $\mathcal{E}_{\text{hybrid}}$, parameters given are: mini-max game tree threshold, minscore, maxscore, starting card endgame are added for the different teams and time is the total for all games. Doppel is Doppelkopf without nines.

| Game | Search Parameters | Time | DeclWon | OppWon |
|---|---|---|---|---|
| Skat | $(21, [0, 120], 7, 100)$ | 110 s | 50 | 50 |
| Belote | $(32, [0, 162], 10, 20)$ | 4.3 s | 82 | 18 |
| Bridge | $(20, [0, 13], 28, 500)$ | 48.77 s | 74 | 26 |
| Spades | $(20, [0, 192], 30, 500)$ | 144 s | 69 | 31 |
| Doppel | $(12, [0, 240], 15, 500)$ | 105 s | 51 | 49 |
| Tarot | $(56, [0, 91], 52, 500)$ | 1 s | 73 | 37 |
| Euchre | $(24, [0, 52], 2, 500)$ | 1 s | 44 | 56 |
| Schafkopf | $(56, [0, 120], 7, 500)$ | 1 s | 67 | 33 |

seconds. To compute the game value of Hearts is computationally far more challenging, as besides being negated its cost function is not monotone. The accurate game value of Belote is available in less than a second.

**Table 3.** Open-card solver efficiency for first card in a random deal, with Θ being the exploration depth in the Minimax search tree. Doppel is Doppelkopf without nines.

| Game | Θ | Time | Nodes | Val | Θ | Time | Nodes | Val | Θ | Time | Nodes | Val |
|---|---|---|---|---|---|---|---|---|---|---|---|---|
| Skat | 32 | 6 s | 25467877 | 44 | 32 | 6 s | 25467877 | 44 | 32 | 6 s | 25467877 | 44 |
| Belote | 32 | .02 s | 139443 | 130 | 32 | .02 s | 139443 | 130 | 32 | .02 s | 139443 | 130 |
| Bridge | 36 | 19 s | 59082808 | 4 | 44 | 29 s | 88142774 | 4 | 52 | 32 s | 96538375 | 4 |
| Hearts | 16 | 2 s | 8736217 | 0 | 20 | 48 s | 107374348 | 0 | 24 | 8 m 47 s | 1096719725 | 0 |
| Spades | 36 | 37 s | 102654703 | 132 | 44 | 1 m 19 s | 237445864 | 132 | 52 | 2 m 26 s | 352232675 | 142 |
| Doppel | 24 | 8.9 s | 36750735 | 96 | 32 | 12.4 s | 50683171 | 96 | 40 | 14.1 s | 60018422 | 96 |
| Tarot | 32 | .6 s | 2169402 | 160 | 40 | 41.3 s | 75813106 | 146 | 48 | 39 m 57 s | 3367581704 | 130 |

## 6.2 Competitive Play

We played all players interactively. For casual games, the playing strength is acceptable and the response time small enough for swift play. According to our subjective judgement, many good cards were found for the AI players. By exploiting expert player knowledge in a database of million recorded games and some further playing rules of a professional player in Skat, we can even reach a level of top human play. For a larger controlled set of experiments with humans, however, we would need a general graphical user interface, which by the support of different games proves tricky, as lots of the existing interfaces only support one type of game.

We played 100 Belote games with ensemble $\mathcal{E}^{\mathrm{OPT}}_{\mathrm{PIMC}}$ (using the card corresponding to the optimal score) against $\mathcal{E}_{\mathrm{hybrid}}$, where $\mathcal{E}^{\mathrm{OPT}}_{\mathrm{PIMC}}$ won $54 : 46$. In 100 games of Skat (with its strong knowledge-based player implementation inherited from [12]) $\mathcal{E}_{\mathrm{hybrid}}$ won all games against $\mathcal{E}^{\mathrm{OPT}}_{\mathrm{PIMC}}$, while in Spades $\mathcal{E}^{\mathrm{OPT}}_{\mathrm{PIMC}}$ won 19:7 against $\mathcal{E}_{\mathrm{hybrid}}$. For Doppelkopf we played 10 games with the ensemble $\mathcal{E}^{\mathrm{DEC}}_{\alpha\mu}$ (using the card with the highest score corresponding to the decision vector) against $\mathcal{E}_{\mathrm{hybrid}}$, where $\mathcal{E}_{\mathrm{hybrid}}$ won $9 : 1$.

Furthermore, we cross-compared against a general AI [4] that incorporates no additional knowledge besides the rules of the games, the card deck and the scoring system. It supports Bridge and Belote and features PIMC and $\alpha\mu$ as search engines for incomplete-information search during trick-taking. The system lacks bidding, so the comparison is limited to a small number of games it can play (clubs in a fixed contract). In initial experiments in Belote over the same 100 games the alternate player achieved 8399 points as a declarer, while we achieved only 6823 points as a declarer, leading to an advantage of $9.7\%$ of the maximal score in average per game.

## 7 Conclusion and Outlook

We saw a generalization of a state-of-the-art domain-dependent card player to a general card game AI. The research objective was to provide a benchmark for the development of incomplete information algorithms, as well as an efficient framework for the implementation of trick-taking games, while at the same time providing an adequate playing strength that allows casual, interactive human vs AI play. The user, who implements a new card game has to provide the rules of the games, the card encoding, an adequate bidding system, and, optionally, expert rules for preferred card recommendations.

This work has implications not only for research on advanced multi-agent game tree search in games with randomness and incomplete information, coalition, and cooperation. It also has an expected impact in practice, as it provides AIs for mobile applications or web services for playing card games in case of insufficient number of players, or for Internet outages. Considering the tremendous rise of online play in Chess over the last few years, interactive card game players are wanted by the individual card playing game communities. There are many servers and applications available, but the playing strength is usually at most moderate.

**Acknowledgment.** The presented work has been supported by the Czech Science Foundation (GAČR) under the research project numbers 24-12046S and 22-30043S.

## References

1. Allis, L.V.: A knowledge-based approach to Connect-Four. The game is solved: White wins. Master's thesis, Vrije Univeriteit, The Netherlands (1998)
2. Billings, D., Papp, D., Schaeffer, J., Szafron, D.: Opponent modeling in Poker. In: Mostow, J., Rich, C. (eds.). In: AAAI, pp. 493–499 (1998)
3. Blanvillain, X.A.P.: Oware is strongly solved. In: Browne, C., Kishimoto, A., Schaeffer, J. (eds.) CG 2022. LNCS, vol. 13865, pp. 87–96. Springer, Cham (2022). https://doi.org/10.1007/978-3-031-34017-8_8

4. Bounan, S., Edelkamp, S.: Trick costs for $\alpha\mu$ and new relatives. In: Barneva, R.P., Brimkov, V.E., Gentile, C., Pacchiano, A. (eds.) Artificial Intelligence and Image Analysis. IWCIA ISAIM 2024. LNCS, vol. 14494, pp. 50–67. Springer, Cham (2024). https://doi.org/10.1007/978-3-031-63735-3_3
5. Bowling, M., Burch, N., Johanson, M., Tammelin, O.: Heads-up limit Hold'em Poker is solved. Commun. ACM **60**(11), 81–88 (2017)
6. Cazenave, T., Ventos, V.: The $\alpha\mu$ search algorithm for the game of bridge. In: Cazenave, T., Teytaud, O., Winands, M.H.M. (eds.) MCS 2020. CCIS, vol. 1379, pp. 1–16. Springer, Cham (2021). https://doi.org/10.1007/978-3-030-89453-5_1
7. Cohensius, G., Meir, R., Oved, N., Stern, R.: Bidding in Spades. In: ECAI, pp. 387–394 (2020)
8. Edelkamp, S.: Challenging human supremacy in Skat. In: SOCS, pp. 52–60 (2019)
9. Edelkamp, S.: Dynamic play via suit factorization search in Skat. In: Schmid, U., Klügl, F., Wolter, D. (eds.) KI 2020. LNCS (LNAI), vol. 12325, pp. 18–32. Springer, Cham (2020). https://doi.org/10.1007/978-3-030-58285-2_2
10. Edelkamp, S.: Representing and reducing uncertainty for enumerating the belief space to improve endgame play in Skat. In: ECAI, pp. 395–402 (2020)
11. Edelkamp, S.: Knowledge-based paranoia search. In: Conference on Games, pp. 1–8 (2021)
12. Edelkamp, S.: Improving computer play in Skat with hope cards. In: Browne, C., Kishimoto, A., Schaeffer, J. (eds.) CG 2022. LNCS, vol. 13865, pp. 133–145. Springer, Cham (2022). https://doi.org/10.1007/978-3-031-34017-8_12
13. Ferrucci, D.: Building Watson: an overview of the DeepQA project. AI Mag. **31**, 59–79 (2010)
14. Finnsson, H.: Cadia-player: a general game playing agent. PhD thesis (2007)
15. Frank, I., Basin, D.: Search in games with incomplete information: a case study using bridge card play. Artif. Intell. **100**(1–2), 87–123 (1998)
16. Gasser, R.: Harnessing computational resources for efficient exhaustive search. PhD thesis, ETH Zürich (1995)
17. Ginsberg, M.: Step toward an expert-level bridge-playing program. In: IJCAI, pp. 584–589 (1999)
18. Keller, T., Kupferschmid, S.: Automatic bidding for the game of Skat. In: KI, pp. 95–102 (2008)
19. Kubicek, O., Burch, N., Lisy, V.: Look-ahead search on top of policy networks in imperfect information games (2024). arXiv/2312.15220
20. Kupferschmid, S.: Entwicklung eines Double-Dummy Skat Solvers mit einer Anwendung für verdeckte Skatspiele. Master's thesis, Albert-Ludwigs-Universität Freiburg (2003)
21. Kupferschmid, S., Helmert, M.: A Skat player based on Monte-Carlo simulation. In: van den Herik, H.J., Ciancarini, P., Donkers, H.H.L.M.J. (eds.) CG 2006. LNCS, vol. 4630, pp. 135–147. Springer, Heidelberg (2007). https://doi.org/10.1007/978-3-540-75538-8_12
22. Nowakowski, R.J., Landman, B.M., Luca, F., Nathanson, M.B.,Nešetřil, J., Robertson, A., (eds.): In: Combinatorial Game Theory. De Gruyter: Berlin (2022)
23. Perolat, J., et al.: Mastering the game of Stratego with model-free multiagent reinforcement learning. Science **378**(6623), 990–996 (2022)
24. Schaeffer, J., et al.: Solving checkers. In: IJCAI, pp. 292–297 (2005)
25. Silver, D., et al.: Mastering the game of go with deep neural networks and tree search. Nature **529**, 484–489 (2016)
26. Silver, D., et al.: Mastering Chess and Shogi by self-play with a general reinforcement learning algorithm. Technical Report 1712.018. arxiv (2017)

# Mechanisms for Data Sharing in Collaborative Causal Inference

Björn Filter[1(✉)], Ralf Möller[2], and Özgür Lütfü Özçep[2]

[1] Institute of Information Systems, University of Lübeck, Lübeck, Germany
b.filter@uni-luebeck.de

[2] Institute for Humanities-Centered AI (CHAI), University of Hamburg, Hamburg, Germany
{ralf.moeller,oezguer.oezcep}@uni-hamburg.de

**Abstract.** Collaborative causal inference (CCI) is a federated learning method for pooling data from multiple, often self-interested, parties, to achieve a common learning goal over causal structures, e.g. estimation and optimization of treatment variables in a medical setting. Since obtaining data can be costly for the participants and sharing unique data poses the risk of losing competitive advantages, motivating the participation of all parties through equitable rewards and incentives is necessary. This paper devises an evaluation scheme to measure the value of each party's data contribution to the common learning task, tailored to causal inference's statistical demands, by comparing completed partially directed acyclic graphs (CPDAGs) inferred from observational data contributed by the participants. The Data Valuation Scheme thus obtained can then be used to introduce mechanisms that incentivize the agents to contribute data. It can be leveraged to reward agents fairly, according to the quality of their data, or to maximize all agents' data contributions.

**Keywords:** Collaborative Causal Inference · Data Evaluation · Fairness · Mechanism Design · Data Sharing

## 1 Introduction

Causal inference estimates the causal effect of treatment variables within a particular population, a method widely adopted across various fields. In healthcare, for example, it assesses the effectiveness of pharmaceutical interventions [1] and explores gene effects on phenotypes [2]. It also finds application in policy decision-making [3], recommender systems [4], education [5], advertisement [6], agriculture [7], and numerous other domains.

Various methodologies for causal inference have emerged to analyze interventional data, obtained from controlled experimental trials, or observational data. However, their effectiveness can be hindered by data quality issues, including data sparsity and non-representativeness. For example, patient preferences for hospitals can limit each institution's records, causing biases and data sparsity.


A. Hotho and S. Rudolph (Eds.): KI 2024, LNAI 14992, pp. 86–98, 2024.
https://doi.org/10.1007/978-3-031-70893-0_7

Thus, causal inference efforts by single hospitals risk inaccurate treatment effect estimates, potentially impeding optimal medication prescriptions [8].

Collaborative causal inference (CCI) leverages collective data pooling from diverse agents, such as companies, organizations, or individuals, to tackle data sparsity and non-representativeness challenges. This approach enhances the precision of causal model estimations. Methods such as simple aggregation or multi-source causal inference [9] yield accurate and statistically significant estimates by utilizing data from all participants. However, success depends on the willingness of agents to share data, which is not always guaranteed. Entities often prioritize self-interest, hesitating to share data due to associated collection costs.

As a concrete example consider the case of genetic research for predicting and treating illnesses which relies on substantial amounts of data. Despite decreasing costs, acquiring such data remains expensive [10]. Pooling genome and gene expression data from multiple sources is crucial for accurate models, especially when dealing with biased subpopulations. To achieve this, assessing data quality and incentivizing parties based on it are necessary.

In this paper, we propose a mechanism for CCI aimed at motivating agents to provide maximal amounts of data to gain access to a high-quality model. We introduce a quantitative measure to assess data quality and each party's contribution to the collective model. This valuation scheme is tailored to the unique requirements of statistical causal inference, where, in contrast to the framework proposed by Karimireddy and colleagues [11], no ground truth is available. Unlike the data valuation scheme introduced by Qiao and colleagues [12], our method considers differences in estimated causal graph models and probability distributions for causal effects among contributors' data. The resulting data valuation function incentivizes agents to maximize their data contribution. Furthermore, it can also be used for the allocation of fair incentives based on data quality, ensuring better causal estimators for agents with higher-quality data.

For an implementation of our data valuation scheme see the following URL https://github.com/bjofilter/Data-Valuation-CCI. An extended version of this paper with proofs and an algorithm for the dSID is available under URL https://www.edit.fis.uni-hamburg.de/ws/files/54866098/ki24.pdf.

## 2 Preliminaries

We collect the basic notions and methods for causal structure learning. *Directed acyclic graphs (DAGs)* can model conditional independencies. For a DAG $\mathcal{D} = (\mathbf{V}, \mathbf{E})$ let the vertices $\mathbf{V} = V_1, ..., V_p$ correspond to a $p$-variate random vector $\mathbf{X} \in \mathbb{R}^p$, while the edges $\mathbf{E}$ denote relationships between the variables they connect. A probability function $P$ over $\mathbf{X}$ is called *Markov* relative to a DAG $\mathcal{D}$ if it factorizes as $P(x_1, ..., x_p) = \prod_{i=1}^{p} P(x_i \mid Pa_{X_i}(\mathcal{D}))$, where $Pa_{X_i}(\mathcal{D})$ are the parents of node $X_i$.

In causal inference, a DAG $\mathcal{D}$ is not only used to represent associational but also causal relationships. Here, intuitively, the parents $Pa_{V_i}(\mathcal{D})$ of a node $V_i \in \mathbf{V}$ can be interpreted as the *direct causes* of $V_i$, and the children as the *direct effects* of $V_i$. Such a model gives information about how a variable would change, if its direct causes are changed, for example by external manipulation. To denote such a manipulation, Pearl's *do-operator* can be used: Let $\mathcal{D} = (\mathbf{V}, \mathbf{E})$ be a DAG that is Markov to a probability distribution $P$ over a set of random variables $\mathbf{V} = \{V_1, ..., V_p\}$. Let $V_i, V_j \in \mathbf{V}$ be two distinct variables and $v_i, v_i$ the values of those variables. The *causal effect* of $V_i$ on $V_j$ is denoted as $P(v_j \mid do(V_i = v_i))$, where $do(V_i = v_i)$ or short $do(v_i)$ is an intervention that sets $V_i = v_i$ [13].

Let $\mathbf{v} = (v_1, ..., v_p)$ be a sequence of values of the variables $\mathbf{V}$. $\mathbf{v}$ is consistent with $v_i'$ if $v_i = v_i'$. Given a DAG $\mathcal{D}$, the distribution after an intervention, called *intervention distribution*, can be expressed as follows: $P_{\mathcal{D}}(\mathbf{v} \mid do(V_i = v_i')) = \prod_{j:V_j \neq V_i} P(v_j \mid Pa_{V_j}(\mathcal{D}))$ if $\mathbf{v}$ is consistent with $v_i'$, else set $= 0$.

The *total causal effect* of $V_i$ on $V_j$ is written as $P(V_j \mid do(V_i))$ and is defined as the probability distribution of $V_j$ after the intervention at $V_i$.

A *linear structural equation model (SEM)* $\mathcal{M}$ comprises endogenous variables $\mathbf{V} = \{V_1, ..., V_p\}$, exogenous variables $\mathbf{U} = \{U_1, ..., U_p\}$, and a joint distribution $P_M$ over $\mathbf{U}$, with structural assignments $V_j = \sum_{i<j} c_{ij} V_i + U_j$ (for $j = 1, ..., p$) defining direct causal effects between variables. When dealing with observational data that is assumed to have been created by an underlying SEM, various algorithms exist to estimate the underlying DAG, like for example the PC algorithm [14] or GES (greedy equivalence search) [15]. However, most algorithms for estimating a DAG from observational data cannot distinguish between two DAGs that are Markov equivalent (two DAGs $\mathcal{G}_1$ and $\mathcal{G}_2$ are said to be Markov equivalent if they encode the same set of conditional independencies [16]). Instead, one can only identify the Markov equivalence class (MEC) of DAGs. This is represented by a *Completed Partially Directed Acyclic Graph (CPDAG)*, which has a directed edge between two nodes if such an edge is present in all the DAGs belonging to the MEC and an undirected edge between two nodes if edges in both directions are present among DAGs in the MEC. The set of *consistent DAG extensions* of $\mathcal{C}$, $CE(\mathcal{C})$, consists of all DAGs represented by $\mathcal{C}$. We define the *intervention distribution of causal effects* given a CPDAG $\mathcal{C}$ as follows:

$$P_{\mathcal{C}}(\mathbf{v} \mid do(V_i = v_i')) = \sum_{\mathcal{D} \in CE(\mathcal{C})} \frac{1}{|CE(\mathcal{C})|} P_{\mathcal{D}}(\mathbf{v} \mid do(V_i = v_i')) \quad (1)$$

When a CPDAG $\mathcal{C}$ has been estimated, the IDA algorithm (Intervention-calculus when the DAG is Absent) [17] can be used to estimate the causal effect of a variable $V_i \in \mathbf{V}$ on another variable $V_j \in \mathbf{V}$. This algorithm estimates the multiset $\Theta^{ij}$ of all possible causal effects of $V_i$ on $V_j$ given the MEC described by $\mathcal{C}$, i.e. the effects described by all possible DAGs $\mathcal{D} \in CE(\mathcal{C})$. The basic idea behind the IDA algorithm is to determine all DAGs $\mathcal{D} \in CE(\mathcal{C})$ for a given CPDAG $\mathcal{C}$. In each of these DAGs, the parents of $V_i$ can be used as a so-called *adjustment set* [13] to estimate the *average causal effect (ACE)* $\tau^{ij}(\mathcal{D})$, which is defined as $\tau^{ij}(\mathcal{D}) = E(V_j \mid do(V_i = v_i + 1)) - E(V_j \mid do(V_i = v_i))$, where $E(V_j \mid do(V_i = v_i))$ stands for the expected value of $V_j$ after the intervention $do(v_i)$. The ACE can be estimated by performing the regression of $V_j$ on $V_i$ and

$Pa_{V_i}(\mathcal{D})$, denoted $\beta_{V_i|Pa_{V_i}(\mathcal{D})}$ as follows:

$$\beta_{V_i|Pa_{V_i}(\mathcal{D})} = \begin{cases} 0 & \text{if } V_j \in Pa_{V_i}(\mathcal{D}), \\ \text{coefficient of } V_i \text{ in } V_j \sim V_i + Pa_{V_i}(\mathcal{D}) & \text{if } V_j \notin Pa_{V_i}(\mathcal{D}), \end{cases} \tag{2}$$

where $V_j \sim V_i + Pa_{V_i}(\mathcal{D})$ stands for the linear regression of $V_j$ on $V_i$ and $Pa_{V_i}(\mathcal{D})$ [17]. Thus $\beta_{V_i|Pa_{V_i}(\mathcal{D})}$ describes a Gaussian distribution $\hat{\theta}^{ij}_{Pa_{V_i}(\mathcal{D})}$ consisting of the estimated effect $\hat{\tau}^{ij}_{Pa_{V_i}(\mathcal{D})}$ and the standard error $\hat{\sigma}^{ij}_{Pa_{V_i}(\mathcal{D})}$. Since only the parents of $V_i$ are used to estimate the effect, DAGs in which $V_i$ has the same parents yield the same effect. However, in DAGs with different parents, the estimated causal effect can be different. Thus, a regression has to be performed for each of the possible parent sets of $V_i$ given $\mathcal{C}$. Let $\mathbf{Pa}_{V_i}(\mathcal{C}) = \{Pa_{V_i}(\mathcal{D}) | \mathcal{D} \in CE(\mathcal{C})\}$. The estimated causal effect of $V_i$ on $V_j$ given CPDAG $\mathcal{C}$ is given by the weighted sum over all possible sets of parents for $V_i$:

$$\hat{\theta}^{ij} = \sum_{\mathbf{Z} \in \mathbf{Pa}_{V_i}(\mathcal{C})} \frac{|\{\mathcal{D} \in CE(\mathcal{C}) | Pa_{V_i}(\mathcal{D}) = \mathbf{Z}\}|}{|CE(\mathcal{C})|} \hat{\theta}^{ij}_{\mathbf{Z}}, \tag{3}$$

that is, the mixture model obtained by summing over all effects induced by a parent-set in $\mathbf{Pa}_{V_i}(\mathcal{C})$, divided by the fraction of DAGs in $CE(\mathcal{C})$ in which $V_i$ has that set of parents.

## 3 Collaborative Causal Inference

We assume, that $n$ self-interested agents $N := \{1, ..., n\}$ want to solve a common causal inference problem by estimating a CPDAG and using observational data of $p$ variables, $\mathbf{V} = \{V_1, ..., V_p\}$ to estimate causal effects between variable-pairs $(V_i, V_j) \in \mathbf{V} \times \mathbf{V}$. Furthermore, agents are non-malicious and they may acquire data from different and potentially biased populations, but these data collectively form the common target population of interest. Each agent $i \in N$ can produce data points $D_i$ for a fixed cost $c_i > 0$ per data point. Both single agents as well as coalitions of agents use their available data to infer a CPDAG $\mathcal{C} = (\mathbf{V}, \mathbf{E}_\mathcal{C})$ and then infer the causal effects using the IDA algorithm. We will make the rather strict assumption, that all agents are equally interested in all effects between all possible variable pairs $(V_i, V_j) \in \mathbf{V} \times \mathbf{V}$.

Let $N$ denote the grand coalition containing all agents and let $C \subseteq N$ denote some coalition of agents. $D_C$ denotes the data provided by coalition $C$. Let $\mathcal{E}(D_C) = \mathcal{E}_C = \left(\mathcal{C}_C, \hat{\Theta}_C\right)$ be the estimator computed from $D_C$ as described in Sect. 2. It consists of an estimated CPDAG $\mathcal{C}_C$ and the estimated distributions of causal effects $\hat{\Theta}_C = \{\hat{\theta}^{ij}_C \mid (V_i, V_j) \in \mathbf{V} \times \mathbf{V}\}$. To simplify notation, the subscript $C$ is replaced by index $i$ when $C = \{i\}$. Thus, $D_C = \cup_{i \in C} D_i$.

In the following, it will be necessary for agents and coalitions, to know the quality of their estimators, to calculate their improvement rate, and to allocate fair incentives to agents. Therefore let $v : \mathcal{E}_C \times \mathcal{E}_B \to \mathbb{R}$ be the valuation function for the estimator $\mathcal{E}_C$, which will be defined in the next chapter. It evaluates an estimator against the benchmark estimator $\mathcal{E}_B$ (usually the estimator obtained from the data available to the grand coalition). We can use this to define a reward value for each agent $i \in N$, denoted as $r_i$. The goal is to reward each agent's data contribution with an estimator $\mathcal{E}_i$, such that the quality of $\mathcal{E}_i$ corresponds to the valuation of $i$'s data contribution $D_i$, which will be measured using the function $v$.

## 4 Data Valuation Scheme

We assume coalitions are interested in both the inferred CPDAG $\mathcal{C}$, as well as the estimated distributions of causal effects $\hat{\theta}^{ij}$ for all pairs of variables $(V_i, V_j) \in \mathbf{V} \times \mathbf{V}$. In practice, the ground truth CPDAG and causal effects are unknown quantities and must be inferred from the observed data. The best available estimate for these are the CPDAG and estimated distributions computed using all data available to the grand coalition $N$, due to the asymptotic consistency of the estimators [14]. Thus, we will usually treat the grand coalition estimates of the CPDAG $\mathcal{C}_N$ as the surrogate for the CPDAG of the true underlying DAG, similarly, for all $V_i, V_j \in \mathbf{V} \times \mathbf{V}$, we use $\hat{\theta}^{ij}_N$ (the estimated distributions using the grand coalitions' data) as the surrogate for the ground truth population ACE $\theta^{ij}$.

Different CPDAGs often yield vastly different effects. Suppose we have two CPDAGs, $\mathcal{C}_1$ and $\mathcal{C}_2$, over the same set of variables $\mathbf{V}$, and we are interested in the possible causal effects of $V_i \in \mathbf{V}$ on $V_j \in \mathbf{V}\backslash\{V_i\}$. To infer these, one has to determine the sets $\mathbf{Pa}_{V_i}(\mathcal{C}_1)$ and $\mathbf{Pa}_{V_i}(\mathcal{C}_2)$ and has to use these to calculate the possible causal effects (see Sect. 2).

Therefore, to evaluate the data of a coalition $C \subseteq N$, we propose to first take into account the difference between the CPDAG estimated by that coalition, $\mathcal{C}_C = (V, E_C)$ and the grand coalition estimates of the CPDAG $\mathcal{C}_N = (V, E_N)$. To do this, we introduce a measure extending the measure of *structural intervention distance (SID)* [18]. We will call this the *distribution-SID (dSID)*. It is the number of node-pairs $V_i, V_j$, for which the distribution of causal effects of $V_i$ on $V_j$ given $\mathcal{C}_1$ is different from the distribution given $\mathcal{C}_2$. For a discussion of the SID and other possible distance measures from the literature see Sect. 7 on related work.

**Definition 1 (Distribution Structural Intervention Distance (dSID)).** *Let $\mathbb{G}$ be the space of CPDAGs over $p$ variables. We then define*

$$\begin{aligned} dSID: \quad & \mathbb{G} \times \mathbb{G} \to \mathbb{N} \\ & (\mathcal{C}_1, \mathcal{C}_2) \mapsto |\{(V_i, V_j) \in \mathbf{V} \times \mathbf{V}, V_i \neq V_j \mid \textit{the distribution of causal} \\ & \qquad \textit{effects of } V_i \textit{ on } V_j \textit{ is falsely identified in } \mathcal{C}_1 \textit{ with respect to } \mathcal{C}_2\}| \end{aligned}$$

This leads to the first part of our data valuation measure. Let $\mathcal{E}$ be an estimator, consisting of a CPDAG $\mathcal{C}_\mathcal{E} = (\mathbf{V}, \mathbf{E})$ and the estimated distributions $\hat{\theta}^{ij}_\mathcal{E}$ for all pairs of variables $V_i, V_j \in \mathbf{V} \times \mathbf{V}$. Further, let $\mathcal{B}$ be a benchmark estimator (usually the estimator obtained from dataset $D_N$ of the grand coalition), also consisting of a CPDAG $\mathcal{C}_\mathcal{B} = (\mathbf{V}, \mathbf{E}')$ and distributions $\hat{\theta}^{ij}_\mathcal{B}$. Let

$$v_{dSID}(\mathcal{E}, \mathcal{B}) := -\frac{dSID(\mathcal{C}_\mathcal{E}, \mathcal{C}_\mathcal{B})}{|\mathbf{V}| \cdot |\mathbf{V} - 1|}. \tag{4}$$

The $v_{dSID}$ metric can therefore be seen as the percentage amount of variable pairs, for which the intervention distribution of causal effects differ between $\mathcal{C}_\mathcal{E}$ and $\mathcal{C}_\mathcal{B}$. Its maximum value is 0, if all intervention distributions in the two CPDAGs are identical, and its minimum is $-1$ if all are different.

We will refine the $v_{dSID}$, because even for pairs of variables for which the distribution of causal effects is correctly identified in $\mathcal{C}_\mathcal{E}$ with respect to (wrt) $\mathcal{C}_\mathcal{B}$, the estimated effect distributions $\theta^{ij}_\mathcal{E}$ and $\theta^{ij}_\mathcal{B}$ can differ since they are computed by regression using different datasets. Thus, we adopt the method proposed by Qiao and colleagues [12], using the negative reverse Kullback-Leibler (KL) divergence. It quantifies, how much one probability distribution differs from another probability distribution.

For each pair of variables $V_i, V_j$ where $\mathcal{C}_\mathcal{E}$ and $\mathcal{C}_\mathcal{B}$ identify the same causal effects, the IDA algorithm will yield relatively similar estimated causal effects. In contrast to the approach of Qiao and colleagues [12], however, for a given variable-pair $(V_i, V_j) \in \mathbf{V} \times \mathbf{V}$, we are not dealing with one causal effect described by a Gaussian distribution, but a distribution over several effects. The mixture model given by the sum of these distributions then describes the effect distribution $\hat{\theta}^{ij}_\mathcal{E}$ (see Eq. 3). This can then be compared to the mixture model of possible causal effects from the benchmark estimator $\mathcal{B}$, $\hat{\theta}^{ij}_\mathcal{B}$. Define $\mathbf{A} = \{(V_i, V_j) \in \mathbf{V} \times \mathbf{V}, V_i \neq V_j |$ *the distribution of causal effects from* $V_i$ *to* $V_j$ *is correctly identified in* $\mathcal{C}_\mathcal{E}$ *wrt* $\mathcal{C}_\mathcal{B}\}$. We can now define the second part of our data valuation method:

$$v_{-KL}(\mathcal{E}, \mathcal{B}) = \frac{\sum_{(V_i,V_j)\in\mathbf{A}} -KL\left(\hat{\theta}^{ij}_\mathcal{E} || \hat{\theta}^{ij}_\mathcal{B}\right)}{|\mathbf{A}|}. \tag{5}$$

Lastly, we propose the combination of $v_{dSHD}$ and $v_{-KL}$ as the final data valuation function:

$$v(\mathcal{E}, \mathcal{B}) = v_{dSHD}(\mathcal{E}, \mathcal{B}) + v_{-KL}(\mathcal{E}, \mathcal{B}). \tag{6}$$

When $\mathcal{C}_\mathcal{E}$ and $\mathcal{C}_\mathcal{B}$ identify the same distribution of causal effects for a variable pair $(V_i, V_j) \in \mathbf{V} \times \mathbf{V}$, the underlying data sets must be similar concerning the effects between these variables. Thus the negative KL-divergence between $\hat{\theta}^{ij}_\mathcal{E}$ and $\hat{\theta}^{ij}_\mathcal{B}$ is relatively small. Consequently, $v_{-KL}$-values are typically much smaller than $v_{dSHD}$-values. Because differences in CPDAGs can lead to vastly different causal effects, this is more significant than an effect, that is identically identified by both CPDAGs, but then slightly differently estimated from the regressions given coalitions different data sets. Thus, $v_{dSHD}$ makes up a significantly larger part of our date valuation measure than the relatively small $v_{-KL}$. Intuitively, the $v$ measure can be seen as a measure of the difference of the estimated CPDAGs, degraded by the difference of the estimated effects.

## 5 Modeling an Individual Agent

To understand how agents will behave in a multi-agent setting, we first consider the case of a single agent $i \in N$. In this case, the agent only has access to the data he produces himself and the estimator arising from this data. We will assume that each agent $i$ has a marginal fixed cost $c_i > 0$ for producing a data point. Thus his cost for producing data $D_i$ is $cost_i(D_i) = c_i|D_i|$.

Using data $D_i$, agent $i$ will be able to produce some estimator $\mathcal{E}_i$. However, there is no benchmark to assess the quality of $\mathcal{E}_i$ without access to more data. Thus, agent $i$ can only evaluate the quality of his estimator based on estimators he has produced before, using a subset of his current data. Let $T = \{1, 2, ...\}$ be a series of timestep. For each $t \in T$, let $D_i^t$ be the data produced so far by $i$ at timestep $t$ with $D_i^{t-1} \subseteq D_i^t$ for all $t > 1$. Let $\mathcal{E}\left(D_i^t\right) = \mathcal{E}_i^t$ be the estimator obtained by agent $i$ at time $t$ (using data $D_i^t$).

For the rest of this paper, we will assume that an agent will compare his current estimator to the estimators he produced before, using the current estimator as the benchmark (the current estimator presents his best knowledge of the real distribution underlying the data). The agent will act according to how much his estimator improved from older versions. If the improvement rate is judged as too small by the agent, he will decide that further improvements are not worth the cost and thus stop producing further data. We will assume, that for as long as an agent produces data, he will produce the same amount of data at every timestep. Let $\Delta_i$ denote this amount of data, thus for all $t > 1$, $|D_i^t \backslash D_i^{t-1}| = \Delta_i$. So while producing data, at every timestep $i$ occurs additional cost $c_i \Delta_i$.

$v(\mathcal{E}_i^{t'}, \mathcal{E}_i^t)$ assesses the discrepancy between estimators $\mathcal{E}_i^{t'}$ and $\mathcal{E}_i^t$, with 0 indicating identical CPDAGs and causal effect distributions, and $-1$ indicating completely different CPDAGs. Conversely, $-v(\mathcal{E}_i^{t'}, \mathcal{E}_i^t)$ measures the improvement in estimator quality from $\mathcal{E}_i^{t'}$ to $\mathcal{E}_i^t$, with 0 denoting no improvement and 1 completely different estimators.

While the PC algorithm's asymptotic consistency guarantees eventual convergence to the true underlying distribution with sufficient data [17], noise in early-stage data may not consistently improve estimator quality over time, Thus, $v(\mathcal{E}_i^{t'}, \mathcal{E}_i^t) > v(\mathcal{E}_i^{t'-1}, \mathcal{E}_i^t)$ is not guaranteed for every $t, t' \in T$ with $t > t' > 1$. For this reason, it makes sense for an agent to not only take into account the improvement from the last estimator to the current one, $v(\mathcal{E}_i^{t-1}, \mathcal{E}_i^t)$ but also the average improvement rate from older estimators towards the current one: $\frac{-v(\mathcal{E}_i^{t'}, \mathcal{E}_i^t)}{t-t'}$. With this, we can define the *improvement rate* as follows:

$$im_i(t) := \frac{1}{t-1} \sum_{t' \in \{1,\ldots,t-1\}} \frac{-v(\mathcal{E}_i^{t'}, \mathcal{E}_i^t) + v\left(\mathcal{E}_i^t, \mathcal{E}_i^t\right)}{t-t'}, \tag{7}$$

This can be seen as the average rate of improvement in model quality towards $\mathcal{E}_i^t$ per timestep, with $im_i(t) \in [0, 1]$ for all $t$. We assume that $i$ will gain some utility $u_i$ from the estimator improvement. We will assume that the utility increases linearly with increasing estimator quality. From this utility, we subtract the cost of producing $\Delta_i$ data points and thus get agent $i$'s utility at time $t$ by:

$$u_i(t) := im_i(t) - c_i\Delta_i \tag{8}$$

Finally, we will assume that agent $i$ will continue to produce data as long as his utility is not shrinking and stop producing once his utility is decreasing. That is, at the first $t \in T$, for which $im_i(t) - c_i\Delta_i < 0$, $i$ with stop producing data. Then the behavior of a single agent will be as follows:

**Definition 2 (Optimal Data Production).** *Let $T = \{1, 2, ...\}$ be an infinite series of timestamps and $i \in N$ an agent. Let $t_i^{s-opt} \in T$ be the unique timestamp, such that for all $t < t_i^{s-opt}$ $im_i(t) - c_i\Delta_i \geq 0$ and $im_i(t_i^{s-opt}) - c_i\Delta_i < 0$. Then the* optimal data production *for $i$ is $D_i^{t_i^{s-opt}}$. For producing this data, $i$ will incur cost $c_i|D_i^{t_i^{s-opt}}| = c_i\Delta_i t_{opt}$.*

Since $i$ stops producing further data after $t_i^{s-opt}$, we have $D_i^{t_i'} = D_i^{t_i^{s-opt}}$ for all $t_i' > t_i^{s-opt}$.

## 6 Modeling Multiple Agents

We will now study how agents behave in a collaborative setting as described in Sect. 3, where multiple agents produce data and the resulting model can be shared. In this framework, a server mediates the interaction between agents. First, the server publishes the mechanism to the agents, then, at each timestep $t$, each agent $i$ contributes some data $D_i^t$ to the server and finally, the server in turn rewards the agents with an estimator $\mathcal{E}_i^t$ of a certain quality. The server also provides agents with the opportunity to check the quality of any estimator they possess, that is, at time $t$ the server computes $v(\mathcal{E}_i^{t'}, \mathcal{E}_N^t)$ for any estimator $\mathcal{E}_i^{t'}$ which $i$ received at some time $t' \leq t$. This way, $i$ can always check the improvement rate of models he received so far, compared to the best available estimator at the time.

For time steps $T = \{1, 2, ...\}$ let $\mathbf{D}^t = \{D_1^t, ..., D_n^t\}$ be the data provided by agents 1 to $n$ at time t. A *mechanism* is formalized as a function $\mathcal{M}(\mathbf{D}) : D_1 \times ... \times D_n \rightarrow \mathcal{E}_1 \times ... \times \mathcal{E}_n$, which maps the agents' contributions to estimators they receive from the mechanism. At $t$, so far each agent $i$ generated and transmitted data $D_i^t$ to the server. For this, he is rewarded with an estimator $\mathcal{E}_i^t$ of quality $v(\mathcal{E}_i^t, \mathcal{E}_N^t)$. From now on, $\mathcal{E}_i^t$ will denote the estimator $i$ receives at $t$ from the server, while $\tilde{\mathcal{E}}_i^t$ will denote the estimator obtained only from $i$'s data $D_i^t$. A requirement for such a mechanism is, for it to be feasible: A mechanism which returns an estimator $\mathcal{E}_i^t$ to agent $i$ is said to be feasible if for any $i \in N$ and any $D_i^t$, it satisfies $v(\mathcal{E}_i^t, \mathcal{E}_N^t) \leq v(\mathcal{E}_N^t, \mathcal{E}_N^t)$. Feasibility is necessary to make sure that the mechanism does not return an estimator that is better than the estimator obtained by pooling all available data, $\mathcal{E}_i^t$. But since $\mathcal{E}_i^t$ is always our benchmark to which every other estimator is compared, by definition for any estimator $v(\mathcal{E}', \mathcal{E}_N^t) \leq 1$, while $v(\mathcal{E}_N^t, \mathcal{E}_N^t) = 1$, thus feasibility always holds.

Another requirement for our mechanism is individual rationality (IR), which is defined as follows: Given data contributions $\mathbf{D}^t$ by the agents, the mechanism provides an estimator $\mathcal{E}_i^t$ to agent $i$. Such a mechanism $\mathcal{M}$ is said to satisfy IR if for any agent $i \in N$ and any contribution $\mathbf{D}^t$, it holds that $v(\mathcal{E}_i^t, \mathcal{E}_N^t) \geq v(\tilde{\mathcal{E}}_i^t, \mathcal{E}_N^t)$. Thus individual rationality guarantees that no agent will obtain a worse estimator than they would have obtained if they were working independently. IR is necessary to guarantee that rational agents participate in the mechanism.

### 6.1 Standard Collaborative Setting

We first examine the most basic setting for collaborative causal inference. Here, each agent immediately has access to the model of the grand coalition, obtained from all data available to the mechanism at time $t$, $\mathcal{E}_N^t$. Thus at all $t \in T$ and for all $i \in N$: $[\mathcal{M}(\mathbf{D}^t)]_i = \mathcal{E}_N^t$. This mechanism is feasible and also satisfies individual rationality since $v(\mathcal{E}_N^t, \mathcal{E}_N^t) \geq v(\tilde{\mathcal{E}}_i^t, \mathcal{E}_N^t)$. We will assume that agents will behave as they would if they were to act on their own. That is, at each time step an agent $i \in N$ will produce and contribute some data to the mechanism and then evaluate, how much the estimator of the grand coalition has improved through their data contribution. If this improvement makes up for the cost $i$ incurred, he will continue to produce data, until the improvement does not outweigh their cost anymore. When first entering the mechanism, $i$ will contribute initial data $D_i^1$, since he does not know yet, how much this data will improve the grand coalition's estimator. After that, at each timestep $t > 1$, $i$ will compute the improvement rate $im_i(t)$ according to Eq. 7, but now comparing estimators of the grand coalitions at times $t' < t$ to the current estimator $\mathcal{E}_N^t$.

Again, $i$'s utility is calculated as $u_i(t) := \sum_{t' \in \{1,\dots,t\}} (im_i(t) - c_i \Delta_i)$, and agent $i$ will continue to produce data as long as this is not decreasing and stop producing once it decreases. Thus the behavior of an agent $i$ in the standard collaborative setting will be as described in Definition 2. However, we will denote the optimal time for agent $i$ to stop producing further data in the collaborative setting by $t_i^{opt}$, as opposed to $t_i^{s-opt}$ in the single agent setting.

### 6.2 Data Maximizing Mechanism

We will now present a mechanism that maximizes the data provided by all agents $i \in N$. A mechanism $\mathcal{M}$ is *data-maximizing* given costs $\mathbf{c} = \{c_1, ..., c_n\}$ if it maximizes the data collected at equilibrium. Since at each timestep $i$ produces a fixed amount of data $\Delta_i$, the total amount of data provided by agent $i$ is $t_i^{opt} \Delta_i$

**Definition 3 (Data Maximization).** *A mechanism $\hat{\mathcal{M}}$ is data-maximizing if $\hat{\mathcal{M}} \in \arg\max_{\mathcal{M}} \sum_{i \in N} \left[t_i^{opt} \Delta_i\right]$, i.e., if it maximizes the amount of data collected, subject to $\hat{\mathcal{M}}$ being feasible and satisfying IR.*

To maximize contributions, we need to make sure agents produce data for as long as possible. To achieve this, we will reward them at every time step with an estimator which is just good enough for them not to achieve a decrease in utility.

We will do this for as long as we can provide such an estimator, given access to all the data contributed by the grand coalition. Following the design described by Karimireddy and colleagues [11], we propose the following mechanism. Now

$$im_i(t) = \frac{1}{t-1} \sum_{t' \in \{1,\dots,t-1\}} \frac{-v\left(\mathcal{E}_i^{t'}, \mathcal{E}_N^t\right) + v\left(\mathcal{E}_i^t, \mathcal{E}_N^t\right)}{t-t'}, \tag{9}$$

as in Eq. 7, $im_i(t)$ measures the improvement towards the quality of the last model $i$ has received, $\mathcal{E}_i^t$, however, this does not have to be the best possible model $\mathcal{E}_N^t$. Now we can formulate our data-maximizing mechanism:

$$[\mathcal{M}(\mathbf{D}^t)]_i = \begin{cases} \tilde{\mathcal{E}}_i^t & \text{if } t < t_i^{s-opt} \\ \mathcal{E}_i^t, \text{such that } im_i(t) = c_i \Delta_i + \epsilon & \text{else, if such } \mathcal{E}_i^t \text{ exists} \\ \mathcal{E}_N & \text{else.} \end{cases} \tag{10}$$

Even without external incentivization, $i$ will continue to produce data until $t_i^{s-opt}$. From here on, $i$ will receive a model $\mathcal{E}_i^t$ such that the improvement rate just makes up for his incurred cost, so that his utility does not decrease. At some point, $i$ will receive the best possible model $\mathcal{E}_N^t$ or something very close to this, so it is no longer possible to keep the average improvement rate over $c_i\Delta_i$. Now, $i$ will receive the best possible model and no further incentivization is possible. This leads to the result expressed in the following theorem.

**Theorem 1 (Data maximization with known costs).** *The mechanism $\mathcal{M}$ defined by Eq. 10 is data-maximizing for $\epsilon \to 0^+$. A rational agent $i$ will contribute $\Delta_i t_i^{opt}$ data points where $\Delta_i t_i^{opt} \geq \Delta_i t_i^{s-opt}$, yielding a total of $\sum_{j\in N} \Delta_j t_j^{opt}$ data points.*

### 6.3 Achieving Fairness

While the mechanism described above maximizes the amount of data extracted from each agent, it does not satisfy fairness. In parallel to work done by Qiao and colleagues [12], we can use our data valuation function $v(\mathcal{E}, \mathcal{B})$, to design a reward scheme $r_i^t$ for every agent $i \in N$ at each time $t \in T$, which fulfills the following fairness conditions defined in the work of Qiao and colleagues [12]. At each $t \in T$, let $v_\emptyset^t := min_{i\in N} v(\mathcal{E}_i^t, \mathcal{E}_N^t)$ for the empty coalition $C = \emptyset$. Then:

- **(F1) Uselessness:** Agent $i$ should receive a valueless reward if his data does not improve the estimator of any other coalitions.
- **(F2) Symmetry:** If two agents yield the same improvement for all other coalitions, then they should receive equally valuable estimates as rewards.
- **(F3) Strict Desirability:** If data from agent $i$ strictly improves the estimate for at least one coalition more than that of agent $j$, but the reverse is not true, then agent $i$ should receive a more valuable reward than agent $j$.

- **(F4) Monotonicity:** For an agent $i$, if his dataset $D_i$ strictly improves the estimate for at least one coalition more compared to that of another dataset $D_{i'}$, but the reverse is not true, then sharing $D_i$ should give agent $i$ more valuable reward than sharing $D_{i'}$.

At each $t \in T$, let $v_\emptyset^t := min_{i \in N} v(\mathcal{E}_i^t, \mathcal{E}_N^t)$ for the empty coalition $C = \emptyset$. The marginal contribution $m_i^t(T)$ of the data of agent $i$ to a coalition $T \subseteq N$ at time $t$ is then defined as $m_i^t(T) := v(\mathcal{E}_{T \cup \{i\}}^t, \mathcal{E}_N^t) - v(\mathcal{E}_T^t, \mathcal{E}_N^t)$, with $\mathcal{E}_N^t$ being the estimator obtained from all data available to the grand coalition $N$. As proposed by Qiao and colleagues, the Shapley Value (SV) can then be calculated as $\phi_i^t := \frac{1}{|N|!} \sum_{T \subseteq N \setminus \{i\}} |T|! \cdot (|N| - |T| - 1)! \cdot m_i^t(T)$. Using this, the modified $\rho$-Shapley fair reward value is defined by

$$r_i^t := max \left\{ v\left(\mathcal{E}_i^t, \mathcal{E}_N^t\right) - v_\emptyset^t, -v_\emptyset^t \cdot \left(\frac{\phi_i^t}{\phi^{t*}}\right)^\rho \right\}, \tag{11}$$

with scaling factor $\rho \in (0, 1]$ and $\phi^{t*} := max_{i \in N} \phi_i^t$. Using this keeps rewards lower than when using the standard SV while still ensuring the fairness conditions [12]. Thus, agents can be incentivized to produce data over a longer period than by using the standard SV. For $\rho \to 0$, $r_i^t \to 0$, thus all agents receive the same reward. The larger $\rho$, the larger the differences between agent's rewards, and thus the more fairness is achieved. A detailed discussion on the effect of the $\rho$ parameter has been done by Qiao and colleagues [12].

We can use this reward scheme, to ensure that agents with more valuable data receive a better estimator than other agents. For this, when computing the modified $\rho$-Shapley fair reward value $r_i^t$, we need to set $\rho \leq \min_{i \in N} log(1 - v\left(\mathcal{E}_i^t, \mathcal{E}_N^t\right) / v_\emptyset^t) / log(\phi_i^t / \phi^{t*})$ [12]. Now we can define our fair mechanism as:

$$[\mathcal{M}_{fair}(\mathbf{D}^t)]_i = \begin{cases} \tilde{\mathcal{E}_i^t} & \text{if } t < t_i^{s-opt} \\ \mathcal{E}_i^t, \text{such that } im_i(t) = c_i \Delta_i + r_i^t & \text{else, if such } \mathcal{E}_i^t \text{ exists} \\ \mathcal{E}_N & \text{else.} \end{cases} \tag{12}$$

**Proposition 1** *$\mathcal{M}_{fair}$ ensures fairness since agents with more valuable data receive a better estimator [12]. The agent with the least valuable data receives an estimator which just ensures that his improvement rate balances his costs, agents with more valuable data receive estimators with larger improvement rates. This, in turn, means that $\mathcal{M}_{fair}$ is not data maximizing anymore, because agents with data of high value receive better estimators faster than they would under $\mathcal{M}$.*

This proposition follows directly from the construction of the reward scheme and the fact that $\mathcal{M}$, as defined in Eq. 10, is data maximizing with $\mathcal{M} \neq \mathcal{M}_{fair}$.

## 7 Related Work

The data valuation scheme presented in this paper is inspired by the scheme presented by Qiao and colleagues [12], in particular w.r.t. the use of negative reverse KL divergence. However, they only performed one regression to estimate one effect, they did not estimate a CPDAG and did not estimate the causal effects between all pairs of variables.

For this reason, a measure for distances between CPDAGs was needed. One possible measure here would be the structural hamming distance (SHD). However, as described by Peters and Bühlmann [18], the SHD does not necessarily describe well, how different two graphs are regarding the causal effects they decode. They therefore introduce the structural intervention distance (SID), which, given two DAGs $\mathcal{H}$ and $\mathcal{G}$ over the same vertices $\mathbf{V}$, describes how many intervention distributions between variable pairs are falsely estimated by $\mathcal{H}$ wrt $\mathcal{G}$. The SID can be applied to CPDAGs as well, here however it returns a range of the SIDs between all possible DAG extensions of the CPDAGs, whereas here a definite measure was needed.

A data maximizing mechanism for federated learning was first presented by Karimireddy and colleagues [11]. They assumed, however, that each data point provided by any agent had the same value and that the quality of a model could be directly computed. In our setting of collaborative causal inference, these assumptions do not hold anymore, therefore the data valuation scheme was introduced, both to measure the value of an agent's data as well as to measure the quality of a given estimator.

## 8 Conclusion and Future Work

This paper presents novel mechanisms for CCI aimed at motivating the engagement of self-interested agents through the provision of superior causal effect estimators compared to those agents that could develop independently. Our initial mechanism incentivizes all participating agents to maximize data contribution, thereby facilitating the construction of a comprehensive model based on extensive data and expected to yield high accuracy. Building upon this, our second mechanism assigns better estimators to agents supplying higher-quality data, ensuring a fair reward allocation given individual contributions.

A limitation of our mechanism is the assumption, that data production costs $c_i$ are known, prompting a future research direction towards maximizing data despite unknown $c_i$'s. Furthermore, the presumption of honesty and non-malicious behavior of agents may not hold in practice, with some parties potentially exploiting the transparent framework for personal gain or harm to others. Consequently, a more robust mechanism resilient to such behavior would be preferable.

**Acknowledgements.** Björn Filter acknowledges funding for project FPOplus by BMDV/TüV Rheinland Consulting GmbH.

## References


1. Shi, J., Norgeot, B.: Learning causal effects from observational data in healthcare: a review and summary. Front. Med. **9**, 864882 (2022)
2. Hu, P., Jiao, R., Jin, L., Xiong, M.: Application of causal inference to genomic analysis: advances in methodology. Front. Genet. **9**, 238 (2018)
3. Yao, L., Chu, Z., Li, S., Li, Y., Gao, J., Zhang, A.: A survey on causal inference. ACM Trans. Knowl. Discov. Data (TKDD) **15**(5), 1–46 (2021)
4. Liang, D., Charlin, L., Blei, D.M.: Causal inference for recommendation. In: Causation: Foundation to Application, Workshop at UAI. AUAI (2016)
5. Cordero, J.M., Cristóbal, V., Santín, D.: Causal inference on education policies: a survey of empirical studies using PISA, TIMSS and PIRLS. J. Econ. Surv. **32**(3), 878–915 (2018)
6. Sun, W., Wang, P., Yin, D., Yang, J., Chang, Y.: Causal inference via sparse additive models with application to online advertising. In: Proceedings of the AAAI Conference on Artificial Intelligence, vol. 29 (2015)
7. Singer, R.S., Williams-Nguyen, J.: Human health impacts of antibiotic use in agriculture: a push for improved causal inference. Curr. Opin. Microbiol. **19**, 1–8 (2014)
8. Masic, I., Miokovic, M., Muhamedagic, B.: Evidence based medicine-new approaches and challenges. Acta Informatica Med. **16**(4), 219 (2008)
9. Bareinboim, E., Pearl, J.: Causal inference and the data-fusion problem. Proc. Natl. Acad. Sci. **113**(27), 7345–7352 (2016)
10. Pareek, C.S., Smoczynski, R., Tretyn, A.: Sequencing technologies and genome sequencing. J. Appl. Genet. **52**, 413–435 (2011)
11. Karimireddy, S.P., Guo, W., Jordan, M.I.: Mechanisms that incentivize data sharing in federated learning. Papers 2207.04557. arXiv.org (2022). https://ideas.repec.org/p/arx/papers/2207.04557.html
12. Qiao, R., Xu, X., Low, B.K.H.: Collaborative causal inference with fair incentives. In: Krause, A., Brunskill, E., Cho, K., Engelhardt, B., Sabato, S., Scarlett, J. (eds.) Proceedings of the 40th International Conference on Machine Learning. Proceedings of Machine Learning Research, vol. 202, pp. 28300–28320. PMLR (2023). https://proceedings.mlr.press/v202/qiao23a.html
13. Pearl, J.: Causality, 2nd edn. Cambridge University Press, Cambridge (2009)
14. Kalisch, M., Bühlman, P.: Estimating high-dimensional directed acyclic graphs with the PC-algorithm. J. Mach. Learn. Res. **8**(3), 613–636 (2007)
15. Chickering, D.M.: Optimal structure identification with greedy search. J. Mach. Learn. Res. **3**, 507–554 (2002)
16. Chickering, D.M.: Learning equivalence classes of Bayesian networks structures. CoRR abs/1302.3566 (2013). http://arxiv.org/abs/1302.3566
17. Maathuis, M.H., Kalisch, M., Bühlmann, P.: Estimating high-dimensional intervention effects from observational data. Ann. Stat. **37**(6A), 3133–3164 (2009)
18. Peters, J., Bühlmann, P.: Structural intervention distance for evaluating causal graphs. Neural Comput. **27**(3), 771–799 (2015)

# SaVeWoT: Scripting and Verifying Web of Things Systems and Their Effects on the Physical World

Justus Fries[1(✉)], Michael Freund[2], and Andreas Harth[2,3]

[1] Technical University of Munich, Munich, Germany
fries@in.tum.de
[2] Fraunhofer Institute for Integrated Circuits IIS, Nuremberg, Germany
[3] Friedrich-Alexander-Universität Erlangen-Nürnberg, Nuremberg, Germany

**Abstract.** We introduce SaVeWoT (Scripting and Verifying Web of Things Systems), an approach for designing, formally verifying, and deploying decentralized control systems based on the W3C WoT. SaVeWoT consists of two main parts: the SaVeWoT language and the SaVeWoT compiler. The SaVeWoT language models devices (i.e., Things), controllers that orchestrate Things, virtual composite Things (i.e., subsystems consisting of multiple Things), interactions between these components, and their effects on the physical world. The SaVeWoT compiler uses Thing Descriptions (TDs) and SaVeWoT behavior descriptions along with correctness specifications in Linear-time Temporal Logic (LTL) to automatically generate a Promela model, which is validated using the SPIN model checker. We demonstrate the feasibility of the SaVeWoT approach by verifying a conveyor belt system as an example and conducting an empirical evaluation.

**Keywords:** Web of Things · Static Verification · Safety

## 1 Introduction

The Web of Things (WoT) and the WoT Thing Description (TD) ontology enable interoperable interaction with Internet of Things (IoT) devices (Things). When building networked automation and control systems, e.g., in manufacturing [1] or building automation [2], interoperability enables the use of devices from various vendors and manufacturers. Interoperability is helpful for deployment and during the development and design phase of a system, as engineers can compare and use different components (e.g., devices, subsystems, and solutions) from various vendors. Because controllers interact with Things according to a specification developed by an engineer [2], integrating Things requires not only a description of the Things but also of the interactions between controllers and Things (i.e., all system components). These interactions can become complex as the system grows and cause unintended but subtle effects or bugs. With SaVeWoT we enable


A. Hotho and S. Rudolph (Eds.): KI 2024, LNAI 14992, pp. 99–113, 2024.
https://doi.org/10.1007/978-3-031-70893-0_8

engineers to find such unintended effects via model checking [3]. We focus on physical world effects and want to integrate SaVeWoT into the design phase of system development to find bugs as early as possible.

Integrating model checking into any design process leads to two main challenges: Ease of use and choosing the right level of abstraction for the model to check [3]. Existing approaches to model checking WoT systems [4] solve these issues by relying on rule-based programming and introducing strong assumptions (discussed in Sect. 3.2) about how Things interact. For SaVeWoT, we introduce and use a scripting language based on JavaScript. SaVeWoT eases the assumptions, which enables SaVeWoT to integrate the physical world into model checking through descriptions of behavior via scripts.

Additionally, previous work on model checking in IoT and WoT has primarily focused on the Trigger-Action Programming (TAP) where TAP rules are specified independently and trigger based on events to execute actions. While TAP rules are useful for smart home scenarios, the W3C WoT covers various domains, and the WoT Scripting API [5] is based on JavaScript and not rules. In addition, many previous approaches did not consider the physical world as a part of the system and instead formally verified the state of Things.

Our approach, SaVeWoT, enables model checking WoT systems using WoT TDs and scripts. Central to our approach is a model of the physical world described using ontologies [6], specifically the TD ontology [7] as well as the Sensor, Observation, Sampler, and Actuator (SOSA) and Semantic Sensor Network (SSN) ontologies [8]. Things sense or actuate so-called Features of Interest (FoIs) of the physical world, e.g., the temperature in a room. Such a model of the physical world is machine-readable, usable for inference or reasoning, and can be integrated with other physical world models. For SaVeWoT we define composite Things [9,10] as Things that control other Things, i.e., subsystems of the WoT system. Concurrency in WoT systems is introduced via concurrently running controllers.

We define a formal language based on JavaScript and the WoT Scripting API that is executable and designed for translation to formally verifiable models. When describing behavior in scripts, interactions with the physical world are expressed via reading (e.g., sensing the temperature) and writing (e.g., setting the temperature on a thermostat) variables representing FoIs.

After introducing our approach using an example (Sect. 2), we provide related work (Sect. 3). The contributions of this paper are:

- A formal WoT system model (Sect. 4) that explicitly facilitates formal verification of physical world effects.
- An executable, formally verifiable language (Sect. 5) based on W3C WoT standards.
- A proof-of-concept implementation (Sect. 6) and an evaluation that illustrates the feasibility of our approach (Sect. 7).

Finally, Sect. 8 summarizes the most important aspects of our approach and highlights future work.

## 2 SaVeWoT by Example

We now illustrate our approach with an example WoT system for a conveyor belt that moves objects on the belt in one direction, see Fig. 1 (1). The system consists of the following components (Fig. 1 (2)): a motor drives the conveyor belt, two photoelectric sensors detect objects on the belt, and a controller. The controller orchestrates the motor to move objects, e.g., work pieces, from the start to the end of the belt. The physical world is modeled using two FoIs that are sensed by the respective sensors. The first sensor measures the physical space at the start of the conveyor belt (`objectAtStartFoI`), while the second sensor measures the physical space at the end of the belt (`objectAtEndFoI`)

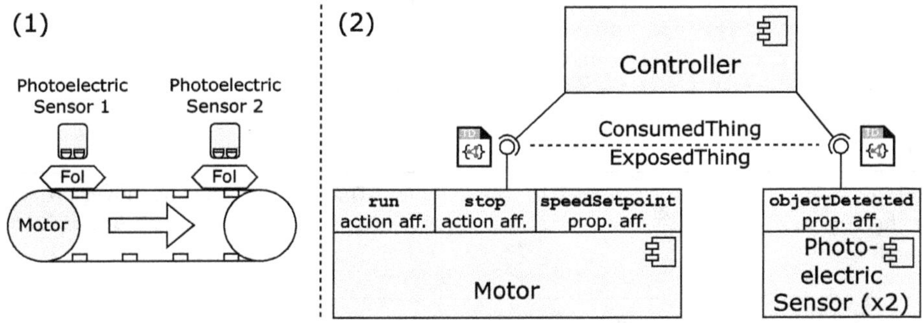


**Fig. 1.** Physical layout with Features of Interest (1) and components and interfaces (2) of the example WoT system. We refer to the FoI measured by "Sensor 1" as `objectAtStartFoI`, while the FoI measured by "Sensor 2" is called `objectAtEndFoI`.

### 2.1 Web of Things by Example

In the context of WoT, the motor and photoelectric sensors are Things described with a TD. The motor has three interaction affordances: the action affordances `run` and `stop`, and an integer property affordance for reading and writing the `speedSetpoint`. If the `speedSetpoint` is not set to a value greater than zero, the motor does not run. The sensor has a read-only boolean property affordance indicating whether the sensor detected an object or not, called `objectDetected`. Both photoelectric sensors have an `objectDetected` property affordance. However, each sensor measures its respective FoI. The relations between interaction affordances and FoIs are part of the TDs of the sensors and actuators.

The WoT Scripting API defines `ConsumedThing` and `ExposedThing` interfaces. A controller script is written against `ConsumedThing` interface methods based on the interaction affordances in a TD. On the other hand, the behavior of a Thing is described by a set of handlers as defined by the `ExposedThing` interface. Each handler describes the behavior of the Thing for a specific interaction affordance (e.g., the action `run`) and operation (in this example: invoke, read or write).

The script for controller behavior is shown in Listing 2.1. If an object is at the start of the belt, the controller runs the motor to move the object. Once the object is at the end, the motor is stopped. Listing 2.2 shows the script for the Thing behavior. The motor moving an object on the belt is modeled through the FoIs. The motor affects the physical world in a single step, i.e., an object is either at the start or end of the belt. As a result, stopping the motor does not affect any FoI.

```
while (true) {
   objectAtStart = sensorStart.readProperty(objectDetected);
   objectAtEnd = sensorEnd.readProperty(objectDetected);
   if (objectAtStart && !objectAtEnd) {
      motor.invokeAction(run);
   } else {
      motor.invokeAction(stop);
}}}
```

**Listing 2.1.** Controller script based on the `ConsumedThing` interface that moves objects on the conveyor belt. This controller script has a bug: It does not set the `speedtSetpoint`.

```
motor.setActionHandler(run, () => {
   if (speedSetpoint > 0 && objectAtStartFoI) {
      objectAtStartFoI = false; objectAtEndFoI = true;
   } else {
      if (speedSetpoint > 0 && objectAtEndFoI) {
         objectAtEndFoI = false;
}}});
motor.setActionHandler(stop, () => {});
motor.setPropertyReadHandler(speedSetpoint, () => {
   return speedSetpoint; });
motor.setPropertyWriteHandler(speedSetpoint, (arg) => {
   speedSetpoint = arg; });
sensorStart.setPropertyReadHandler(objectDetected, () => {
   return objectAtStartFoI; });
sensorEnd.setPropertyReadHandler(objectDetected, () => {
   return objectAtEndFoI; });
```

**Listing 2.2.** Thing script with motor and sensor handlers based on the `ExposedThing` interface.

## 2.2 Model Checking by Example

SaVeWoT is designed to reuse the syntax of existing W3C WoT standards while ensuring formally verifiable execution semantics. For each handler (Listing 2.2) and the controller (Listing 2.1), the SaVeWoT compiler adds a process to the SPIN model, and for each FoI, a variable is introduced. The controller can interact with Things using channels (see Sect. 4.1), and standalone Things can change the physical world by changing FoI variables. The FoI variable for `objectAtStartFoI` is initialized as true and `objectAtEndFoI` as false to represent that there is an object at the start.

The example controller script contains a bug that causes the WoT system to not move objects: the motor driving the belt must have its speed set point set to a value $> 0$ to move any object on the conveyor belt, which the controller does not do. With SaVeWoT and model checking, an engineer can specify the requirement "the conveyor belt must eventually move objects to the end of the belt," using the "existence" LTL pattern [11] "`objectAtEndFoI` must be true at least once" (see Sect. 4.2). Checking the LTL pattern on the SPIN model shows that the program does not satisfy the pattern.

## 3 Related Work

We group previous work into behavioral extensions to WoT and formal verification, including translation approaches, of control, IoT, and WoT systems.

### 3.1 W3C Web of Things Behavior

Several proposed extensions for WoT describe behavior through annotations. Such extensions include permitted sequences of interactions [12], JSON-based descriptions corresponding to UML sequence diagrams that represent mashups [13], effects of action affordances on property affordances [14], effects expressed as JavaScript embedded in TDs [15], and extending TDs with physics simulations [16]. Generally, these approaches just represent applications built on top of WoT interfaces and do not support any form of formal verification or mechanisms to enforce restrictions or requirements.

### 3.2 Formal Verification

Recently, multiple tools and approaches have been proposed where a programming language is used to generate both models and deployments. Lin *et al.* [3] propose model checking models generated from Lingua Franca, a coordination language for Cyber-Physical Systems. Similarly, for distributed systems, Hackett *et al.* [17] present the language MPCal and the compiler PGo to translate TLA+ models [18] to Go. Lastly, Lattuada *et al.* [19] enable formal verification of a subset of the Rust language for systems programming.

Related work on formally verifying control systems is primarily based on Programmable Logic Controllers (PLCs). Approaches with a focus on usability using LTL patterns [20], on modeling communication link failures [21], or on verifying communication time constraints [22] have been proposed. An early case study for model checking a production cell [23,24] modeled a production cell without concrete PLCs.

Previous work [25–29] on model checking IoT systems focused on TAP (also called Event-Condition-Action (ECA)) rules and interactions between rules/programs due to a focus on smart home IoT. The general goal is to verify that rules do not cause bad situations or interfere with each other. Some approaches focus purely on the interactions of rules, while others incorporate models for IoT devices, traces of the system, or timed automata. Instead of TAP rules, we use a DSL based on JavaScript to ensure compatibility with existing WoT standards.

Krishna *et al.* [4] model check WoT systems based on a predecessor of the W3C WoT. The authors create models of Things from just TDs using the following assumptions: Actions directly cause the emission of corresponding events, actions modify the properties of a Thing, and Things cannot be composed to create composite Things. SaVeWoT, in contrast, does not restrict how interaction affordances relate to each other and instead relies on the descriptions of Thing behavior in scripts.

# 4 Modeling Things and the Physical World

We first introduce the components of a WoT system and next show how the architectural composition of Things is represented in SaVeWoT. We explain how the physical world, communication and behavior are described in SaVeWoT and finally show how these previous design decisions enable formally verifying effects on the physical world via LTL formulas.

## 4.1 WoT System Components

In SaVeWoT, a WoT system consists of Things and controllers. Formally, we define a set of Things $\mathsf{T}$ and a set of controllers $\mathsf{C}$. Each Thing can be uniquely identified by an IRI. A Thing can have property affordances $\mathsf{P}$, action affordances $\mathsf{A}$, and event affordances $\mathsf{E}$. Conceptually, interaction affordances are uniquely identifiable individuals and belong to exactly one Thing, while a Thing may have multiple affordances: $\mathsf{pAff} : \mathsf{P} \rightarrow \mathsf{T}$, $\mathsf{aAff} : \mathsf{A} \rightarrow \mathsf{T}$ and $\mathsf{eAff} : \mathsf{E} \rightarrow \mathsf{T}$.

Property affordances are either read-only, write-only or read-write. We introduce a set of possible read-write combinations $\mathsf{RW} = \{r, w, rw\}$ and assign one to each property affordance: $\mathsf{propRW} : \mathsf{P} \rightarrow \mathsf{RW}$. Reading a property returns data to the controller synchronously, while writing pushes data to the Thing.

Action affordances are invoked by controllers. When invoking an action affordance, a controller can provide input parameters and receive output data.

Finally, event affordances enable a controller to subscribe to receive notifications from a Thing. The Thing can then emit events that contain data to the controller.

**Composition.** SaVeWoT enables the design of subsystems through the composition of Things. An example of such a subsystem is the initial conveyor belt example. If a conveyor belt is also modeled as a Thing with interaction affordances, the belt is a composite Thing that controls a motor. Composite

Things are Things that control other Things. Relationships between Things can be expressed using the SSN ontology through the hosts relation.

**Restrictions Based on W3C Implementation Report TDs.** For the development of SaVeWoT, we surveyed the TDs from the W3C WoT TD implementation report[1] using SPARQL queries to find an expressive and often used subset of interaction affordances and data types (i.e., data schemas) to limit the state space of WoT system models. The surveyed TDs contain 90 Things with 368 property affordances, 66 action affordances and five event affordances. Due to the low prevalence of event affordances, the initial version of SaVeWoT does not support event notifications and instead focuses on the more relevant property and action affordances.

**Modeling the Physical World and Standalone Things.** The initial conveyor belt example showed how FoIs can be used. FoIs can cover a wide variety of aspects of the physical world: FoIs can represent a window, a room in a house, or the house itself or the physical space deployed in front of a sensor[2]. Running a conveyor belt's motor moves objects on the belt toward the sensor. We model that sensors measure the results of actuations by introducing global variables for FoIs in the system model. Actuators can change a FoI variable's value, and sensors read the value. How the actuator changes the value is described in a script.

```
@prefix td: <https://www.w3.org/2019/wot/td#> .
@prefix sosa: <http://www.w3.org/ns/sosa/> .
@prefix ssn: <http://www.w3.org/ns/ssn/> .
_:objectAtEndFoI a sosa:FeatureOfInterest .
[ a td:PropertyAffordance ;
    ssn:forProperty [ a sosa:ObservableProperty ;
                ssn:isPropertyOf _:objectAtEndFoI ]].
[ a td:ActionAffordance ;
    ssn:forProperty [ a sosa:ActuatableProperty ;
                ssn:isPropertyOf _:objectAtEndFoI ]].
```

**Listing 4.1.** TD (in Turtle) with a FoI and two interaction affordances (based on [30]).

Formally, we introduce a set $\mathsf{F}$ of all FoIs. The relation between FoIs and Things is based on previous work by Charpenay and Käbisch [30] and the W3C WoT TD standard [7, sec. 7.1] that aligns SOSA/SSN with the TD ontology. Listing 4.1 shows an RDF example of the FoIs of a `objectDetected` property affordance and a `run` action affordance in a conveyor belt system. An interaction affordance is related to an observable (for sensors) or actuatable (for actuators) property (*a different concept than property affordances*). A property is associated

[1] https://w3c.github.io/wot-thing-description/testing/report.html; available as a single JSON-LD file at https://www.vcharpenay.link/talks/td-sem-interop.html.

[2] See https://www.w3.org/TR/vocab-ssn/#SOSASample.

with (`ssn:isPropertyOf`) a FoI. For our formalization, we ignore properties and instead introduce a direct relation from affordances to FoIs: $hasFoI \subseteq (\mathsf{P} \cup \mathsf{A} \cup \mathsf{E}) \times \mathsf{F}$.

**Interactions as Channel Communication.** Verifiable models consist of sequential processes communicating with each other using channels, which are FIFO queues containing messages [31]. Due to our focus on property and action affordances, we can model interaction affordances using a request-response pattern. When a controller uses an interaction affordance, in the SPIN model, the controller sends a request message on a channel and immediately waits for a response message.

In the translation to SPIN, we introduce a channel for each interaction affordance operation of each Thing and translate controllers and each affordance handler to processes that interact via channels. As a result, different handlers of a Thing can be executed concurrently. However, one controller can use only one handler at a time.

### 4.2 Specifying Required Physical Effects

With SaVeWoT we aim to enable formal verification of effects on the physical world. For formal verification, correctness properties that the system must fulfill, i.e., requirements, are specified in a temporal logic. LTL has emerged as one of the main temporal logics for model checking [32].

LTL is based on propositional logic. Thus, LTL formulas contain atomic propositions (APs) that are either *true* or *false* in a system model's states. Our approach aims to verify effects on the physical world, which we achieve by introducing FoIs as global variables. Since FoIs are variables, they can be used in LTL formulas in value comparisons. For example, a presence sensor measuring occupancy of a physical space can be expressed as the AP `objectDetected=true`.

Additionally, LTL formulas consist of the operators from propositional logic (e.g., $\rightarrow$, $\neg$, $\vee$, $\wedge$) as connectives between APs, and temporal operators can be used as connectives to refer to time. The relevant temporal operators are: "Always" ($\Box\varphi$) means the system model always has to satisfy $\varphi$, "Eventually" ($\Diamond\varphi$) means the system model has to satisfy $\varphi$ now or at some point in the future, and "Until" ($\varphi_1 \,\mathcal{U}\, \varphi_2$) means the model has to satisfy $\varphi_1$ until the model satisfies $\varphi_2$.

- Eventually there is an object at the end of the conveyor belt (*existence pattern*): $\Diamond$`objectAtEndFoI`
- An object on a conveyor belt cannot be at two locations (FoIs) at the same time (*absence pattern*): $\Box\neg$(`objectAtStartFoI` $\wedge$ `objectAtEndFoI`)
- An object progresses from the start to the end of the belt (*response pattern*): $\Box$(`objectAtStartFoI` $\rightarrow$ $\Diamond$`objectAtEndFoI`)
- Trivial LTL formula that is satisfied by any system model (used in Sect. 7): $\Box\, true$

– After an object is at the start of a belt, the object must be detected at the end next (used in Sect. 7): $\square(\texttt{objectAtStartFoI} \rightarrow \texttt{objectAtStartFoI}\,\mathcal{U}\,(\neg\texttt{objectAtStartFoI} \land (\neg\texttt{objectAtStartFoI}\,\mathcal{U}\,\texttt{objectAtEndFoI})))$

LTL formulas, including nested formulas, are generally difficult to grasp for non-experts [33]. A solution could be LTL patterns or templates that enable non-experts to specify correctness properties [34].

## 5 Behavior Language

Now we present the formal grammar of the SaVeWoT language. The language is a subset of JavaScript combined with the WoT Scripting API. The grammar has C-style control flow with `if` and `while` statements. We split the grammar into two parts, the grammar for controllers and the grammar for Things.

### 5.1 Controller Scripts

```
ControllerScript ::= statement+
statement ::= ifStmt | whileStmt | exprStmt | assignStmt | interactAffStmt
ifStmt ::= "if" parenExpr blockStmt ("else" blockStmt)?
whileStmt ::= "while" parenExpr blockStmt
exprStmt ::= expression";"
assignStmt ::= idOrField "=" (interactAffExpr | expression)";"
interactAffStmt ::= ID "." (invActExprStmt | writePropStmt)";"
parenExpr ::= "(" expression ")"
blockStmt ::= "{" statement* "}"
interactAffExpr ::= ID "." (invActExprStmt | readPropExpr)
expression ::= term (relationExpr | binOpExpr)? | "!"term
invActExprStmt ::= "invokeAction(" ID ("," term)? ")"
writePropStmt ::= "writeProperty(" ID "," term ")"
readPropExpr ::= "readProperty(" ID ")"
relationExpr ::= ("<" | ">" | "==" | "≤" | "≥" | "≠" | "&&" | "||") term
binOpExpr ::= ("+" | "−" | "×" | "÷") term
term ::= idOrField | INT | STRING | BOOL | parenExpr
idOrField ::= ID | ID "." ID
```

**Listing 5.1.** Simplified controller grammar based on the `ConsumedThing` interface for client interactions.

Each controller script represents the behavior of one controller. Listing 5.1 shows the formal grammar of the language for controller scripts in W3C EBNF[3]. The WoT Scripting API `ConsumedThing` interface is integrated by using expressions (**Expr*) for interaction affordances with output and by using statements (**Stmt*) for affordances without output. Statements are nonterminals that do not produce any output and represent an instruction to do something, while expressions are evaluated and stand for a value that is returned. For easier translation

[3] https://www.w3.org/TR/2010/REC-xquery-20101214/#EBNFNotation.

while ensuring correct operational semantics, the production rules restrict interaction affordance expressions to only appear on the right-hand side of assignment statements and not, e.g., within an `if` statement's condition.

### 5.2 Thing Scripts

```
ThingScript ::= (handler)+
handler ::= ID "." (invActHandler| writePropHandler
                   | readPropHandler)";"
invActHandler ::= "setActionHandler(" ID
                  ", (" ID? ") => " (blockStmt| blockStmtWithRet) ")"
writePropHandler ::= "setPropertyWriteHandler(" ID
                  ", (" ID ") => " blockStmt ")"
readPropHandler ::= "setPropertyReadHandler(" ID
                  ", () => " blockStmtWithRet ")"
blockStmtWithRet ::= "{" statement* returnStmt "}"
returnStmt ::= "return" expression";"
```

**Listing 5.2.** Abbreviated Thing grammar based on the `ExposedThing` interface for server interactions.

Listing 5.2 shows the formal grammar of the language for Thing scripts. A handler program is a set of handlers for one or more Things. Handlers implement the `ExposedThing` interface. A handler can act as a controller because the handler bodies consist of *statements* from the controller script syntax (Listing 5.1), which enables composite Things. Like the controller grammar, the Thing grammar enforces that reading a property affordance returns a value but does not accept any argument, and vice versa for writing property affordances.

## 6 Implementation of the Compiler

We implement a proof-of-concept (PoC) of SaVeWoT in JavaScript using the Chevrotain parser library[4]. Our executable scripts use the WoT Scripting API implementation `node-wot`[5]. As a target model checker, we chose SPIN [35] as Fu *et al.* [36] already showed that SPIN can verify various Web Services based on behavior descriptions. The translation workflow from TDs and scripts to models can be seen in Fig. 2. The visitors iterate over the syntax trees and emit partial models, which are assembled to a PROMELA model by the linker. The visitors and the linker use the sets and relations used to formalize WoT systems (extracted from TDs using SPARQL) to link all behavior descriptions and variables in the model together.

[4] https://chevrotain.io.

[5] https://github.com/eclipse-thingweb/node-wot.

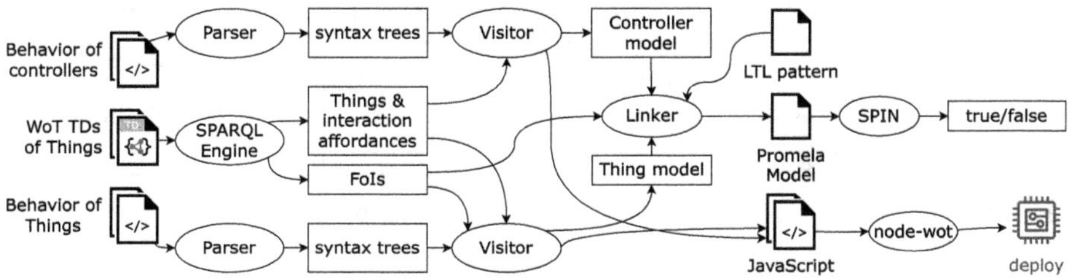


**Fig. 2.** Dataflow diagram showing the translation of TDs, scripts and correctness requirements to the SPIN model checker.

The input language of SPIN, PROMELA, supports channel and C-style control flow (`if` and `while`), which makes the translation from the SaVeWoT language to PROMELA straightforward. The PoC handles the differences in blocking semantics and variable scope between JavaScript and PROMELA. The translation to JavaScript avoids the event loop (JavaScript's built-in concurrency mechanism) because its operational semantics are more complicated [37] than the operational semantics of PROMELA [38]. We instead facilitate concurrency by creating separate scripts and using mutex locks during translation by the visitors.

## 7 Evaluation

To study the feasibility of our approach for large control systems with many components, we evaluate PROMELA models translated by our PoC on consumer hardware with an Intel i5-8295U CPU and 16 GB of RAM. We consider a series of conveyor belts orchestrated by a controller. Each belt has a motor and an ultrasonic sensor, which interact with the physical environment via a FoI. The controller implements a control loop for each belt. The belt's motor runs until a certain distance is observed, after which the motor stops. We evaluate the number of states, verification time, and verification memory consumption as measured by SPIN for PROMELA models.

First we increase the number of belts from one to 58, which increases the number of Things (two per belt) from two to 116 respectively. Figure 3 (left) shows the results. The number of states is linear in the number of Things. The initial model with two Things has 76 states, which is increased to 4180 states for 116 Things. The memory consumption appears almost linear, starting at 128.73 MB for one belt and increasing to 134.39 MB for 116 Things. The verification time exhibits a step-wise pattern. SPIN measures zero ms for up to 30 Things and 60 ms for 116 Things.

In a second step, we verify the system with 28 belts (the highest number of belts with a verification time of 10 ms) for increasingly large conjunctive LTL formulas on the FoIs of each belt. We generate the verifier for up to 18 belt motors in the LTL formula. For LTL formulas longer than 18 belts SPIN takes multiple hours to generate a verifier, independently of the verification time.

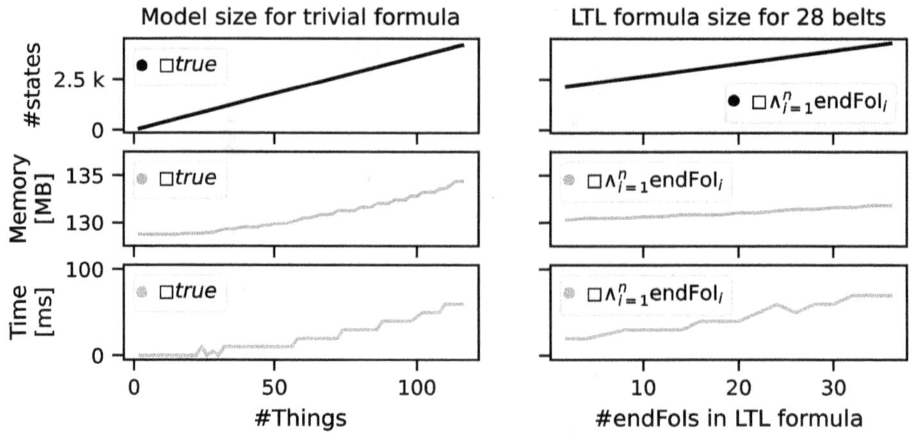


**Fig. 3.** Number of states, memory consumption and verification time for increasing model and LTL formula sizes.

Figure 3 (right) shows that the state space is linear in the number of belts (up to 4288 states), while the memory footprint increases linearly (up to 131.86 MB). The verification time is between 20 and 70 ms.

# 8 Conclusion and Future Work

We presented SaVeWoT, an approach to formally verify that WoT systems fulfill requirements that specify desired or undesired physical effects. The main components of SaVeWoT are a model of the physical world based on SOSA/SSN and WoT, which enables expressing requirements on physical effects in LTL, and a language that can be translated to models and deployments of WoT systems. We evaluate our approach based on the model checker SPIN on consumer hardware. Our evaluation shows that SaVeWoT is feasible for verifying large WoT systems on consumer hardware. As a result, SaVeWoT can be used to check whether various subsystems, when integrated, exhibit the intended effects on the physical world.

Because we inherit various limitations from translating to SPIN, we plan to explore different tools like UPPAAL, NUXMV or LTL satisfiability checking tools like BLACK [39] in the future. We expect these tools to provide better performance because they rely on modern and performant satisfiability solvers. Further, we aim to integrate event affordances into SaVeWoT and want to incorporate graph-structured process representations [40] in the future.

**Acknowledgements.** This work was funded by the German Federal Ministry for Economic Affairs and Climate Action (BMWK) through the Antrieb 4.0 project (Grant No. 13IK015B).

## References


1. Thuluva, A.S., Anicic, D., Rudolph, S., Adikari, M.: Semantic Node-RED for rapid development of interoperable industrial IoT applications. Semant. Web **11**(6), 949–975 (2020). https://doi.org/10.3233/SW-200405
2. Ramanathan, G., Husmann, M., Mayer, S.: Interoperability vs. tradition: benefits and challenges of web of things in building automation. In: IoT 2021: 11th International Conference on the Internet of Things, 8–12 November 2021, pp. 57–63. ACM (2021). https://doi.org/10.1145/3494322.3494330
3. Lin, S., et al.: Towards building verifiable CPS using lingua franca. ACM Trans. Embed. Comput. Syst. **22**(5s) (2023). https://doi.org/10.1145/3609134
4. Krishna, A., Le Pallec, M., Mateescu, R., Salaün, G.: Design and deployment of expressive and correct web of things applications. ACM Trans. Internet Things **3**(1) (2021). https://doi.org/10.1145/3475964
5. Kis, Z., Peintner, D., Aguzzi, C., Hund, J., Nimura, K.: Web of Things (WoT) Scripting API. Working group note, W3C (2020). https://www.w3.org/TR/2020/NOTE-wot-scripting-api-20201124/
6. Cena, F., Haller, A., Lefrançois, M.: Semantics in the edge: sensors and actuators in the web of linked data and things. Semant. Web **11**(4), 571–580 (2020). https://doi.org/10.3233/SW-200379
7. Käbisch, S., Kamiya, T., McCool, M., Charpenay, V., Kovatsch, M.: Web of Things (WoT) Thing Description. Recommendation, W3C (2020). https://www.w3.org/TR/2020/REC-wot-thing-description-20200409/
8. Haller, A., et al.: The modular SSN ontology: a joint W3C and OGC standard specifying the semantics of sensors, observations, sampling, and actuation. Semant. Web **10**(1), 9–32 (2019)
9. Freund, M., Fries, J., Dorsch, R., Schiller, P., Harth, A.: WoT2Pod: an architecture enabling an edge-to-cloud continuum. In: Proceedings of the 13th International Conference on the Internet of Things, IoT 2023, pp. 42–49. Association for Computing Machinery, New York (2024). https://doi.org/10.1145/3627050.3627063
10. Kovatsch, M., Matsukura, R., Lagally, M., Kawaguchi, T., Toumura, K., Kajimoto, K.: Web of Things (WoT) Architecture. Recommendation, W3C (2020). https://www.w3.org/TR/2020/REC-wot-architecture-20200409/
11. Giacomo, G.D., Masellis, R.D., Montali, M.: Reasoning on LTL on finite traces: insensitivity to infiniteness. In: Proceedings of the Twenty-Eighth AAAI Conference on Artificial Intelligence, Québec City, Québec, Canada, 27–31 July 2014, pp. 1027–1033. AAAI Press (2014). https://doi.org/10.1609/aaai.v28i1.8872
12. Korkan, E., Kaebisch, S., Kovatsch, M., Steinhorst, S.: Safe interoperability for web of things devices and systems. In: Kazmierski, T.J., Steinhorst, S., Große, D. (eds.) Languages, Design Methods, and Tools for Electronic System Design. LNEE, vol. 611, pp. 47–69. Springer, Cham (2020). https://doi.org/10.1007/978-3-030-31585-6_3
13. Kast, A., Korkan, E., Käbisch, S., Steinhorst, S.: Web of things system description for representation of mashups. In: IEEE International Conference on Omni-Layer Intelligent Systems, COINS 2020, 31 August–2 September 2020, pp. 1–8 (2020).https://doi.org/10.1109/COINS49042.2020.9191677
14. Salama, F., Korkan, E., Käbisch, S., Steinhorst, S.: Towards a behavioral description of cyber-physical systems using the thing description. In: Proceedings of the 2021 Workshop on Descriptive Approaches to IoT Security, Network, and Application Configuration, DAI-SNAC 2021, pp. 6–9. Association for Computing Machinery (2021). https://doi.org/10.1145/3488661.3494030

15. Mena, M., Criado, J., Iribarne, L., Corral, A.: Defining interactions of WoT servients with causality relations. In: Proceedings of the 13th International Conference on Management of Digital EcoSystems, MEDES 2021, pp. 112–119. Association for Computing Machinery (2021). https://doi.org/10.1145/3444757.3485102
16. Salama, F., Tsirkunenko, A., Korkan, E., Käbisch, S., Steinhorst, S.: WoT-PhyngSim: integrating physics simulations with IoT digital twins using the web of things. In: IEEE International Conference on Omni-Layer Intelligent Systems, COINS 2023, 23–25 July 2023, pp. 1–8 (2023). https://doi.org/10.1109/COINS57856.2023.10189326
17. Hackett, F., Hosseini, S., Costa, R., Do, M., Beschastnikh, I.: Compiling distributed system models with PGo. In: Proceedings of the 28th ACM International Conference on Architectural Support for Programming Languages and Operating Systems, ASPLOS 2023, 25–29 March 2023, vol. 2, pp. 159–175. ACM (2023). https://doi.org/10.1145/3575693.3575695
18. Lamport, L.: Specifying Systems. Addison-Wesley, The TLA+ Language and Tools for Hardware and Software Engineers (2002)
19. Lattuada, A., et al.: Verus: verifying rust programs using linear ghost types. Proc. ACM Program. Lang. **7**(OOPSLA1) (2023). https://doi.org/10.1145/3586037
20. Adiego, B.F., Darvas, D., Tournier, J., Viñuela, E.B., Suárez, V.M.G.: Bringing automated model checking to PLC program development - a CERN case study. In: 12th International Workshop on Discrete Event Systems, WODES 2014, 14–16 May 2014, pp. 394–399. International Federation of Automatic Control (2014). https://doi.org/10.3182/20140514-3-FR-4046.00051
21. Lesi, V., Jakovljevic, Z., Pajic, M.: Reliable industrial IoT-based distributed automation. In: Proceedings of the International Conference on Internet of Things Design and Implementation, IoTDI 2019, Montreal, QC, Canada, 15–18 April 2019, pp. 94–105. ACM (2019). https://doi.org/10.1145/3302505.3310072
22. Vogel-Heuser, B., Folmer, J., Frey, G., Liu, L., Hermanns, H., Hartmanns, A.: Modeling of networked automation systems for simulation and model checking of time behavior. In: International Multi-Conference on Systems, Signals & Devices, SSD 2012, 20–23 March 2012, pp. 1–5. IEEE (2012). https://doi.org/10.1109/SSD.2012.6197943
23. Lewerentz, C., Lindner, T.: Formal Development of Reactive Systems: Case Study Production Cell. LNCS, vol. 891. Springer, Heidelberg (1995). https://doi.org/10.1007/3-540-58867-1
24. Paun, D.O., Marsha, C., Biechele, B.: Production cell revisited. In: Proceedings of SPIN 1998 (1998)
25. Zhang, L., He, W., Martinez, J.J., Brackenbury, N., Lu, S., Ur, B.: AutoTap: synthesizing and repairing trigger-action programs using LTL properties. In: Proceedings of the 41st International Conference on Software Engineering, ICSE 2019, Montreal, QC, Canada, 25–31 May 2019, pp. 281–291. IEEE/ACM (2019). https://doi.org/10.1109/ICSE.2019.00043
26. Trimananda, R., Aqajari, S.A.H., Chuang, J., Demsky, B., Xu, G.H., Lu, S.: Understanding and automatically detecting conflicting interactions between smart home IoT applications. In: Proceedings of the 28th ACM Joint Meeting on European Software Engineering Conference and Symposium on the Foundations of Software Engineering, ESEC/FSE 2020, pp. 1215–1227. Association for Computing Machinery (2020). https://doi.org/10.1145/3368089.3409682
27. Yu, Y., Liu, J.: TAPInspector: safety and liveness verification of concurrent trigger-action IoT systems. IEEE Trans. Inf. Forensics Secur. **17**, 3773–3788 (2022). https://doi.org/10.1109/TIFS.2022.3214084

28. Kashaf, A., Sekar, V., Agarwal, Y.: Protecting smart homes from unintended application actions. In: 13th ACM/IEEE International Conference on Cyber-Physical Systems, ICCPS 2022, Milano, Italy, 4–6 May 2022, pp. 270–281 (2022). https://doi.org/10.1109/ICCPS54341.2022.00031
29. Alhanahnah, M., Stevens, C., Chen, B., Yan, Q., Bagheri, H.: IoTCom: dissecting interaction threats in IoT systems. IEEE Trans. Softw. Eng. **49**(4), 1523–1539 (2023). https://doi.org/10.1109/TSE.2022.3179294
30. Charpenay, V., Käbisch, S.: On modeling the physical world as a collection of things: the W3C thing description ontology. In: Harth, A., et al. (eds.) ESWC 2020. LNCS, vol. 12123, pp. 599–615. Springer, Cham (2020). https://doi.org/10.1007/978-3-030-49461-2_35
31. Brand, D., Zafiropulo, P.: On communicating finite-state machines. J. ACM **30**(2), 323–342 (1983). https://doi.org/10.1145/322374.322380
32. Biere, A., Artho, C., Schuppan, V.: Liveness checking as safety checking. Electron. Notes Theor. Comput. Sci. **66**(2), 160–177 (2002). https://doi.org/10.1016/S1571-0661(04)80410-9. fMICS 2002, 7th International ERCIM Workshop in Formal Methods for Industrial Critical Systems
33. Greenman, B., Saarinen, S., Nelson, T., Krishnamurthi, S.: Little tricky logic: misconceptions in the understanding of LTL. Art Sci. Eng. Program. **7**(2) (2023). https://doi.org/10.22152/programming-journal.org/2023/7/7
34. Dwyer, M.B., Avrunin, G.S., Corbett, J.C.: Patterns in property specifications for finite-state verification. In: Proceedings of the 21st International Conference on Software Engineering, ICSE 1999, pp. 411–420. Association for Computing Machinery, New York (1999). https://doi.org/10.1145/302405.302672
35. Holzmann, G.J.: The SPIN Model Checker - Primer and Reference Manual. Addison-Wesley (2004)
36. Fu, X., Bultan, T., Su, J.: Analysis of interacting BPEL web services. In: Proceedings of the 13th International Conference on World Wide Web, WWW 2004, 17–20 May 2004, pp. 621–630. ACM, New York (2004). https://doi.org/10.1145/988672.988756
37. Loring, M.C., Marron, M., Leijen, D.: Semantics of asynchronous JavaScript. In: Proceedings of the 13th ACM SIGPLAN International Symposium on on Dynamic Languages, DLS 2017, pp. 51–62. Association for Computing Machinery (2017). https://doi.org/10.1145/3133841.3133846
38. Baier, C., Katoen, J.: Principles of Model Checking. MIT Press, Cambridge (2008)
39. Geatti, L., Gigante, N., Montanari, A.: BLACK: a fast, flexible and reliable LTL satisfiability checker. In: Proceedings of the 3rd Workshop on Artificial Intelligence and Formal Verification, Logic, Automata, and Synthesis hosted by the Twelfth International Symposium on Games, Automata, Logics, and Formal Verification (GandALF 2021), Padua, Italy, 22 September 2021. CEUR Workshop Proceedings, vol. 2987, pp. 7–12. CEUR-WS.org (2021). https://ceur-ws.org/Vol-2987/paper2.pdf
40. Harth, A., Käfer, T., Rula, A., Calbimonte, J.P., Kamburjan, E., Giese, M.: Towards representing processes and reasoning with process descriptions on the web. Trans. Graph Data Knowl. **2**(1), 1:1–1:32 (2024). https://doi.org/10.4230/TGDK.2.1.1

# Active Learning in Multi-label Classification of Bioacoustic Data

Hannes Kath[1,2(✉)], Thiago S. Gouvêa[1], and Daniel Sonntag[1,2]

[1] German Research Center for Artificial Intelligence (DFKI), Oldenburg, Germany
{hannes.kath,thiago.gouvea,daniel.sonntag}@dfki.de

[2] Applied Artificial Intelligence (AAI), University of Oldenburg, Oldenburg, Germany

**Abstract.** Passive Acoustic Monitoring (PAM) has become a key technology in wildlife monitoring, providing vast amounts of acoustic data. The recording process naturally generates multi-label datasets; however, due to the significant annotation time required, most available datasets use exclusive labels. While active learning (AL) has shown the potential to speed up the annotation process of multi-label PAM data, it lacks standardized performance metrics across experimental setups. We present a novel performance metric for AL, the 'speedup factor', which remains constant across experimental setups. It quantifies the fraction of samples required by an AL strategy compared to random sampling to achieve equivalent model performance. Using two multi-label PAM datasets, we investigate the effects of class sparsity, ceiling performance, number of classes, and different AL strategies on AL performance. Our results show that AL performance is superior on datasets with sparser classes, lower ceiling performance, fewer classes, and when using uncertainty sampling strategies.

**Keywords:** Passive Acoustic Monitoring · Multi-label data · Active Learning · Transfer Learning · Speedup Factor

## 1 Introduction

***Passive acoustic monitoring (PAM)*** has emerged as a powerful technology for wildlife monitoring, allowing researchers and biodiversity managers to gather extensive acoustic data without disturbing natural habitats [21,22]. PAM systems continuously record sounds from various environments, offering valuable insights into animal behavior, species richness, and ecosystem health, with important applications in ecosystem management, rapid assessments of biodiversity [20], and basic research [17]. However, effectively utilizing this vast amount of data for sound event detection poses significant challenges due to the need for annotated data to train machine learning models. The low quality of automatically generated annotations for PAM datasets often requires manual annotation.

***Multi-label PAM Datasets.*** Because PAM datasets are typically recorded without controlling the sources of sound events, PAM recordings are naturally


A. Hotho and S. Rudolph (Eds.): KI 2024, LNAI 14992, pp. 114–127, 2024.
https://doi.org/10.1007/978-3-031-70893-0_9

multi-label datasets, meaning that multiple labels can be assigned to a single point in time. However, there is a lack of availability of multi-label PAM datasets. The main reasons for this scarcity are that only 21% of the data examined in the literature is published [1], and that manually labelling a PAM dataset with multi-label annotations can be extremely time-consuming and costly, typically requiring annotation times of 10 to 15 times the duration of the audio [12]. For this reason, the vast majority of PAM datasets are annotated with exclusive labels, neglecting the multi-label aspect. Few research efforts are dedicated to the development of multi-label annotation tools for PAM datasets [8,12].

***Active learning (AL)*** is a promising research direction for accelerating the annotation and analysis of large multi-label PAM datasets [9]. AL provides an iterative strategic approach to prioritize the most informative samples for annotation. Unlike random sampling of data points, AL algorithms intelligently select samples that accelerate model convergence (see Fig. 1). AL strategies can be categorized into prediction-based, data-based and model-based strategies [24]. Prediction-based strategies use model outputs to select the next batch of samples, e.g., uncertainty sampling prioritizes samples close to the decision boundary. Data-based strategies select samples based on the internal structure of the data, e.g., diversity sampling selects batches that cover the entire input space. Model-based strategies select samples based on the change in the model, e.g., prioritizing samples with high model influence regardless of the label. While model-based strategies directly aim to improve model performance, they are rarely used in practice due to the high computational cost of training the model for all samples in all label combinations [14]. Multi-label AL is not a new task, yet this research field has gained popularity only recently [26]. Various approaches have been proposed to the problem of selecting samples while optimizing for multiple classes [13,16,26], with the practice of pooling results from established strategies applied to binary classifiers being used as a baseline [14]. The evaluation of AL performance is primarily done by visually comparing the learning curves of the applied sampling strategies [15,19]. While this approach is feasible for a few and clearly separated curves, it complicates the comparison of AL strategies across datasets, experiments, and projects [15]. Evaluation metrics provide statistical justification by quantifying features of the learning curve [15]. The most popular evaluation metrics used in the literature for (multi-label) AL strategies are accuracy [2,23] or multi-class accuracy [10], sensitivity and specificity [11,23], and area under the curve [10,25] or multi-class area under the receiver operating characteristic curve [5,11]. These methods offer numerically comparable results under equal experimental setups, but they are limited by their variation with different numbers of training samples, leading to incomparability when different experimental setups are used.

***Research Contributions.*** We propose a quantitative performance metric for AL termed the speedup factor. The metric represents the fraction of samples required for an AL strategy to achieve the same model performance as when using random sampling. The speedup factor remains constant regardless of the number of training samples, providing a comparable score even across experimental setups. Furthermore, driven by the quest to accelerate the annotation of multi-label PAM datasets, we investigate the performance of AL on a synthetically generated variant of the Watkins Marine Mammal Sound Database (WMMD) [18], as well as on AnuraSet [3], a multi-label PAM dataset of reasonable size. Multi-label datasets often exhibit significant class imbalance, where the amount of samples containing a particular class, termed **positive samples**, varies significantly across classes. We examine the correlation between AL performance and the number of positive samples. Since neural models typically perform best on balanced datasets with 50% positive samples [6], and random sampling on average selects the same fraction of positive samples as in the unlabelled set, we hypothesize that AL performance decreases as the fraction of positive samples in the dataset increases. Conversely, we expect to see an improvement in performance for sparse datasets where random sampling has difficulty selecting positive samples. The maximum performance, termed **ceiling performance**, of a neural model for a given dataset depends on several factors, including model architecture, class similarity, class sparsity, and feature characteristics. We investigate the correlation between AL performance and ceiling performance. Since classes with high ceiling performance have more room for improvement through AL strategies, while classes with low ceiling performance may already be facing challenges and could benefit from AL, we do not hypothesize a clear trend in this regard. The number of classes in multi-label datasets is not fixed, but open-ended. We investigate the correlation between the performance of AL and the number of classes. Since AL strategies are forced to optimize for multiple goals when the dataset contains more classes, we hypothesize that AL performance decreases as the number of classes increases. Due to the high computational cost associated with model-based AL strategies and the large number of experiments, we limit our analysis to one uncertainty-based method and one diversity-based method. Diversity strategies rely on the internal structure of the data and do not consider information from previously sampled data. As a result, they face challenges such as repeatedly sampling from the same clusters and sampling from clusters irrelevant to the classification task [14]. Uncertainty strategies aim to refine the decision boundary by selecting samples close to it, but they can also face challenges such as getting trapped by sampling from small clusters where the decision boundary is unclear, rather than establishing a reliable decision boundary for the entire space [14]. Because diversity sampling strategies select data regardless of model performance and the classes being optimized for, we hypothesize that uncertainty sampling outperforms diversity sampling. The present study tests the following four hypotheses on two multi-label PAM datasets:

- H1 AL performance decreases as the number of positive samples in the dataset increases.
- H2 AL performance changes as the ceiling performance changes.
- H3 AL performance decreases as the number of classes increases.
- H4 AL performance decreases when using diversity sampling strategies instead of uncertainty sampling strategies.

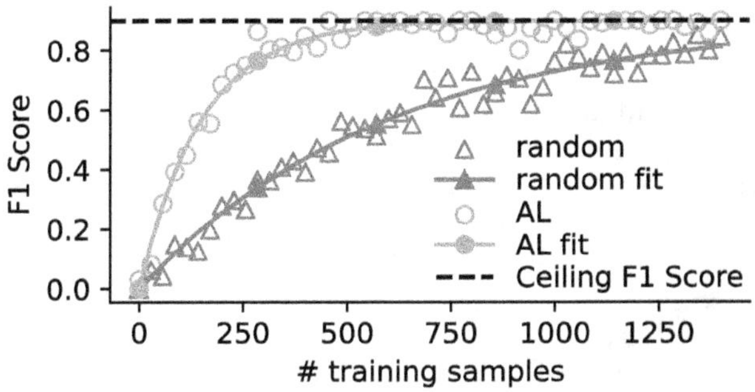


**Fig. 1.** Schematic representation of the learning curve for a neural model employing both random sampling and active learning (AL). The curve is fitted using the function $P = a\left(1 - e^{-\frac{x}{b}}\right)$, where the ceiling F1 score is $a = 0.9$ and the parameterized learning rates are $b_{random} = 600$ and $b_{AL} = 150$.

# 2 Methodology

## 2.1 Active Learning Validation Metric

The learning curves of neural models typically exhibit three prominent features: performance starting at 0 for 0 training samples, a general increase with the addition of training data, and an asymptotic shape with the ceiling performance as the limit. Therefore, the learning curves of neural models can be effectively modeled using the equation

$$P = a\left(1 - e^{-\frac{x}{b}}\right), \tag{1}$$

where $P$ is the model performance, $a$ is the ceiling performance, $b$ is the parameter that represents the learning rate, and $x$ is the number of training samples.

Active learning algorithms are designed to identify and select relevant samples from the unlabelled dataset. As a result, they reduce the number of samples required to achieve a predetermined level of model performance. Figure 1 shows a schematic comparison of learning curves using random sampling and AL. Given the assumption that increasing the number of training samples improves performance, $a$ corresponds to the performance of the model when trained on the entire dataset. The parameter $b$ is calculated from Eq. (1) using the least squares method, using the samples obtained from the experimental data. Therefore, $a$ is constant for all sampling strategies, while $b$ varies depending on the sample selection method. We define the speedup factor as an evaluation metric for AL, representing the fraction of samples required to achieve the same level of performance using AL as compared to random sampling. The speedup factor (S) serves as validation metric for assessing the efficiency of active learning. Inverting Eq. (1) and assuming that $a_{AL} = a_{random}$, the speedup factor is calculated with Eq. (2).

The speedup factor remains constant regardless of the number of training samples; for example, a factor of 0.4 means that the AL strategy requires, on average, 40% of the amount of data that random sampling requires to achieve the same performance. A lower speedup factor indicates that AL requires fewer data samples to achieve comparable performance, while a speedup factor greater than 1 indicates that random sampling is more efficient than AL.

$$S = \frac{x_{AL}}{x_{random}} = \frac{-\ln\left(1 - \frac{P}{a_{AL}}\right) \cdot b_{AL}}{-\ln\left(1 - \frac{P}{a_{random}}\right) \cdot b_{random}} = \frac{b_{AL}}{b_{random}} \quad (2)$$

### 2.2 Active Learning Strategies

To test the AL performance, we opted for simple but interpretable and well-established AL strategies to test hypotheses H1-H4 [14].

We use 'ratio' as the AL uncertainty sampling strategy [14]. The uncertainty score $\Phi$ for a binary classifier is calculated using the equation $\Phi_{bi}(y) = \frac{0.5-|y-0.5|}{0.5+|y-0.5|}$, where $y \in [0;1]$ represents the output of the model prediction. Multi-label classifiers with $n$ species use $n$ binary classifiers, hence $n$ uncertainty scores per sample are computed. To obtain a meaningful single uncertainty score for a sample, we use the maximum of the computed uncertainty scores [9].

Following [14], we implement k-means clustering using the Euclidean distance measure as diversity sampling strategy. Within each cluster, we select the centroid (the sample with the smallest distance to the cluster centre), an outlier (the sample farthest from the nearest cluster centre) and three random samples. The number of clusters is inversely determined; e.g., to annotate 20 samples at a rate of 5 samples per cluster, we use 4 clusters [14].

### 2.3 Experimental Setup

***Datasets.*** In this study we take advantage of AnuraSet, a recently released real-world benchmark multi-label PAM dataset containing 27 h of audio plus manually created expert annotations for 42 species of anurans (frogs and toads) from two different biomes [3]. To the best of our knowledge, this is currently the only freely available multi-label PAM dataset of reasonable size. The original authors divide the one-minute audio files recorded in four different areas into segments of three seconds each, with an overlap of two seconds. This segmentation approach resulted in 58 three-second audio files per minute, increasing the dataset to 77 h of audio. The sample rate is 22.05 kHz. The authors provide a training/evaluation split for all files. The AnuraSet dataset has a high degree of class imbalance among its classes. We used the evaluation set as defined by the original authors. Additionally, we synthesized another multi-label dataset using underwater sound[1] and inserting events from the WMMD [18] at random positions, including the possibility of overlapping events. We segmented the data

[1] https://www.youtube.com/watch?v=sCc3UtzZDEo.

into three-second files. From the total duration of 7.8 h of audio with 40 highly unbalanced classes, the evaluation set (1.6 h) includes 20% of the events from each class.

***Pre-processing.*** Recent studies show that the use of BirdNet [7], a neural model trained on vocal bird recordings, is an effective approach for embedding PAM datasets [4,9]. As a pre-processing step, all three-second files are resampled to 48 kHz and embedded using the penultimate layer of BirdNet 2.4. This layer has an embedding size of 1024, which has shown superior performance [9].

***Training.*** Since the goal of this study is not to construct a state-of-the-art classifier, but to investigate the dependencies of different factors on AL performance, a linear (multi-label) classifier is trained to evaluate the performance of active learning. The resulting architecture consists of a single fully connected layer with an equal number of output nodes as there are species in the dataset. Because the classifier comprises only a single layer without shared weights, it facilitates the assignment of results to different causes. Each output node indicates the presence or absence of a particular species and is independent of the other output nodes. A binary cross-entropy loss function and logistic activation are used since we train a multi-label classifier. The classifiers are trained on frozen BirdNet embeddings (no fine-tuning) for a maximum of 1 000 epochs. Early stopping criteria are based on validation loss, with a minimum delta of 0.1 and a patience of 10 epochs, with reinstatement of the best weights.

***Active learning*** experiments use an initial random selection of 20 samples. Using the sampling strategies random, ratio max, and clustering, 20 samples are added per iteration until a total of 1400 samples (70 iterations) is reached.

***Evaluation.*** All experiments are conducted with 5 random seeds to ensure robustness and reliability of the results. All results are evaluated on the respective evaluation set, ensuring that this set is never used for training purposes. After each sample selection iteration, all selected samples are used to train a linear classifier, which is then evaluated on the evaluation set. Given the highly imbalanced classes, we use the macro F1 score as score aggregation metric for experiments with multi-label datasets to assign equal importance to all species. To compute the speedup factor, we calculate the mean across random seeds of the true positive, false positive and false negative values to compute the mean macro F1 score for the currently used AL sampling strategy and for random sampling. Using the macro F1 score as learning curve, the $b$-values and the respective speedup factor are computed. If the speedup factor cannot be computed, e.g. if the class is extremely sparse and no positive samples are selected for the training set over all iterations, the results are discarded. To compute the ceiling performance, we use all available data from the training split to train the model and average the result over the independent runs.

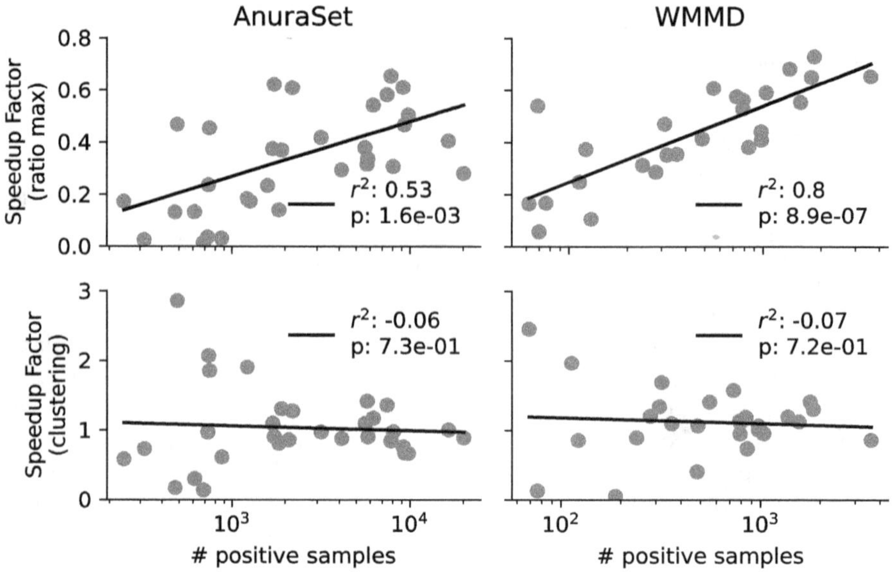


**Fig. 2.** Speedup factor of the single-label datasets over the number of positive samples of the respective class. AnuraSet (left) and WMMD (right) with the active learning strategies ratio max (top) and clustering (bottom).

## 3 Experiments

The following experiments investigate the influence of the number of positive samples, the ceiling performance, the number of classes, and the active learning strategy on the performance of active learning, measured by the speedup factor.

### 3.1 Single-Label Data

To isolate the variables of interest, we first evaluate hypotheses H1, H2 and H4 on single-label dataets generated by iteratively selecting a single class.

To test hypothesis H1, Fig. 2 shows the speedup factor for the single-class datasets of AnuraSet and WMMD over the total number of positive samples for the respective class used. Using ratio max, Fig. 2 (top) shows a strong positive correlation between the speedup factor and the number of positive samples. The speedup factor remains consistently below 1. Conversely, using clustering, Fig. 2 (bottom) shows no significant correlation. The speedup factor is around 1. Figure 3 provides a detailed overview of the results for AnuraSet, showing the learning curves for both a rare and a frequent species over the AL sampling strategies ratio max and clustering. As can be seen in the figures, clustering does not improve model performance compared to random sampling, while ratio max improves model performance, showing higher improvement for rare classes.

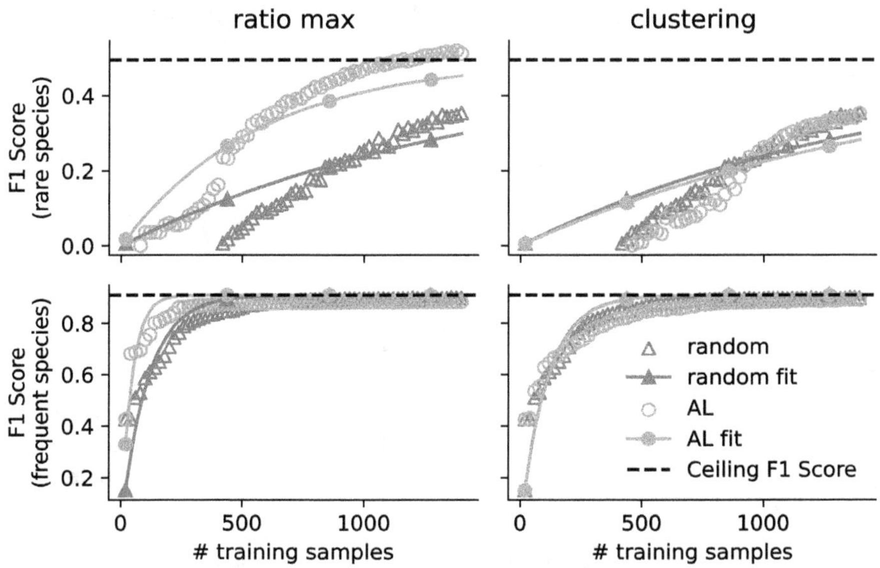


**Fig. 3.** AnuraSet: Learning curves for two single-label datasets. Positive samples: 1.8%, (class ELABIC, top row); 17.4%, (class BOABIS, bottom row). Speedup factors for active learning (AL) strategies: ratio max (left col): 0.3 (ELABIC), 0.5 (BOABIS); clustering (right col): 1.0 (both classes).

Figure 4 (left) shows for AnuraSet for all 3 sampling strategies the fraction of positive samples in the training set over the total amount of training data for a rare class and a frequent class. After 70 iterations, the number of positive samples in the training set for the ratio max sampling strategy is 50.0% for the rare dataset and 49.5% for the frequent dataset. In contrast, for random sampling and clustering, the proportion remains approximately at the default level. Figure 4 (right) shows similar trends for the WMMD dataset, with the difference that for the rare species the fraction for ratio max decreases after a certain number of iterations.

These results indicate that ratio max is effective as an AL strategy for PAM datasets, while clustering does not provide significant benefits. Therefore in the following we present and discuss the results of the AL strategy ratio max.

To test hypothesis H2, Fig. 5 (left) shows the ceiling performance for the single-class datasets over the total number of positive samples for the respective class used. A strong positive correlation between the speedup factor and the number of positive samples is recognisable. Observing a strong correlation between the ceiling performance and the number of positive samples (Fig. 5 (left)) and a strong correlation between the speedup factor and the number of positive samples (Fig. 2 (top)), we also observe a strong positive correlation of the speedup factor and the ceiling performance (Fig. 5 (right)).

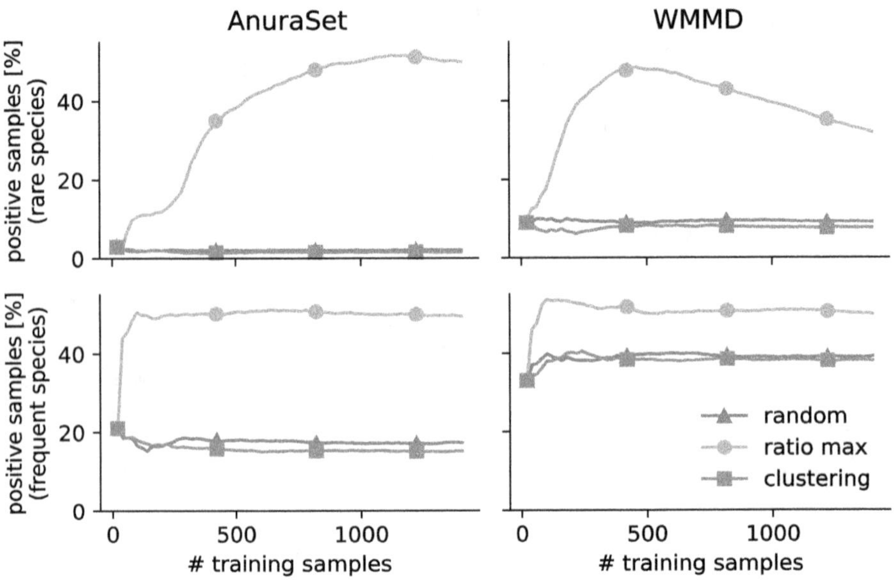


**Fig. 4.** Fraction of positive samples in the training set selected by the sampling strategies random, ratio max and clustering over the total amount of training samples for two single-label datasets from AnuraSet: one with a low number of positive samples in the dataset (left, class ELABIC, 1705 positive samples (1.8%)) and the other with a high number of positive samples in the dataset (right, class BOABIS, 16524 positive samples (17.4%)).

### 3.2 Multi-label Data

To test hypothesis H3, we create multi-label datasets.

To isolate the variable of interest (number of classes), we first create datasets that include classes with similar amounts of positive samples, ceiling performances and speedup factors. Figure 6 (top) shows a strong positive correlation between the speedup factor and the number of classes. Figure 7 shows the fraction of positive samples in the training set selected by the sampling strategy ratio max over the total amount of training samples for multi-label datasets for one class per dataset. The proportion of positive samples for a single class decreases as the number of classes optimised increases. Figure 7 (right) shows that the proportion of selected positive samples begins to decrease after several iterations for the WMMD dataset.

Finally, we used all classes for our analysis. To minimize the variance within the selected datasets regarding the number of positive samples, we first sorted all classes based on their respective number of positive samples. Subsequently, we generated datasets containing 1 to 40 (42) classes. For $n$ classes, we utilized all combinations of classes with neighboring numbers of positive samples, resulting in 40 (42) $- n + 1$ combinations. Figure 6 (bottom) shows the speedup factor over the number of classes with all used combinations for the sampling strategy

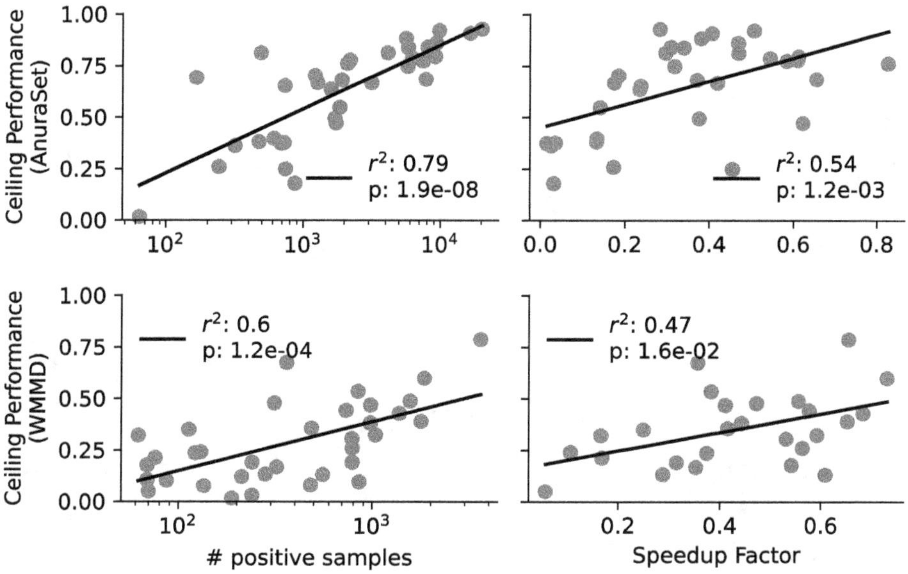


**Fig. 5.** Ceiling performance of the single-label datasets of AnuraSet (top) and WMMD (bottom) over the number of positive samples of the respective class (left) and the speedup factor (right) for the active learning strategy ratio max.

ratio max. The average number of positive samples across all classes used for each dataset is color-coded. There is a strong positive correlation between the number of classes and the speedup factor, where the speedup factor tends to be lower with fewer classes and approaches 1 with more classes. There is also a clear positive correlation between the number of positive samples in the dataset and the speedup factor, with the speedup factor being lower for datasets with fewer positive samples and higher for datasets with more positive samples. While for some combinations the speedup factor exceeds 1, indicating that random sampling outperforms AL, for the majority of combinations it remains below 1, indicating that AL is more efficient in those cases.

## 4 Discussion

In this study, we introduced a quantitative performance measure for AL, the speedup factor, which is independent of the number of training samples and states the fraction of samples needed to reach the same level of performance using AL compared to random sampling. Motivated by the desire to develop a methodology capable of annotating and interpreting large volumes of multi-label PAM data, we investigate the influence of several factors on AL performance. These factors include the number of positive samples, the ceiling performance, the number of classes, and the AL sampling strategy.

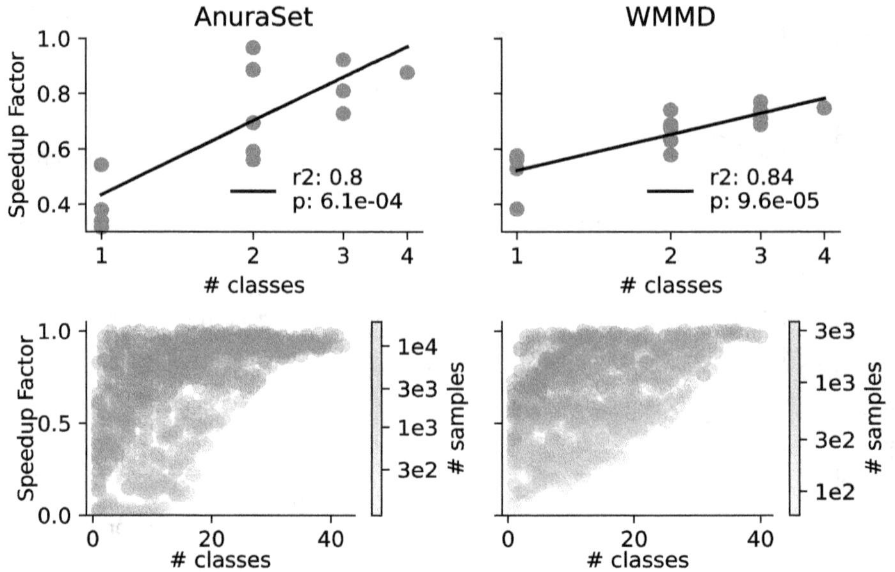


**Fig. 6.** Active learning (ratio max) performance over the used number of classes for multi-label datasets generated with AnuraSet (left) and WMMD (right). Top row: 4 classes with similar properties (left: classes: BOAALB, SCIPER, DENNAN, PHYCUV, # positive samples: $5917 \pm 309$, ceiling performance: $\mu = 0.82$, $\sigma^2 = 0.003$, speedup factor: $\mu = 0.46$, $\sigma^2 = 0.012$; right: classes: FalseKillerWhale, SpinnerDolphin, NorthernRightWhale, BowheadWhale, # positive samples: $787 \pm 56$, ceiling performance: $\mu = 0.43$, $\sigma^2 = 0.105$, speedup factor: $\mu = 0.51$, $\sigma^2 = 0.006$). All combinations are used. Bottom row: All classes. For each number of classes, all combinations with neighboring numbers of positive samples are used. The mean number of positive samples is color-coded.

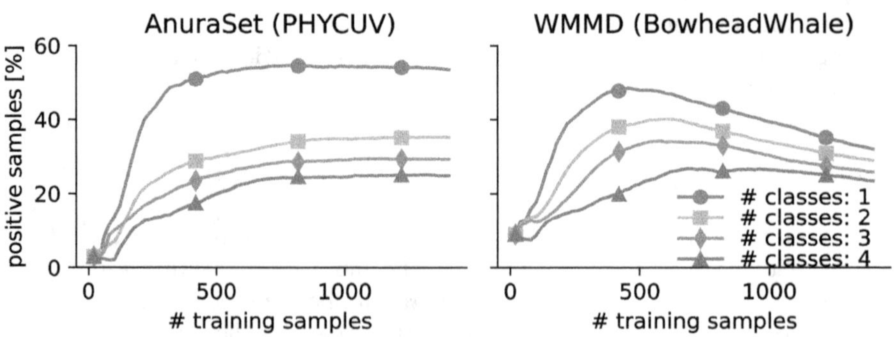


**Fig. 7.** Fraction of positive samples in the training set selected by the sampling strategy ratio max over the total amount of training samples for multi-label datasets for one class. For 2 and 3 classes, where 3 combinations are possible, the curves are averaged over these combinations.

***H1.*** We hypothesised, that AL performance decreases as the number of positive samples in the dataset increases. Our results shown in Figs. 2 and 3 support this hypothesis for single-class datasets, showing a strong positive correlation between the number of positive samples and the speedup factor. Figure 4 shows that random sampling maintains an equivalent proportion of positive samples in the training set as in the original unlabelled dataset. By selecting samples close to the decision boundary, the AL method ratio max is trained to select approximately 50% positive samples. Since a model tends to perform better with a more balanced dataset, the effect of AL is more pronounced for sparse classes where the fraction of positive samples selected by random sampling is very small. Figure 6 (bottom) shows the same trend for multi-label datasets, where the color indicates that the speedup factor decreases for sparser multi-label datasets. Using the sampling strategy ratio max on the smaller dataset WMMD with sparse classes results in a decrease in the number of unlabelled positive samples over iterations, resulting in a decreasing proportion of positive samples within the training set, as shown in Fig. 4 (top right) and Fig. 7 (right).

***H2.*** We hypothesised, that AL performance changes as the ceiling performance changes. While Fig. 5 (right) shows a decrease in AL performance with increasing ceiling performance, Fig. 5 (left) also shows a strong correlation between the number of positive samples and ceiling performance. Therefore, we conclude that the ceiling performance does not directly influence the AL performance, but rather the number of positive samples has a strong influence on both the AL performance and the ceiling performance.

***H3.*** We hypothesised, that AL performance decreases as the number of classes increases. Holding the variables of positive samples, ceiling performance, and speedup factors constant for the single-label datasets, Fig. 6 (top) illustrates that the performance of AL decreases as the number of classes increases. As shown in Fig. 7, this phenomenon occurs because the selected fraction of positive samples decreases when optimizing for more classes. Figure 6 (bottom) shows consistent results even when considering different numbers of positive samples, ceiling performances, and speedup factors, asymptotically approaching a speedup factor of 1 as the number of classes increases.

***H4.*** We hypothesised, that AL performance decreases when using diversity sampling strategies instead of uncertainty sampling strategies. Figures 2 and 3 show that for single-label datasets, ratio max significantly outperforms random sampling, while clustering exhibit an average speedup factor of 1, indicating a learning curve close to random sampling. Figure 4 reveals that the number of selected positive samples using clustering is even slightly lower than that for random sampling. In addition, we conducted all experiments using diversity sampling. The results indicate that the correlation between the speedup factor and both maximum performance and the number of classes suggests no performance advantage when using clustering compared to random sampling.

Regarding the limitations of our elaboration, while the speedup factor effectively captures the AL performance, it could be better adapted to the learning curves by allowing a shift along the x-axis (see Fig. 3). For this study, we limited our analysis to two well-established, computationally tractable AL strategies in the domain of PAM datasets. Future research include investigating these claims for multi-label datasets from other domains, such as image, text, and tabular data, as well as exploring more sophisticated AL strategies.

**Acknowledgements.** This research is part of the Computational Sustainability & Technology project area (https://cst.dfki.de/), and has been supported by the Ministry for Science and Culture of Lower Saxony (MWK), the Endowed Chair of Applied Artificial Intelligence, Oldenburg University, and DFKI.

## References

1. Baker, E., Vincent, S.: A deafening silence: a lack of data and reproducibility in published bioacoustics research? Biodivers. Data J. **7**, e36783 (2019). https://doi.org/10.3897/BDJ.7.e36783
2. Boney, R., Ilin, A.: Semi-supervised and active few-shot learning with prototypical networks. arXiv preprint arXiv:1711.10856 (2017)
3. Cañas, J., Toro-Gómez, M., Sugai, L., et al.: A dataset for benchmarking neotropical anuran calls identification in passive acoustic monitoring. Sci. Data **10**(1), 771 (2023). https://doi.org/10.1038/s41597-023-02666-2
4. Ghani, B., Denton, T., Kahl, S., Klinck, H.: Global birdsong embeddings enable superior transfer learning for bioacoustic classification. Sci. Rep. **13**(1), 22876 (2023)
5. Hand, D.J., Till, R.J.: A simple generalisation of the area under the roc curve for multiple class classification problems. Mach. Learn. **45**, 171–186 (2001)
6. He, H., Garcia, E.A.: Learning from imbalanced data. IEEE Trans. Knowl. Data Eng. **21**(9), 1263–1284 (2009). https://doi.org/10.1109/TKDE.2008.239
7. Kahl, S., Wood, C.M., Eibl, M., Klinck, H.: BirdNET: a deep learning solution for avian diversity monitoring. Ecol. Inform. **61**, 101236 (2021). https://doi.org/10.1016/j.ecoinf.2021.101236
8. Kath, H., Gouvêa, T., Sonntag, D.: A Human-in-the-loop tool for annotating passive acoustic monitoring datasets. In: Proceedings of the Thirty-Second International Joint Conference on Artificial Intelligence, Macau, SAR China, pp. 7140–7144 (2023). https://doi.org/10.24963/ijcai.2023/835
9. Kath, H., Serafini, P.P., Campos, I.B., Gouvêa, T.S., Sonntag, D.: Leveraging transfer learning and active learning for sound event detection in passive acoustic monitoring of wildlife. In: 3rd Annual AAAI Workshop on AI to Accelerate Science and Engineering (AI2ASE) (2024)
10. Lampert, C.H., Nickisch, H., Harmeling, S.: Attribute-based classification for zero-shot visual object categorization. IEEE Trans. Pattern Anal. Mach. Intell. **36**(3), 453–465 (2014). https://doi.org/10.1109/TPAMI.2013.140
11. Liu, W., Zhang, H., Ding, Z., Liu, Q., Zhu, C.: A comprehensive active learning method for multiclass imbalanced data streams with concept drift. Knowl.-Based Syst. **215**, 106778 (2021). https://doi.org/10.1016/j.knosys.2021.106778
12. Lüers, B., Serafini, P.P., Campos, I.B., Gouvêa, T.S., Sonntag, D.: BirdNET-annotator: AI-assisted strong labelling of bird sound datasets. In: 3rd Annual AAAI Workshop on AI to Accelerate Science and Engineering (AI2ASE), Vancouver, Canada (2024)

13. Möllenbrok, L., Sumbul, G., Demir, B.: Deep active learning for multi-label classification of remote sensing images. IEEE Geosci. Remote Sens. Lett. **20**, 1–5 (2023). https://doi.org/10.1109/LGRS.2023.3305647
14. Monarch, R.: Human-in-the-Loop Machine Learning: Active Learning and Annotation for Human-Centered AI. Simon and Schuster (2021)
15. Pupo, O.G.R., Altalhi, A.H., Ventura, S.: Statistical comparisons of active learning strategies over multiple datasets. Knowl. Based Syst. **145**, 274–288 (2018). https://doi.org/10.1016/J.KNOSYS.2018.01.033
16. Reichart, R., Tomanek, K., Hahn, U., Rappoport, A.: Multi-task active learning for linguistic annotations. In: McKeown, K.R., Moore, J.D., Teufel, S., Allan, J., Furui, S. (eds.) ACL 2008, Proceedings of the 46th Annual Meeting of the Association for Computational Linguistics, 15–20 June 2008, Columbus, Ohio, USA, pp. 861–869. The Association for Computer Linguistics (2008)
17. Ross, S., O'Connell, D., Deichmann, J., et al.: Passive acoustic monitoring provides a fresh perspective on fundamental ecological questions. Funct. Ecol. **37**(4), 959–975 (2023). https://doi.org/10.1111/1365-2435.14275
18. Sayigh, L., et al.: The Watkins marine mammal sound database: An online, freely accessible resource. In: Proceedings of Meetings on Acoustics, vol. 27, p. 040013 (2016). https://doi.org/10.1121/2.0000358
19. Settles, B.: Active Learning. Synthesis Lectures on Artificial Intelligence and Machine Learning. Morgan & Claypool Publishers (2012). https://doi.org/10.2200/S00429ED1V01Y201207AIM018
20. Sueur, J., Pavoine, S., Hamerlynck, O., Duvail, S.: Rapid acoustic survey for biodiversity appraisal. PLoS One **3**(12), e4065 (2008). https://doi.org/10.1371/journal.pone.0004065
21. Sugai, L., Llusia, D.: Bioacoustic time capsules: using acoustic monitoring to document biodiversity. Ecol. Ind. **99**, 149–152 (2019). https://doi.org/10.1016/j.ecolind.2018.12.021
22. Sugai, L., Silva, T., Ribeiro, J., Llusia, D.: Terrestrial passive acoustic monitoring: review and perspectives. Bioscience **69**(1), 15–25 (2019). https://doi.org/10.1093/biosci/biy147
23. Tharwat, A., Schenck, W.: Balancing exploration and exploitation: a novel active learner for imbalanced data. Knowl.-Based Syst. **210**, 106500 (2020). https://doi.org/10.1016/j.knosys.2020.106500
24. Tharwat, A., Schenck, W.: A survey on active learning: state-of-the-art, practical challenges and research directions. Mathematics **11**(4) (2023). https://doi.org/10.3390/math11040820
25. Vasilakes, J., Rizvi, R., Melton, G.B., Pakhomov, S., Zhang, R.: Evaluating active learning methods for annotating semantic predications. JAMIA Open **1**(2), 275–282 (2018)
26. Wu, J., et al.: Multi-label active learning algorithms for image classification: overview and future promise. ACM Comput. Surv. **53**(2) (2020). https://doi.org/10.1145/3379504

# Quantifying the Trade-Offs Between Dimensions of Trustworthy AI - An Empirical Study on Fairness, Explainability, Privacy, and Robustness

Nils Kemmerzell(✉) and Annika Schreiner

Friedrich-Alexander-Universität Erlangen-Nürnberg, Erlangen, Germany
{nils.kemmerzell,annika.schreiner}@fau.de

**Abstract.** Trustworthy AI encompasses various requirements for AI systems, including explainability, fairness, privacy, and robustness. Addressing these dimensions concurrently is challenging due to inherent tensions and trade-offs between them. Current research highlights these trade-offs, focusing on specific interactions, but comprehensive and systematic evaluations remain insufficient. This study aims to enhance the understanding of trade-offs among explainability, fairness, privacy, and robustness in AI. By conducting extensive experiments in the domain of image classification, it quantitatively assesses how methods to improve one requirement impact others. More specifically, it explores different training adaptations to enhance each requirement and measures their effects on the others on four datasets for gender classification. The experiments revealed that the Local Gradient Alignment method improved explainability and robustness but introduced trade-offs in fairness, privacy, and accuracy. Fairness-focused training adaptations only enhanced fairness for the most biased models. For all other cases, fairness, explainability and robustness are reduced. Differential Privacy improved privacy but compromised explainability, fairness, and accuracy, with varied impacts on robustness. Data augmentation techniques enhanced robustness, explainability and accuracy with minor trade-offs in privacy and fairness.

**Keywords:** machine learning · trustworthiness · interactions · training

## 1 Introduction

Artificial Intelligence (AI) is rapidly expanding across numerous sectors, presenting both benefits and challenges. As AI sees broader real-world application, ensuring its trustworthiness becomes crucial [30]. Trustworthiness is a construct that takes into account different requirements on AI [41,48,51]. The requirements of explainability, fairness, privacy, and robustness, which we refer to as dimensions of trustworthiness, are particularly commonly addressed. However, it can be challenging to target these dimensions simultaneously as previous research


A. Hotho and S. Rudolph (Eds.): KI 2024, LNAI 14992, pp. 128–146, 2024.
https://doi.org/10.1007/978-3-031-70893-0_10

has acknowledged that tensions exist between them [48,86]. This may lead to inevitable trade-offs between the dimensions [41], which may prevent trustworthiness from being achieved by improving individual dimensions in isolation [51].

Recent research has begun to explore these trade-offs. For instance, [82] reviewed trade-offs between privacy and fairness, and explainability and robustness, while [98] addressed the link between fairness and adversarial robustness. [15] investigated the connection between fairness, privacy, and accuracy, and [33] examined the relations among accuracy, robustness, fairness, and privacy. However, research on the trade-offs among different dimensions is still in an early stage and not all trade-offs have been fully investigated, yet [54]. Additionally, empirical evidence on this issue has been scarce.

This work aims to contribute to a better understanding of the trade-offs between explainability, fairness, privacy and robustness. To this end, we conduct an extensive experimental study in the domain of image classification in which we investigate how methods for improving one dimension affect other dimensions. For example, differential privacy attempts to improve the privacy of machine learning (ML) models, but it is unclear how this affects fairness, explainability and robustness.

This study examines trustworthiness in AI within the context of facial analysis. AI models have exhibited biases and fairness issues in this domain, highlighting the need for rigorous evaluation to ensure equitable outcomes for all users [12,37,97]. Furthermore, facial analysis poses significant regulatory and technical challenges surrounding privacy, consent, and surveillance, making it an ideal domain for exploring the complexities of AI trustworthiness [28].

We train several ResNet models on four datasets for gender classification, using various approaches to enhance one of the four dimensions by adjusting the training algorithm and measuring its impact on the other three. We evaluate the results using quantitative metrics for explainability, fairness, privacy, and robustness.

Our experiments showed that Local Gradient Alignment (LGA) [45] enhances explainability and robustness but has trade-offs in fairness, privacy, and accuracy. Training adaptations to improve fairness showed mixed results, enhancing fairness only in the LFWA+ dataset but generally decreasing explainability and robustness, with minor effects on privacy and accuracy. Differential Privacy (DP) significantly improved privacy, particularly against model extraction attacks, but at the expense of explainability, fairness, and accuracy, with mixed impacts on robustness. Data augmentation techniques improved robustness against distribution shifts and explainability, with minor trade-offs in privacy and fairness, and generally positive effects on accuracy. We discuss our results in the context of related work, contributing to a better understanding of the trade-offs between explainability, fairness, privacy, and robustness in ML models.

## 2 Fundamentals

*Explainability.* Explainability in ML refers to an ML model's capacity to explain its decision-making process [2]. Models are often classified as white-box or black-box based on their interpretability [14]. Black-box models like deep neural networks necessitate post-hoc methods for explanations [2]. Feature attribution techniques are a type of post-hoc explanation method used to assign importance to input features for model outputs [8]. However, it has been identified, that they are sensitive to small input changes [32]. To enhance their robustness and therefore improve explainability, several approaches have been introduced [20,45].

*Fairness.* The concept of fairness in ML aims to ensure that decisions made by algorithms are not biased toward specific individuals or population groups [59]. Various definitions of fairness exist, which can be categorized into group-based and individual-based approaches. Group fairness, on which we focus in this study, is based on the concept that certain groups, defined by sensitive attributes, should be equally likely to be classified into a particular category [16]. Approaches attempting to improve fairness can intervene at different levels of the ML modelling process [16]. Pre-processing methods manipulate the training data, e.g. through reweighting [92]. In-processing methods incorporate fairness metrics into the model optimization functions, e.g. through constraints [44]. Post-processing methods transform the actual model output [71].

*Privacy.* Privacy in ML entails the protection of sensitive personal data used for model training and inference [23]. The integrity of ML applications is threatened not only by direct information exposure such as data leaks, but also by indirect threats that target the ML models themselves [61]. A breach occurs when training data or model parameters can be deduced from the model via various attack methods, such as membership inference, model inversion, or model extraction [3]. In the context of ML, differential privacy (DP) [26] refers to techniques that enable the training of models in such a way that they do not reveal sensitive information about individual data points, thus protecting their privacy [1,64].

*Robustness.* Robustness ensures the performance of ML models in a variety of circumstances. In practice, ML models may encounter data containing various types of perturbations misleading the model, which can be naturally occurring or intentionally crafted [52]. Robustness to distribution shift refers to the ability of an ML model to perform well when the data distributions change over time [89]. Robustness to adversarial attacks refers to the ability of a model to resist being misled by intentionally crafted small perturbations of the input data [13]. To defend against adversarial attacks, several techniques are suggested, e.g. adversarial training [58]. Research on distribution shift concentrates on data augmentation mechanisms to improve robustness [40].

## 3 Methodology

In this section, we describe our methodology for the following experiments, focusing on training setups and the evaluation of explainability, fairness, privacy and robustness.

### 3.1 Data and Model

For our study, we utilize the publicly available datasets *UTKFace* [104], *CelebA* [56], *FairFace* [47], and *LFWA+* [56], each comprising facial images along with various attributes. These datasets were chosen for their differences in size and attribute distribution. We focus on *gender* as the target variable and *race* or *young* (in the case of CelebA) as binary sensitive attributes. The LFWA+ and UTKFace datasets, containing 13,233 and 23,705 samples, respectively, are much smaller compared to FairFace and CelebA, which have 108,501 and 202,599 samples. Among these, LFWA+ exhibits the most pronounced imbalance in demographic categories, followed by CelebA, while FairFace shows imbalance only in the sensitive attribute, and UTKFace displays minimal biases. Our chosen model architecture is a ResNet-50 model pre-trained on ImageNet, a common setup for image classification.

### 3.2 Training

We partition each dataset into training, validation, and test sets, using a 60-20-20 split. To create equal evaluation settings, we balance the test set by adjusting the distribution of target and sensitive attributes. For optimization, we employ the cross-entropy loss function, a batch size of 128, the Adam optimizer with a learning rate of 1e-3 and use early stopping to prevent overfitting. Each experiment is repeated 5 times. On top of this basic procedure, we implement different adaptations of the training to improve one of the dimensions.

*Explainability.* In our experiments, we leverage *Local Gradient Alignment* (LGA) [45] to enhance the quality of saliency maps. It has been identified, that they are sensitive to small input changes [20,32], which can often be attributed to non-smooth decision boundaries present in neural networks [45]. LGA addresses this issue by adding a constraint to the loss function which smoothens the gradient and thereby increases the robustness of the saliency map. We investigate two unique settings, in line with the recommended hyperparameters [45]. Specifically, we choose $\lambda_{cos} \in \{0.1, 1\}$ and $\lambda_{L2} = 1$. A larger $\lambda$ is linked to a stronger emphasis on the regularization term in the loss function.

*Fairness.* Intending to mitigate bias in ML models, we employ two distinct approaches: *Weighting* and *Constraint Optimization.* Adapting the implementation of [92], we allocate weights to the training samples, based on the frequency of their target and attribute class occurrences. This improves fairness by ensuring minority classes are appropriately represented by receiving stronger weights, which in turn reduces biases in the model's predictions. The Constraint Optimization integrates a fairness metric as a constraint into the classifier's loss function, which creates a secondary optimization target next to model accuracy during the training process [16]. We used the *fairtorch* framework [31], focusing on the equalized odds metric and two different weight terms $\alpha = 1$ and $\alpha = 100$ for the constraint.

*Privacy.* To improve model privacy, we employ *Differential Privacy* (DP), specifically through the implementation of Differential Private Stochastic Gradient Descent [1]. This technique improves privacy by adding noise to the gradients during the training process, which ensures that the contribution of any single data point to the overall model is indistinguishable from random noise. This protects individual data points from being inferred or re-identified within the dataset. In this context, the privacy budget parameter $\epsilon$ quantifies the added noise. We experiment with three different $\epsilon$ settings, 1, 10, and 100, representing high, medium, and low levels of noise addition, respectively.

*Robustness.* To improve the robustness during training, we employ two data augmentation algorithms: *Augmix* and *Patch Gaussian Augmentation.* Data augmentation improves model robustness by increasing the diversity of training data, which helps the model generalize better to unseen data. Augmix combines stochastically sampled augmentations of the same input image while applying a Jensen-Shannon divergence consistency loss [40]. Patch Gaussian augmentation focuses on adding noise to randomly selected patches within the input image [57].

### 3.3 Evaluation

We evaluate the trustworthiness of ML models by utilizing a thorough collection of 20 quantitative metrics that incorporate various aspects of fairness, robustness, privacy, and explainability, relying on the framework proposed in [76]. To make the results more comparable across the four dimensions, we normalize each metric to a range of zero to ten. As an indicator of the model utility, we also introduce a normalized accuracy score using the same scale. To reduce the complexity of this extensive set of metrics, we introduce scores for each dimension averaging the scores of the individual metrics. This enables a more simplified overview of the results, giving initial insights on the nature of the trade-offs. However, we note that these scores do not make a detailed analysis of individual metrics obsolete but should rather be seen as a helpful addition. Hence, we provide fully detailed result tables including all metrics in the appendix[1].

*Explainability.* We evaluate explainability using a two-stage setup. Initially, we generate saliency maps via the *Integrated Gradients* technique [83]. Subsequently, we employ quantitative metrics to examine the characteristics of these maps [39]. We evaluate four key metrics: *Robustness* is defined as the degree to which an explanation remains consistent in the face of minor alterations to the input data [100]. *Faithfulness* describes the precision of the importance scores produced by the saliency function in capturing the relevance of each input to the model's prediction [10]. *Randomization* compares the explanations derived from model inputs with those obtained from a randomized logit [81]. The *Complexity* of an explanation is indicated by the number of input features needed to elucidate the model's decision-making process; ideally, explanations should be succinct, focusing only on pertinent features [17].

[1] https://github.com/Nilskem/trustworthiness_trade-offs.

*Fairness.* Regarding the evaluation of fairness, mostly group-fairness metrics are used in current literature, comparing a model's performance on subgroups defined by a sensitive attribute [16]. We compute the widely used metrics of *Demographic Parity (DemP) difference* and *Equalized Odds (EOd) difference.* DemP requires that the probability of being classified with the positive label is equal across groups [9]. EOd defines a model as fair if True Positive Rates (TPR) and False Negative Rates are equal across groups [36]. Additionally, we calculate the value difference between subgroups regarding typical classification metrics, namely *Accuracy*, *Precision*, *Recall/TPR* and *False Positive Rate (FPR).* The resulting set of six metrics only represents a small partition, yet a selection of frequently addressed metrics, of the numerous options to evaluate fairness [91].

*Privacy.* We concentrate on evaluating privacy in terms of ML models' vulnerability to specific attacks [3,61,70]. Our investigation quantifies this vulnerability by focusing on two principal types of attacks: *Membership Inference* [79] and *Model Extraction* [87]. For both, we assess two scenarios, depending on assumptions on the attacker's access to training data. In the first scenario, the attacker does not have access to the original training data but instead uses a shadow dataset which follows a similar distribution. In the second scenario, the adversary gains parts of the data used to train the model, which should improve the attack chances. We use the *Roc-auc-score* to evaluate the effectiveness of the membership inference attacks, while we rely on *Accuracy* for the model extraction attacks.

*Robustness.* Regarding robustness, we aim to measure the models' robustness under circumstances where image samples deviate from the training data. Hereby, we cover two different, frequently described situations: Distribution shifts and adversarial attacks. The former can be evaluated by creating a shifted test set and measuring the model's performance on this corrupted set compared to the original test set [84]. We utilize a random combination of *pytorch* transformations, i.e. colour jitter, Gaussian blur, sharpness, and contrast adjustments, to generate realistic augmented samples. Inspired by [89], we then calculate the model's *Accuracy*, the *Roc-auc-Score*, and the *Brier Score* on the shifted and the original test set, and compute the absolute difference between the results.

On the one hand, the model's robustness against adversarial attacks can be measured by confronting the model with adversarial samples created by a certain attack and assessing its accuracy [11] compared to the accuracy of the original test set. We reduce ourselves to two commonly known $L_\infty$-Norm adversarial attacks due to the generally high computational effort: The *Fast Gradient Sign Method (FGSM)* [34] and *Projected Gradient Descent (PGD)* [58]. Besides, we measure an attack-independent metric, the *Loss Sensitivity* computes the effect of each data sample on the average loss [5].

*Significance Test.* Finally, we apply the Wilcoxon signed-rank test to compare the training adaptations with the baseline model [67]. We use one-sided tests for each metric individually to infer whether the scores for the training adaptations are significantly different from the baseline model. Due to the extensive computational requirements (e.g., Patch Gaussian augmentation took up to 22 h per run) and the total number of configurations (44), repeating experiments more frequently was unfeasible. Thus, significance levels should be interpreted cautiously as p-values tend to be higher due to fewer repetitions.

# 4 Results

In this section, we present the results of our experiments and discuss them with regard to previous findings in the literature. The extended result tables including all metrics can be found in the appendix.

## 4.1 Explainability

**Table 1.** Trade-Offs of Improving Explainability

| Dataset | Method | Explainability | Fairness | Privacy | Robustness | Accuracy |
|---|---|---|---|---|---|---|
| UTKFace | Baseline | 6.21 | 9.73 | 5.85 | 6.66 | 9.54 |
| | $\lambda_{cos} = 0.1$ | 8.51* | 9.63 | 6.03* | 6.78 | 9.21* |
| | $\lambda_{cos} = 1$ | 8.39* | 9.34* | 6.11* | 6.95 | 9.04* |
| CelebA | Baseline | 5.45 | 9.48 | 5.26 | 6.47 | 9.74 |
| | $\lambda_{cos} = 0.1$ | 7.87* | 9.51 | 5.29 | 6.33 | 9.68* |
| | $\lambda_{cos} = 1$ | 6.68* | 8.36* | 7.39* | 7.04 | 7.49* |
| Fairface | Baseline | 5.99 | 9.69 | 7.20 | 6.91 | 9.09 |
| | $\lambda_{cos} = 0.1$ | 7.29* | 8.72* | 8.45 | 8.29* | 6.65* |
| | $\lambda_{cos} = 1$ | 6.00 | 8.85* | 8.98* | 8.83* | 5.99* |
| LFWA+ | Baseline | 5.42 | 6.59 | 6.73 | 4.41 | 8.86 |
| | $\lambda_{cos} = 0.1$ | 7.85* | 7.39 | 6.40* | 5.73 | 9.12* |
| | $\lambda_{cos} = 1$ | 7.68* | 7.27 | 6.57 | 6.10* | 8.89 |

Significance is denoted as follows: *: $p \leq 0.05$

In our experiments, LGA effectively enhanced explainability in image classification by increasing explanation robustness and faithfulness across all datasets. This supports existing research on gradient regularization techniques improving model explainability [4, 63, 73]. Unlike prior studies that relied on qualitative evaluations, our findings provide quantitative evidence of LGA's benefits. However, the literature on gradient regularization's effect on other trustworthiness dimensions, apart from adversarial robustness, is limited (Table 1).

LGA's impact on fairness showed mixed results. While it decreased fairness in the UTKFace, CelebA, and FairFace datasets, its mean values hint towards an opposite effect on the LFWA+ dataset. This improvement might be due to LGA's ability to smooth decision boundaries, reducing overfitting by promoting the learning of more robust and generalizable features.

LGA also had mixed effects on privacy risks. It improved resistance to membership inference attacks in UTKFace and CelebA but decreased it in FairFace and LFWA+. Prior research suggests regularization can enhance privacy [42,79], which can explain the former findings. For model extraction attacks, stronger constraints generally improved resistance, correlating with decreased model accuracy [46].

LGA positively impacted model robustness, particularly regarding distribution shifts, with significant improvements on the LFWA+ dataset. However, its effect on adversarial attacks was mixed and dependent on regularization strength. Weaker regularization reduced robustness in CelebA and LFWA+ but improved it in UTKFace and FairFace. Stronger regularization generally enhanced robustness, aligning with previous research on gradient regularization and adversarial robustness [4,73].

LGA had a mixed impact on model accuracy. Generally, regularization decreased accuracy, especially with stronger constraints. However, an exception was observed in the LFWA+ dataset with $\lambda_{cos} = 0.1$, where LGA improved performance. This finding warrants further investigation to explore performance and fairness improvements on biased datasets.

### 4.2 Fairness

**Table 2.** Trade-offs of Improving Fairness

| Dataset | Method | Explainability | Fairness | Privacy | Robustness | Accuracy |
|---|---|---|---|---|---|---|
| UTKFace | Baseline | 6.21 | 9.73 | 5.85 | 6.66 | 9.54 |
| | Weighting | 5.97* | 9.58 | 5.94 | 6.63 | 9.50 |
| | CO ($\alpha = 1$) | 5.74* | 9.61 | 5.95 | 6.44 | 9.44 |
| | CO ($\alpha = 100$) | 5.59* | 9.65 | 5.96* | 4.81* | 9.32 |
| CelebA | Baseline | 5.45 | 9.48 | 5.26 | 6.47 | 9.74 |
| | Weighting | 5.36 | 9.39 | 5.26 | 6.24 | 9.74 |
| | CO ($\alpha = 1$) | 5.39 | 9.37 | 5.20* | 6.24 | 9.76 |
| | CO ($\alpha = 100$) | 5.38 | 9.45 | 5.27 | 5.66* | 9.68* |
| FairFace | Baseline | 5.99 | 9.69 | 7.20 | 6.91 | 9.09 |
| | Weighting | 6.10* | 9.57 | 7.13* | 6.91 | 9.07 |
| | CO ($\alpha = 1$) | 6.00 | 9.64 | 7.13* | 6.75 | 9.03* |
| | CO ($\alpha = 100$) | 5.97 | 9.64 | 7.10* | 6.67* | 8.96* |
| LFWA+ | Baseline | 5.42 | 6.59 | 6.73 | 4.41 | 8.86 |
| | Weighting | 5.58 | 6.70 | 7.07* | 5.26 | 8.71 |
| | CO ($\alpha = 1$) | 5.22* | 6.72 | 6.72 | 4.85 | 8.64 |
| | CO ($\alpha = 100$) | 5.57 | 9.87* | 9.01* | 8.17* | 5.59* |

Significance is denoted as follows: *: $p \leq 0.05$

Training adaptations aimed at improving fairness yielded mixed results, significantly improving the Fairness Score only for the LFWA+ dataset with the CO setup and $\alpha = 100$. This might be due to the stronger bias in LFWA+ compared to other datasets. Overall, this setup behaved differently from other training methods and datasets (Table 2).

Generally, training adaptations negatively affected explainability, especially for UTKFace and LFWA+, with minor changes for CelebA. The reductions were mainly in the Faithfulness and Complexity metrics. While the literature has highlighted fairness concerns in explanations [7,25], the impact of fairness measures on explainability has not been studied yet.

The effects on privacy metrics were mixed and small. The impact on membership inference attacks was mostly insignificant and negative in three settings, aligning with literature suggesting fairness constraints can increase privacy risks by forcing models to memorize training data for underprivileged subgroups [19]. While the literature has not addressed fairness in model extraction attacks, our results showed small improvements with fair training methods, except for LFWA+ with $\alpha = 100$, where resistance against model extraction increased significantly, likely due to lower model accuracy [46].

Overall, fairness methods negatively impacted model robustness. Robustness against distribution shifts decreased significantly, especially on UTKFace. Existing literature discusses maintaining fairness under distribution shifts [21,77], but not the impact of fairness constraints on accuracy under these shifts. Regarding adversarial attacks, bias mitigation techniques may increase susceptibility [18,43,72]. Our results support this, showing significant reductions in loss sensitivity, but CO training adaptations increased robustness on all datasets. Robustness was measured as the difference between original and adversarial test sets; thus, lower benign accuracy with the same adversarial accuracy increases robustness, as seen in CelebA and FairFace. For LFWA+ and UTKFace, CO showed higher adversarial accuracy and lower benign accuracy, increasing the Robustness Score.

Finally, model accuracies decreased significantly with the CO setup, especially with $\alpha = 100$. This trade-off between fairness and accuracy aligns with previous studies [27,60,74], though it is not universally accepted as inherent [22,94].

### 4.3 Privacy

When implementing DP, we observed significant increases in the Privacy Score across all datasets and $\epsilon$ values, with the largest increases for lower $\epsilon$ values, indicating enhanced privacy protection. Resistance to membership inference attacks generally improved, aligning with previous findings [38,42,53]. Only one attack setting on FairFace showed a significant decrease due to the baseline model's perfect score, likely influenced by early stopping to mitigate overfitting [79,101] and FairFace's diverse training data [79]. Furthermore, our experiments demonstrated DP's effectiveness against model extraction attacks across all datasets. This effect is less addressed in the literature, but [53] reported mixed results of DP-SGD on model extraction. Improved resistance could be due to reduced model accuracy caused by DP, as higher accuracy makes models more susceptible to extraction attacks [35,53] (Table 3).

However, these improvements came at a significant cost to other trustworthiness requirements and model accuracy. Lower $\epsilon$ values exacerbated trade-offs in

**Table 3.** Trade-offs of Improving Privacy

| Dataset | Method | Explainability | Fairness | Privacy | Robustness | Accuracy |
|---|---|---|---|---|---|---|
| UTKFace | Baseline | 6.21 | 9.73 | 5.85 | 6.66 | 9.54 |
| | $\epsilon = 10$ | 3.56* | 9.27 | 7.29* | 7.28 | 7.70* |
| | $\epsilon = 100$ | 3.48* | 9.40* | 6.95* | 6.42 | 8.03* |
| CelebA | Baseline | 5.45 | 9.48 | 5.26 | 6.47 | 9.74 |
| | $\epsilon = 1$ | 3.36* | 7.55* | 6.55* | 3.23* | 8.23* |
| | $\epsilon = 10$ | 3.45* | 8.34* | 5.80* | 2.19* | 8.89* |
| | $\epsilon = 100$ | 3.35* | 8.55* | 5.63* | 4.21* | 9.12* |
| FairFace | Baseline | 5.99 | 9.69 | 7.20 | 6.91 | 9.09 |
| | $\epsilon = 1$ | 3.40* | 9.29 | 9.16* | 8.89* | 5.83* |
| | $\epsilon = 10$ | 3.56* | 8.86 | 8.82* | 8.71* | 6.21* |
| | $\epsilon = 100$ | 3.46* | 9.20* | 8.05* | 7.39 | 7.00* |

Non-converging models on UTKFace and LFWA+ are excluded from the table.
Significance is denoted as follows: *: $p \leq 0.05$

other metrics, and noise added during training prevented convergence on UTKFace and LFWA+. DP negatively impacted saliency map quality, diminishing explanation robustness and faithfulness compared to the baseline. While some studies suggest DP might enhance saliency map clarity [62], others argue it degrades interpretability and fidelity [75]. Our findings align with the latter.

DP also negatively impacted fairness across datasets, with lower $\epsilon$ values exacerbating declines in fairness metrics, potentially intensifying biases and harming lower-representation subgroups [6,29,88]. This supports the idea that achieving both equalized odds and differential privacy could be fundamentally incompatible [24].

At the intersection of DP and robustness, we observed mixed outcomes. DP models generally underperformed in robustness to distribution shifts [90]. However, DP improved resistance to FGSM and PGD attacks, supported by theoretical links between DP and adversarial robustness [50,66], although this contrasts with findings showing increased accuracy differences between benign and adversarial samples [90]. The attack-agnostic loss sensitivity metric indicated a decline in model robustness compared to baseline models, suggesting an ambiguous trade-off between privacy and robustness.

Regarding model performance, DP caused significant drawbacks, including up to a 33% point reduction in accuracy and potential training convergence issues, especially in smaller datasets like LFWA+ and UTKFace, aligning with previous research highlighting DP's detrimental effect on accuracy [1,6,65].

### 4.4 Robustness

**Table 4.** Trade-offs of Improving Robustness

| Dataset | Method | Explainability | Fairness | Privacy | Robustness | Accuracy |
|---|---|---|---|---|---|---|
| UTKFace | Baseline | 6.21 | 9.73 | 5.85 | 6.66 | 9.54 |
| | Augmix | 5.78* | 9.76 | 5.88 | 7.02 | 9.60 |
| | PG | 7.68* | 9.70 | 5.82 | 6.17 | 9.54 |
| CelebA | Baseline | 5.45 | 9.48 | 5.26 | 6.47 | 9.74 |
| | Augmix | 5.78* | 9.33 | 5.27 | 6.80 | 9.76* |
| | PG | 7.09* | 9.41* | 5.23 | 6.49 | 9.76* |
| FairFace | Baseline | 5.99 | 9.69 | 7.20 | 6.91 | 9.09 |
| | Augmix | 6.26* | 9.82 | 6.98* | 7.03 | 9.22* |
| | PG | 7.76* | 9.68 | 7.06* | 6.80 | 9.07 |
| LFWA+ | Baseline | 5.42 | 6.59 | 6.73 | 4.41 | 8.86 |
| | Augmix | 5.59 | 8.40* | 6.58 | 6.60* | 8.70 |
| | PG | 7.06* | 7.01 | 6.63 | 6.26 | 8.59* |

Significance is denoted as follows: *: $p \leq 0.05$

In our experiments, we applied two data augmentation techniques to improve robustness. These techniques effectively enhanced robustness against distribution shifts, consistent with previous studies [55,95,99]. However, their impact on adversarial robustness was inconclusive, depending on the augmentation algorithm, dataset, and robustness metric, aligning with prior research [68,69] (Table 4).

Data augmentation significantly enhanced explainability, suggesting a synergy between robustness and explainability. PG improved the Explainability Score across all datasets by enhancing the robustness and faithfulness of saliency maps. Augmix had less pronounced effects, with mixed improvements. Literature on this issue is sparse, but our findings indicate that the effect of data augmentation on explainability depends on the saliency function, type of augmentation, and dataset [85,96].

We observed significant effects on fairness in two scenarios. PG slightly decreased fairness on CelebA, while Augmix greatly improved it on LFWA+, a dataset with a lower fairness baseline suggesting that data augmentation might be suitable for increasing model fairness on small datasets by artificially increasing dataset size and improving generalization. Previous publications support this idea by highlighting overlapping goals between fairness and generalization under out-of-distribution data [78,93,103].

Data augmentation had a minor impact on resistance to privacy attacks, with mixed results for membership inference attacks and a tendency towards reduced model extraction attacks. Literature findings are non-uniform, suggesting that the impact depends on the specific augmentation implementation and dataset [38,49,102]. Enhanced model generalization, assumed to improve with data augmentation, may increase vulnerability to model extraction attacks, which our results tend to support [53].

Finally, data augmentations generally improved model accuracy, with one exception showing negative effects. This relationship is well-documented in the literature [40,80,105], and our findings align with these conclusions.

## 5 Limitations and Transferability

While we have aimed to offer insights into the trade-offs of the trustworthiness of ML models in this study, it is crucial to acknowledge the inherent limitations of our approach and consider the transferability of our results. This research narrows its focus to four dimensions evaluated through specific metrics, encompassing only a small part of the wider range of metrics in this domain. The aggregation of metrics into scores enables a concise and helpful overview of the trade-offs between the dimensions. However, they entail the risk of concealing opposing results of individual metrics. Therefore, these scores do not replace a deep analysis of the underlying metrics which we provide in the appendix. One notable constraint when evaluating fairness was our reliance on annotations for sensitive attributes, potentially limiting the applicability of our approach to datasets without such annotations. The transferability of our experiments is contingent on the task and data type. While our results provide directions for other use cases in the domain of image classification, they cannot replace an application-specific analysis. While our general methodology can be transferred to other data and tasks, not all training adaptations might be feasible outside of image classification. Furthermore, adaptations of metrics and models might be necessary to suit other tasks and data. For instance, image segmentation requires different model architectures and performance metrics, tree-based models in tabular data settings complicate the use of certain gradient-driven metrics and training techniques, and aligning evaluation metrics in the realm of text data poses challenges regarding fairness and explainability.

## 6 Conclusion

This study explored the interplay between explainability, fairness, privacy, and robustness in ML models, particularly within the domain of facial analysis. By conducting extensive experiments and employing various methodologies to enhance each dimension individually, we gained valuable insights into the trade-offs involved. Hereby, we addressed novel interactions and confirm previous research through empirical evidence. Our findings underscore the complexity of achieving trustworthiness in AI systems, as improvements in one dimension often entail compromises in others.

**Disclosure of Interests.** The authors have no competing interests to declare that are relevant to the content of this article.

## References


1. Abadi, M., et al.: Deep learning with differential privacy. In: Proceedings of the 2016 ACM SIGSAC Conference on Computer and Communications Security, CCS 2016, pp. 308–318. Association for Computing Machinery, New York (2016). https://doi.org/10.1145/2976749.2978318
2. Adadi, A., Berrada, M.: Peeking inside the black-box: a survey on explainable artificial intelligence (XAI). IEEE Access **6**, 52138–52160 (2018). https://doi.org/10.1109/ACCESS.2018.2870052
3. Al-Rubaie, M., Chang, J.M.: Privacy-preserving machine learning: threats and solutions. IEEE Secur. Priv. **17**(2), 49–58 (2019). https://doi.org/10.1109/MSEC.2018.2888775
4. Anil, C., Lucas, J., Grosse, R.: Sorting out Lipschitz function approximation. In: Chaudhuri, K., Salakhutdinov, R. (eds.) Proceedings of the 36th International Conference on Machine Learning. Proceedings of Machine Learning Research, vol. 97, pp. 291–301. PMLR (2019)
5. Arpit, D., et al.: A closer look at memorization in deep networks. In: Precup, D., Teh, Y.W. (eds.) Proceedings of the 34th International Conference on Machine Learning. Proceedings of Machine Learning Research, vol. 70, pp. 233–242. PMLR (2017). https://proceedings.mlr.press/v70/arpit17a.html
6. Bagdasaryan, E., Poursaeed, O., Shmatikov, V.: Differential privacy has disparate impact on model accuracy. In: Wallach, H., Larochelle, H., Beygelzimer, A., d'Alché-Buc, F., Fox, E., Garnett, R. (eds.) Advances in Neural Information Processing Systems, vol. 32. Curran Associates, Inc (2019)
7. Balagopalan, A., Zhang, H., Hamidieh, K., Hartvigsen, T., Rudzicz, F., Ghassemi, M.: The road to explainability is paved with bias: measuring the fairness of explanations (2022). https://doi.org/10.1145/3531146.3533179
8. Barredo Arrieta, A., et al.: Explainable artificial intelligence (XAI): concepts, taxonomies, opportunities and challenges toward responsible AI. Inf. Fusion **58**, 82–115 (2020). https://doi.org/10.1016/j.inffus.2019.12.012, https://www.sciencedirect.com/science/article/pii/S1566253519308103
9. Berk, R., Heidari, H., Jabbari, S., Kearns, M., Roth, A.: Fairness in criminal justice risk assessments: the state of the art. Sociol. Methods Res. **50**(1), 3–44 (2021) https://doi.org/10.1177/0049124118782533, https://journals.sagepub.com/doi/pdf/10.1177/0049124118782533
10. Bhatt, U., et al.: Explainable machine learning in deployment. In: Hildebrandt, M. (ed.) Proceedings of the 2020 Conference on Fairness, Accountability, and Transparency, pp. 648–657. ACM Digital Library, Association for Computing Machinery, New York (2020). https://doi.org/10.1145/3351095.3375624
11. Brendel, W., Rauber, J., Kümmerer, M., Ustyuzhaninov, I., Bethge, M.: Accurate, reliable and fast robustness evaluation. In: Wallach, H., Larochelle, H., Beygelzimer, A., d'Alché-Buc, F., Fox, E., Garnett, R. (eds.) Advances in Neural Information Processing Systems, vol. 32. Curran Associates, Inc (2019)
12. Buolamwini, J., Gebru, T.: Gender shades: intersectional accuracy disparities in commercial gender classification. In: Conference on Fairness, Accountability and Transparency, pp. 77–91 (2018)
13. Carlini, N., Wagner, D.: Towards evaluating the robustness of neural networks. In: 2017 IEEE Symposium on Security and Privacy (SP), pp. 39–57. IEEE (2017). https://doi.org/10.1109/SP.2017.49

14. Carvalho, D.V., Pereira, E.M., Cardoso, J.S.: Machine learning interpretability: a survey on methods and metrics. Electronics **8**(8), 832 (2019). https://doi.org/10.3390/electronics8080832, https://www.mdpi.com/2079-9292/8/8/832
15. Carvalho, T., Moniz, N., Antunes, L.: A three-way knot: privacy, fairness, and predictive performance dynamics. In: Moniz, N., Vale, Z., Cascalho, J., Silva, C., Sebastião, R. (eds.) EPIA 2023. LNCS, vol. 14115, pp. 55–66. Springer, Cham (2023). https://doi.org/10.1007/978-3-031-49008-8_5
16. Caton, S., Haas, C.: Fairness in machine learning: a survey. arXiv (2020). https://doi.org/10.48550/arXiv.2010.04053
17. Chalasani, P., Chen, J., Chowdhury, A.R., Wu, X., Jha, S.: Concise explanations of neural networks using adversarial training. In: Daume III, H., Singh, A. (eds.) Proceedings of the 37th International Conference on Machine Learning. Proceedings of Machine Learning Research, vol. 119, pp. 1383–1391. PMLR (2020). https://proceedings.mlr.press/v119/chalasani20a.html
18. Chang, H., Nguyen, T.D., Murakonda, S.K., Kazemi, E., Shokri, R.: On adversarial bias and the robustness of fair machine learning. arXiv:2006.08669 (2020)
19. Chang, H., Shokri, R.: On the privacy risks of algorithmic fairness. In: 2021 IEEE European Symposium on Security and Privacy (EuroS&P), pp. 292–303 (2021). https://doi.org/10.1109/EuroSP51992.2021.00028
20. Chen, J., Wu, X., Rastogi, V., Liang, Y., Jha, S.: Robust attribution regularization. In: Wallach, H., Larochelle, H., Beygelzimer, A., d'Alché-Buc, F., Fox, E., Garnett, R. (eds.) Advances in Neural Information Processing Systems, vol. 32. Curran Associates, Inc. (2019)
21. Chen, Y., Raab, R., Wang, J., Liu, Y.: Fairness transferability subject to bounded distribution shift. In: Koyejo, S., Mohamed, S., Agarwal, A., Belgrave, D., Cho, K., Oh, A. (eds.) Advances in Neural Information Processing Systems, vol. 35, pp. 11266–11278. Curran Associates, Inc. (2022)
22. Cooper, A.F., Abrams, E., NA, N.A.: Emergent unfairness in algorithmic fairness-accuracy trade-off research. In: Proceedings of the 2021 AAAI/ACM Conference on AI, Ethics, and Society. ACM, New York (2021). https://doi.org/10.1145/3461702.3462519
23. de Cristofaro, E.: An overview of privacy in machine learning. arXiv:2005.08679 (2020)
24. Cummings, R., Gupta, V., Kimpara, D., Morgenstern, J.: On the compatibility of privacy and fairness. In: Adjunct Publication of the 27th Conference on User Modeling, Adaptation and Personalization, UMAP 2019 Adjunct, pp. 309–315. Association for Computing Machinery, New York (2019). https://doi.org/10.1145/3314183.3323847
25. Dai, J., Upadhyay, S., Aivodji, U., Bach, S.H., Lakkaraju, H.: Fairness via explanation quality: evaluating disparities in the quality of post hoc explanations. In: Proceedings of the 2022 AAAI/ACM Conference on AI, Ethics, and Society, AIES 2022, pp. 203–214. Association for Computing Machinery, New York (2022). https://doi.org/10.1145/3514094.3534159
26. Dwork, C.: Differential privacy: a survey of results. In: International conference on theory and Applications of Models of Computation, pp. 1–19 (2008)
27. Dwork, C., Immorlica, N., Kalai, A.T., Leiserson, M.: Decoupled classifiers for group-fair and efficient machine learning. In: Conference on Fairness, Accountability and Transparency, pp. 119–133 (2018). https://proceedings.mlr.press/v81/dwork18a.html
28. European Commission: Laying down harmonised rules on artificial intelligence and amending certain union legislative acts: Artificial intelligence act

29. Fioretto, F., Tran, C., van Hentenryck, P., Zhu, K.: Differential privacy and fairness in decisions and learning tasks: a survey. In: Proceedings of the Thirty-First International Joint Conference on Artificial Intelligence. International Joint Conferences on Artificial Intelligence Organization (2022). https://doi.org/10.24963/ijcai.2022/766
30. Floridi, L.: Establishing the rules for building trustworthy AI. Nat. Mach. Intell. **1**(6), 261–262 (2019). https://doi.org/10.1038/s42256-019-0055-y
31. Fukuchi, A., Yabe, Y., Sode, M.: FairTorch: PyTorch implementation of parity loss as constraints function to realize the fairness of machine learning (17012023). https://github.com/wbawakate/fairtorch
32. Ghorbani, A., Abid, A., Zou, J.: Interpretation of neural networks is fragile. In: Proceedings of the AAAI Conference on Artificial Intelligence, vol. 33, no. 01, pp. 3681–3688 (2019). https://doi.org/10.1609/aaai.v33i01.33013681
33. Gittens, A., Yener, B., Yung, M.: An adversarial perspective on accuracy, robustness, fairness, and privacy: multilateral-tradeoffs in trustworthy ml. IEEE Access 1 (2022). https://doi.org/10.1109/ACCESS.2022.3218715
34. Goodfellow, I.J., Shlens, J., Szegedy, C.: Explaining and harnessing adversarial examples (2014). https://doi.org/10.48550/arXiv.1412.6572
35. Guo, J., et al.: Isolation and induction: training robust deep neural networks against model stealing attacks. In: Proceedings of the 31st ACM International Conference on Multimedia, MM 2023, pp. 4178–4189. Association for Computing Machinery, New York (2023). https://doi.org/10.1145/3581783.3612092
36. Hardt, M., Price, E., Srebro, N.: Equality of opportunity in supervised learning. In: Lee, D., Sugiyama, M., Luxburg, U., Guyon, I., Garnett, R. (eds.) Advances in Neural Information Processing Systems, vol. 29. Curran Associates, Inc. (2016)
37. Harrison, S., Boxiao, P.: Mitigating bias in facial recognition with FairGAN (2020)
38. He, X., Li, Z., Xu, W., Cornelius, C., Zhang, Y.: Membership-doctor: comprehensive assessment of membership inference against machine learning models (2022)
39. Hedström, A., et al.: Quantus: an explainable AI toolkit for responsible evaluation of neural network explanations. arXiv:2202.06861 (2022)
40. Hendrycks, D., Dietterich, T.: Benchmarking neural network robustness to common corruptions and perturbations (2019). http://arxiv.org/pdf/1903.12261v1
41. High-Level Expert Group on Artificial Intelligence: Ethics guidelines for trustworthy AI: set up by the European Commission (2019)
42. Hu, H., Salcic, Z., Sun, L., Dobbie, G., Yu, P.S., Zhang, X.: Membership inference attacks on machine learning: a survey. ACM Comput. Surv. **54**(11s), 1–37 (2022). https://doi.org/10.1145/3523273
43. Ji, J., et al.: AI alignment: a comprehensive survey. arXiv preprint arXiv:2310.19852 (2023)
44. Jiang, L., Zhang, J., Deng, B.: Robust RGB-D face recognition using attribute-aware loss. IEEE Trans. Pattern Anal. Mach. Intell. **42**(10), 2552–2566 (2020). https://doi.org/10.1109/tpami.2019.2919284
45. Joo, S., Jeong, S., Heo, J., Weller, A., Moon, T.: Towards more robust interpretation via local gradient alignment. In: Proceedings of the AAAI Conference on Artificial Intelligence, vol. 37, no. 7, pp. 8168–8176 (2023). https://doi.org/10.1609/aaai.v37i7.25986, https://ojs.aaai.org/index.php/AAAI/article/view/25986
46. Kariyappa, S., Qureshi, M.K.: Defending against model stealing attacks with adaptive misinformation. In: 2020 IEEE/CVF Conference on Computer Vision and Pattern Recognition (CVPR). IEEE (2020). https://doi.org/10.1109/cvpr42600.2020.00085

47. Karkkainen, K., Joo, J.: Fairface: face attribute dataset for balanced race, gender, and age for bias measurement and mitigation. In: Proceedings of the IEEE/CVF Winter Conference on Applications of Computer Vision, pp. 1548–1558 (2021)
48. Kaur, D., Uslu, S., Rittichier, K.J., Durresi, A.: Trustworthy artificial intelligence: a review. ACM Comput. Surv. **55**(2) (2022). https://doi.org/10.1145/3491209
49. Kaya, Y., Dumitras, T.: When does data augmentation help with membership inference attacks? In: Meila, M., Zhang, T. (eds.) Proceedings of the 38th International Conference on Machine Learning. Proceedings of Machine Learning Research, vol. 139, pp. 5345–5355. PMLR (2021)
50. Lecuyer, M., Atlidakis, V., Geambasu, R., Hsu, D., Jana, S.: Certified robustness to adversarial examples with differential privacy. In: 2019 IEEE Symposium on Security and Privacy (SP), pp. 656–672 (2019). https://doi.org/10.1109/SP.2019.00044
51. Li, B., Qi, P., Liu, B., Di, S., Liu, J., Pei, J., Yi, J., Zhou, B.: Trustworthy AI: from principles to practices. ACM Comput. Surv. **55**(9) (2023). https://doi.org/10.1145/3555803
52. Liu, B., Ding, M., Shaham, S., Rahayu, W., Farokhi, F., Lin, Z.: When machine learning meets privacy: a survey and outlook. ACM Comput. Surv. (CSUR) **54**(2), 1–36 (2021). https://doi.org/10.1145/3436755
53. Liu, H., et al.: Trustworthy AI: a computational perspective. ACM Trans. Intell. Syst. Technol. **14**(1) (2022). https://doi.org/10.1145/3546872
54. Liu, Y., et al.: ML-doctor: Holistic risk assessment of inference attacks against machine learning models. In: 31st USENIX Security Symposium (USENIX Security 2022), pp. 4525–4542. USENIX Association, Boston (2022)
55. Liu, Z., et al: An empirical study on distribution shift robustness from the perspective of pre-training and data augmentation (2022)
56. Liu, Z., Luo, P., Wang, X., Tang, X.: Deep learning face attributes in the wild. In: Proceedings of the IEEE International Conference on Computer Vision (2015)
57. Lopes, R.G., Yin, D., Poole, B., Gilmer, J., Cubuk, E.D.: Improving robustness without sacrificing accuracy with patch gaussian augmentation (2019). arXiv:1906.02611
58. Madry, A., Makelov, A., Schmidt, L., Tsipras, D., Vladu, A.: Towards deep learning models resistant to adversarial attacks (2017). https://doi.org/10.48550/arXiv.1706.06083
59. Mehrabi, N., Morstatter, F., Saxena, N., Lerman, K., Galstyan, A.: A survey on bias and fairness in machine learning. ACM Comput. Surv. **54**(6), 1–35 (2021). https://doi.org/10.1145/3457607
60. Menon, A.K., Williamson, R.C.: The cost of fairness in binary classification. In: Conference on Fairness, Accountability and Transparency, pp. 107–118 (2018). https://proceedings.mlr.press/v81/menon18a.html
61. Mireshghallah, F., Taram, M., Vepakomma, P., Singh, A., Raskar, R., Esmaeilzadeh, H.: Privacy in deep learning: a survey. CoRR abs/2004.12254 (2020)
62. Naidu, R., Priyanshu, A., Kumar, A., Kotti, S., Wang, H., Mireshghallah, F.: When differential privacy meets interpretability: a case study (2021)
63. Noack, A., Ahern, I., Dou, D., Li, B.: An empirical study on the relation between network interpretability and adversarial robustness. SN Comput. Sci. **2**(1), 32 (2021). https://doi.org/10.1007/s42979-020-00390-x
64. Papernot, N., Abadi, M., Erlingsson, Ú., Goodfellow, I., Talwar, K.: Semi-supervised knowledge transfer for deep learning from private training data (2017)

65. Petti, S., Flaxman, A.: Differential privacy in the 2020 us census: what will it do? Quantifying the accuracy/privacy tradeoff. Gates Open Res. **3**, 1722 (2019). https://doi.org/10.12688/gatesopenres.13089.2
66. Pinot, R., Yger, F., Gouy-Pailler, C., Atif, J.: A unified view on differential privacy and robustness to adversarial examples (2019)
67. Rainio, O., Teuho, J., Klén, R.: Evaluation metrics and statistical tests for machine learning. Sci. Rep. **14**, 6086 (2024). https://doi.org/10.1038/s41598-024-56706-x
68. Rebuffi, S.A., Gowal, S., Calian, D.A., Stimberg, F., Wiles, O., Mann, T.A.: Data augmentation can improve robustness. In: Ranzato, M., Beygelzimer, A., Dauphin, Y., Liang, P.S., Wortman Vaughan, J. (eds.) Advances in Neural Information Processing Systems. vol. 34, pp. 29935–29948. Curran Associates, Inc. (2021)
69. Rice, L., Wong, E., Kolter, Z.: Overfitting in adversarially robust deep learning. In: Daume III, H., Singh, A. (eds.) Proceedings of the 37th International Conference on Machine Learning. Proceedings of Machine Learning Research, vol. 119, pp. 8093–8104. PMLR (2020)
70. Rigaki, M., Garcia, S.: A survey of privacy attacks in machine learning. arXiv:2007.07646 (2020)
71. Robinson, J.P., Livitz, G., Henon, Y., Qin, C., Fu, Y., Timoner, S.: Face recognition: too bias, or not too bias? In: 2020 IEEE/CVF Conference on Computer Vision and Pattern Recognition Workshops (CVPRW). IEEE (2020). https://doi.org/10.1109/cvprw50498.2020.00008
72. Roh, Y., Lee, K., Whang, S., Suh, C.: Fr-train: A mutual information-based approach to fair and robust training. In: Daume III, H., Singh, A. (eds.) Proceedings of the 37th International Conference on Machine Learning. Proceedings of Machine Learning Research, vol. 119, pp. 8147–8157. PMLR (2020)
73. Ross, A., Doshi-Velez, F.: Improving the adversarial robustness and interpretability of deep neural networks by regularizing their input gradients. In: Proceedings of the AAAI Conference on Artificial Intelligence, vol. 32, no. 1 (2018). https://doi.org/10.1609/aaai.v32i1.11504
74. Sabato, S., Yom-Tov, E.: Bounding the fairness and accuracy of classifiers from population statistics. In: International Conference on Machine Learning, pp. 8316–8325 (2020)
75. Saifullah, S., Mercier, D., Lucieri, A., Dengel, A., Ahmed, S.: Privacy meets explainability: a comprehensive impact benchmark (2022)
76. Schreiner, A., Kemmerzell, N.: Towards a quantitative evaluation framework for trustworthy AI in facial analysis. In: Bui, T.X. (ed.) 57th Hawaii International Conference on System Sciences, HICSS 2024, Hilton Hawaiian Village Waikiki Beach Resort, Hawaii, USA, 3–6 January 2024, pp. 7821–7830. ScholarSpace (2024). https://hdl.handle.net/10125/107326
77. Schrouff, J., et al.: Maintaining fairness across distribution shift: do we have viable solutions for real-world applications? (2022). https://doi.org/10.48550/arXiv.2202.01034
78. Sharma, S., Henderson, J., Ghosh, J.: CERTIFAI: a common framework to provide explanations and analyse the fairness and robustness of black-box models (2020). https://doi.org/10.1145/3375627.3375812
79. Shokri, R., Stronati, M., Song, C., Shmatikov, V.: Membership inference attacks against machine learning models. In: 2017 IEEE symposium on security and privacy (SP), pp. 3–18 (2017)

80. Shorten, C., Khoshgoftaar, T.M.: A survey on image data augmentation for deep learning. J. Big Data **6**(1), 60 (2019). https://doi.org/10.1186/s40537-019-0197-0
81. Sixt, L., Granz, M., Landgraf, T.: When explanations lie: why many modified BP attributions fail (2019). arXiv:1912.09818
82. Strobel, M., Shokri, R.: Data privacy and trustworthy machine learning. IEEE Secur. Priv. **20**(5), 44–49 (2022). https://doi.org/10.1109/MSEC.2022.3178187
83. Sundararajan, M., Taly, A., Yan, Q.: Axiomatic attribution for deep networks. In: International Conference on Machine Learning, pp. 3319–3328 (2017). https://proceedings.mlr.press/v70/sundararajan17a.html
84. Taori, R., Dave, A., Shankar, V., Carlini, N., Recht, B., Schmidt, L.: Measuring robustness to natural distribution shifts in image classification. In: Larochelle, H., Ranzato, M., Hadsell, R., Balcan, M.F., Lin, H. (eds.) Advances in Neural Information Processing Systems, vol. 33, pp. 18583–18599. Curran Associates, Inc. (2020)
85. Tětková, L., Hansen, L.K.: Robustness of visual explanations to common data augmentation methods. In: Proceedings of the IEEE/CVF Conference on Computer Vision and Pattern Recognition (CVPR) Workshops, pp. 3714–3719 (2023)
86. Thiebes, S., Lins, S., Sunyaev, A.: Trustworthy artificial intelligence. Electron. Mark. **31**(2), 447–464 (2021). https://doi.org/10.1007/s12525-020-00441-4
87. Tramèr, F., Zhang, F., Juels, A., Reiter, M.K., Ristenpart, T.: Stealing machine learning models via prediction APIs. In: 25th USENIX Security Symposium (USENIX Security 2016), pp. 601–618 (2016)
88. Tran, C., Dinh, M., Fioretto, F.: Differentially private empirical risk minimization under the fairness lens. In: Ranzato, M., Beygelzimer, A., Dauphin, Y., Liang, P.S., Wortman Vaughan, J. (eds.) Advances in Neural Information Processing Systems, vol. 34, pp. 27555–27565. Curran Associates, Inc. (2021)
89. Tran, D., et al.: Plex: towards reliability using pretrained large model extensions. arXiv:2207.07411 (2022)
90. Tursynbek, N., Petiushko, A., Oseledets, I.: Robustness threats of differential privacy (2021)
91. Verma, S., Rubin, J.: Fairness definitions explained. In: Brun, Y. (ed.) Proceedings of the International Workshop on Software Fairness. ACM Conferences, pp. 1–7. ACM, New York (2018). https://doi.org/10.1145/3194770.3194776, https://dl.acm.org/doi/pdf/10.1145/3194770.3194776
92. Wang, Z., et al.: Towards fairness in visual recognition: effective strategies for bias mitigation. In: Mortensen, E., Masson-Forsythe, M. (eds.) 2020 IEEE/CVF Conference on Computer Vision and Pattern Recognition, pp. 8916–8925. IEEE, Piscataway (2020)
93. Wang, Z., et al.: Fairness-aware adversarial perturbation towards bias mitigation for deployed deep models. In: Proceedings of the IEEE/CVF Conference on Computer Vision and Pattern Recognition (CVPR), pp. 10379–10388 (2022)
94. Wick, M., Panda, S., Tristan, J.B.: Unlocking fairness: a trade-off revisited (2019)
95. Wiles, O., et al.: A fine-grained analysis on distribution shift. arXiv:2110.11328 (2021)
96. Won, S., Bae, S.H., Kim, S.T.: Analyzing effects of mixed sample data augmentation on model interpretability (2023)
97. Wu, W., Protopapas, P., Yang, Z., Michalatos, P.: Gender classification and bias mitigation in facial images. In: 12th ACM Conference on Web Science. ACM, New York (2020). https://doi.org/10.1145/3394231.3397900

98. Xu, H., Liu, X., Li, Y., Jain, A., Tang, J.: To be robust or to be fair: Towards fairness in adversarial training. In: Meila, M., Zhang, T. (eds.) Proceedings of the 38th International Conference on Machine Learning. Proceedings of Machine Learning Research, vol. 139, pp. 11492–11501. PMLR (2021)
99. Yao, H., Wang, Y., Li, S., Zhang, L., Liang, W., Zou, J., Finn, C.: Improving out-of-distribution robustness via selective augmentation. In: Chaudhuri, K., Jegelka, S., Song, L., Szepesvari, C., Niu, G., Sabato, S. (eds.) Proceedings of the 39th International Conference on Machine Learning. Proceedings of Machine Learning Research, vol. 162, pp. 25407–25437. PMLR (2022)
100. Yeh, C.K., Hsieh, C.Y., Suggala, A., Inouye, D.I., Ravikumar, P.K.: On the (in) fidelity and sensitivity of explanations. In: Advances in Neural Information Processing Systems, vol. 32 (2019)
101. Yeom, S., Giacomelli, I., Fredrikson, M., Jha, S.: Privacy risk in machine learning: analyzing the connection to overfitting. In: 2018 IEEE 31st Computer Security Foundations Symposium (CSF), pp. 268–282 (2018). https://doi.org/10.1109/CSF.2018.00027
102. Yu, D., Zhang, H., Chen, W., Yin, J., Liu, T.Y.: How does data augmentation affect privacy in machine learning? In: Proceedings of the AAAI Conference on Artificial Intelligence, vol. 35, no. 12, pp. 10746–10753 (2021). https://doi.org/10.1609/aaai.v35i12.17284
103. Zhang, Y., Sang, J.: Towards accuracy-fairness paradox: adversarial example-based data augmentation for visual debiasing. In: Proceedings of the 28th ACM International Conference on Multimedia, MM 2020, pp. 4346–4354. Association for Computing Machinery, New York (2020). https://doi.org/10.1145/3394171.3413772
104. Zhang, Z., Song, Y., Qi, H.: Age progression/regression by conditional adversarial autoencoder. In: Proceedings of the IEEE Conference on Computer Vision and Pattern Recognition (CVPR) (2017)
105. Zhong, Z., Zheng, L., Kang, G., Li, S., Yang, Y.: Random erasing data augmentation. In: Proceedings of the AAAI Conference on Artificial Intelligence, vol. 34, no. 07, pp. 13001–13008 (2020). https://doi.org/10.1609/aaai.v34i07.7000, https://ojs.aaai.org/index.php/AAAI/article/view/7000

# Image Dataset Quality Assessment Through Descriptive Out-of-Distribution Detection

Sami Kharma(✉) and Jürgen Großmann

Fraunhofer-Institut für Offene Kommunikationssysteme FOKUS, Berlin, Germany
{sami.kharma,juergen.grossmann}@fokus.fraunhofer.de

**Abstract.** Out-of-distribution detection ensures trustworthiness in machine learning systems by detecting anomalous data points and adjusting confidence in predictions accordingly. However, another key use-case of out-of-distribution detection is the assessment of data quality with respect to a desired distribution or semantic range of data. This work proposes a simple but powerful approach that allows for cleaning of image data based on descriptively defining desired data as well as undesired data. Notably, this method does not require the training of a machine learning model. In addition, this work presents a new image dataset suited for evaluating data cleaning tasks in a way that has practical relevance, and demonstrates satisfactory experimental results.

**Keywords:** Out-of-Distribution Detection · Machine Learning · Data Quality

## 1 Introduction

Out-of-distribution detection (ODD) aims to identify instances of data which are significantly different from a given distribution of data. One essential application of ODD, however, is the assessment of data quality. Irrespective of how data may be used, ODD allows the detection of anomalies. Such anomalous data may be undesirable and can be filtered out. Care must be taken, however, to differentiate between anomalies of the distribution of the given data, and anomalies of the desired distribution of data. Specifically, a data point anomalous in a given dataset may very well be desired, while another anomalous data point is not.

To address this issue, the desired distribution would need to be defined separately and provided to the ODD system. In this work, an approach is presented which allows the definition of desired and undesired data descriptively. These descriptions can be considered classes of data. This approach is using these classes to form a classification problem. By then classifying each element in a dataset, desired and undesired data can be separated. The dataset can then be cleaned by removing elements which have been classified as undesired. Key for this work is the restriction to a generalized pre-trained model to achieve this classification. At no point does the method require training a machine learning model for the particular dataset at hand.


A. Hotho and S. Rudolph (Eds.): KI 2024, LNAI 14992, pp. 147–159, 2024.
https://doi.org/10.1007/978-3-031-70893-0_11

Specifically, the contributions of this work are the presentation and evaluation of a CLIP-based approach to ODD for image data quality assessment and data cleaning which does not require training a task-specific model. The evaluation is aimed at replicating potential real world scenarios in which such methods are desirable, including the creation of a dataset to that end. In doing so, the question of practical feasibility of the method can be answered.

## 2 Motivation and Related Work

The scraping of desired data from public archives or the internet is a common approach for gathering lage datasets. The main drawback of this, however, is the potential lack of quality of scraped data. Specifically, the obtained data may be contaminated with undesirable data points. For example unverified online data may be associated to incorrect keywords. In such a case, a scraped dataset will contain potentially large amounts of falsely labeled or low-quality data points. The focus on leveraging descriptive ODD to this end is a relatively novel direction, in particular when using generalized models such as CLIP [9].

The various specific types of ODD problems and existing state-of-the-art solutions have been analyzed and integrated into a theoretical generalized ODD framework previously [11]. This work will integrate itself in said framework in Sect. 3. Two conclusions from the survey are worth to be highlighted here: First, there is a severe lack of real-world benchmarks and evaluation in the ODD space. Specifically, most existing methods are evaluated on data such as CIFAR. Such datasets are simple and cover only a narrow semantic range. Second, existing work using world models, in particular multi-modal (capable of handling data from both the textual and visual domain) models such as CLIP [9], significantly enhance ODD performance.

The first point is addressed by creating a dataset which represents a usecase in which such methods may be applied. This dataset, named *scrapeface*, contains images scraped online based on various keywords related to the concept of a human face. Unlike other large datasets such as ImageNet, this dataset is created specifically for the purpose of evaluation OOD detectors. Specifically, the OOD data points are as abundant and diverse as is expected in a real-world situation.

Regarding the second highlighted conclusion, vision-language models have been a recent focus of attention due to their ability to bridge the semantic gab between visual and textual modalities. A particularly well known state of the art example of such a model is CLIP [9]. Such cross-modal understanding may prove extremely valuable for assessing and potentially enhancing dataset quality. The survey rightly points out, that using such models results in significantly better ODD performance, beating the previous state-of-the-art such as generalized ODIN [4].

CLIP was also leveraged for ODD by [3] in a way similar to this work. However, [3] uses the CIFAR dataset [5] and does not meaningfully explore the approach further, instead focusing on using a fine-tuned vision transformer, which requires separate training.

A notable existing approach to CLIP-based ODD used maximum concept matching (MCM) [8], leveraging a constructed set of so-called concept vectors. The distance of visual features of a given data point to the closest concept vector is then used for determining the OOD uncertainty. Therefore, the authors are leveraging existing ID examples to generate prototypes which are then used for ODD. In a similar vein to concept vectors, [7] also employs examples to generate partial descriptions that fit ID data. A particularly well performing recent approach is presented in [1]. The authors also generate and use descriptors based on provided ID data. However, the authors also estimate a confidence score for a given set of descriptors to ensure the descriptors are properly representing ID data. Notably, all these approaches use existing example ID data to first construct some representation of what constitutes ID data. An experiment with the goal of comparing the presented methods to the results in [1] is conducted, specifically based on the ImageNet dataset. This comparison, however, comes with various caveats discussed at a later point.

## 3 Generalized ODD

For a complete understanding of the proposed method, it is essential to also be aware of the broader context of ODD. In particular, it is valuable to understand the nature of the task itself this method aims to address. A broad framework for generalized ODD has already been established [11]. The taxonomy of ODD tasks proposed by the author is built on four bases: Distribution shift, in-distribution data type, in-distribution classification and whether a task is transductive or inductive. The problem of dataset quality assessment and cleaning addressed in this work can thus be categorized as the inductive measurement of a semantic shift without ID classification, that is, novelty detection.

Solving the task of novelty detection can be achieved with various fundamentally different approaches. The authors of the previously named survey have organized these approaches into four categories [11]: Density-based, reconstruction-based, distance-based and classification-based methods. The CLIP model as used in this work is a distance-based approach. Solutions for novelty detection typically involve training a task-specific model. This work is set apart in that this is not necessary. The so-called f-AnoGAN approach has been chosen as a comparison in the experiment [10] as a representative of both density-based as well as reconstruction-based approaches.

## 4 CLIP-Based ODD

### 4.1 Method

The CLIP model is capable of embedding both textual as well as visual inputs into a shared feature space. In this space, cosine similarity represents semantic similarity between any two given feature vectors [9]. Due to the domain-agnostic nature of CLIP, it stands to reason that it can be used without any task-specific training for use cases that fall into the convex hull of concepts encountered during pre-training. When assessing the quality of a given dataset, however, task-specific knowledge is introduced. Specifically, that knowledge pertains to what defines desirable and undesirable data. The same holds true for data cleaning, in that low-quality, and as such undesired, data is not just recognized as such but also eliminated from the dataset.

The definition of high-quality and low-quality data and the recognition based on it effectively spans an ODD problem in which CLIP-embedded images or descriptions represent the ID concepts. However, two concepts can be semantically very similar, yet one is desired and the other is not. To alleviate this issue, counterexamples are introduced. This allows measuring similarity to positive examples not nominally, but relatively to negative examples. Consider a given data point $Xx$ and examples (or example descriptions) $x_1^{(p)}, \ldots, x_k^{(p)}, x_1^{(n)}, \ldots, x_l^{(n)}$ where $(p)$ denotes positive and $(n)$ denotes negative examples (or example descriptions). Furthermore, let $E$ denote the appropriate CLIP encoder, that is, either the text encoder or the image encoder and let $\sigma$ be the softmax function. Then the probability of data point $x$ being of the $j$-th class $P(y = j \mid x)$ is

$$\sigma\left(\begin{pmatrix} \mathrm{Cossim}\left(\mathrm{E(x)}, \mathrm{E}\left(\mathrm{x}_1^{(p)}\right)\right) \\ \vdots \\ \mathrm{Cossim}\left(\mathrm{E(x)}, \mathrm{E}\left(\mathrm{x}_k^{(p)}\right)\right) \\ \mathrm{Cossim}\left(\mathrm{E(x)}, \mathrm{E}\left(\mathrm{x}_1^{(n)}\right)\right) \\ \vdots \\ \mathrm{Cossim}\left(\mathrm{E(x)}, \mathrm{E}\left(\mathrm{x}_l^{(n)}\right)\right) \end{pmatrix}\right)_j .$$

Without loss of generality, all positive examples or descriptions are in the lower indices, while the negative ones are found in the higher indices. It follows that

$$P^{(p)}(x) = \sum_{i=1}^{k} P(y = i \mid x)$$

is the overall probability of data point $x$ being classified as adhering to a positive example or description, and thus being high quality or desired. Analogously,

$$P^{(n)}(x) = \sum_{i=k+1}^{k+l} P(y = i \mid x)$$

is the overall probability of data point $x$ being undesired low-quality data.

The underlying multi-class, specifically $(k + l)$-class, classification is thus reduced to binary classification of data being either desired, or undesired. The particular class (high-quality desired or low-quality undesired) is then determined as

$$\underset{c \in \{n,p\}}{\arg\max} P^{(c)}(x).$$

This can then be applied to every element $x$ in a given dataset. All elements classified as undesired may be eliminated to obtain a new dataset which has been cleaned.

### 4.2 Scrapeface Dataset

The goal of this work is to evaluate the feasibility of this approach in a real-world setting. As previously mentioned, there is a severe lack of work using sufficiently realistic OOD datasets. Therefore, to evaluate the method properly, an image dataset containing a mixture of desired and undesired data that represents real-world use-cases must first be acquired. Due to lack of such a dataset, the *scrapeface* dataset was created. The scrapeface dataset consists of pictures scraped from Google Images[1]. Scraping is a common practice for gathering large amounts of data automatically. Therefore this dataset is expected to be a fair representation of how undesired data may be introduced when creating datasets. For this purpose, four keywords or phrases were used: *face*, *human face photo*, *gesicht*, *gesichtsphoto*. This yielded a total of 1881 images[2].

Each image was manually annotated. For the purposes of these experiments, each image is considered either desired (that is, good) or undesired (that is, bad). To ensure consistency, the following clearly defined criteria were used. For an image to be considered good data, it must contain a human-recognizable, and at least partial photograph of a real human face. Otherwise, the image is considered bad data.

In total, out of the 1881 images precisely 675 were good data and 1206 were bad data. This means the portion of good data is approximately 35.9%.

### 4.3 Experimental Setup

In the context of the scrapeface dataset, the positive description *a photo of a real human face*, as well as the negative descriptions *a drawing*, *an emoji* and *a cartoon* are used. These are common concepts associated with images of a human face. While semantically similar, these negative descriptions indeed represent OOD data.

The method as detailed in Sect. 4.1 is applied to each element in the scrapeface dataset and data is classified as either desirable or undesirable based on the resulting probabilities. The experiment is conducted using two different definitions of ID, namely textual descriptions of ID data and examples of ID and OOD data themselves.

For descriptions, the experiment is performed twice. First, in experiment 1, both the ID and OOD description(s) are directly descriptive. In experiment 2, the OOD data is instead described though a natural language negation of the ID description. This means that the description *not a photo of a real human face*

---

[1] Google Image Scraper repository: https://github.com/ohyicong/Google-Image-Scraper.

[2] Repository: https://scm.cms.hu-berlin.de/kharmasa/descriptive-odd.

is used for OOD instead. Furthermore, experiment 3 evaluates the CLIP-based approach by instead using ID and OOD examples directly as descriptions instead (as apposed to the textual descriptions previously).

For comparison, the f-AnoGAN based approach is used [10] in experiment 4. While that method is only one of many which could be used, it suffices as a basic reference for other approaches. A key difference is the need for training and the fact that it is not distance (or similarity) based, but rather density- and reconstruction-based. For a realistic use-case, training with known ID data from the dataset under test, namely scrapeface, is not appropriate. Instead, a dataset of known high-quality ID data for the task has been used. Specifically, a subset of a dataset containing real and fake faces was used[3]. Training data constituted precisely 1081 images of real faces. This data is ID, because it aligns precisely with the ID definition provided in Sect. 4.2.

Lastly, to facilitate a comparison to the results in [1], experiment 5 is devised. For this purpose, subsets of the ImageNet dataset are constructed, specifically ImageNet-10 (ID) and ImageNet-20 (OOD). The specific subclasses used are identical to those detailed in [1]. The class names of the total of 30 classes (10 ID, 20 OOD) are used as descriptors directly. This poses a noteworthy problem for comparing the results to the example-based approach presented in [1] discussed later.

We will test whether the presented CLIP-based approach performs significantly better than the presented f-AnoGAN based approach. This will be tested based on the observations of Experiment 1, experiment 2, experiment 3 and Experiment 4. Furthermore we will test whether the presented CLIP-based approach can compete with existing state-of-the-art methods. For this purpose, Experiment 5 is conducted. Experiment 2 will furthermore test whether description with natural language negation significantly hurt the performance of the presented CLIP-based approach.

### 4.4 Results and Discussion

**Experiment 1: Scrapeface with Direct Descriptions.** The first experiment involves applying the methodology using the positive description *a photo of a real human face*, as well as the negative descriptions *a drawing*, *an emoji* and *a cartoon*. A good immediate indicator for the methods ability to make high confidence decision (irrespective of its correctness) is the probability distribution of the ID prediction $P^{(p)}$. This is visualized in Sect. 4.4.

The bimodal nature of the distribution is clearly visible. For any data point $x$ it holds that $P^{(n)}(x) = 1 - P^{(p)}(x)$. It is therefore clear that in many cases the model is either confident that $x$ is ID (high $P^{(p)}(x)$) or it is confident that $x$ is OOD (low $P^{(p)}(x)$, implying high $P^{(n)}(x)$).

To gauge the efficacy of the method, however, high confidence results do not suffice. As such, the confusion matrix is shown in Sect. 4.4. False positives represent OOD data points which were incorrectly considered ID. This data

[3] Dataset: https://www.kaggle.com/datasets/xhlulu/140k-real-and-fake-faces.

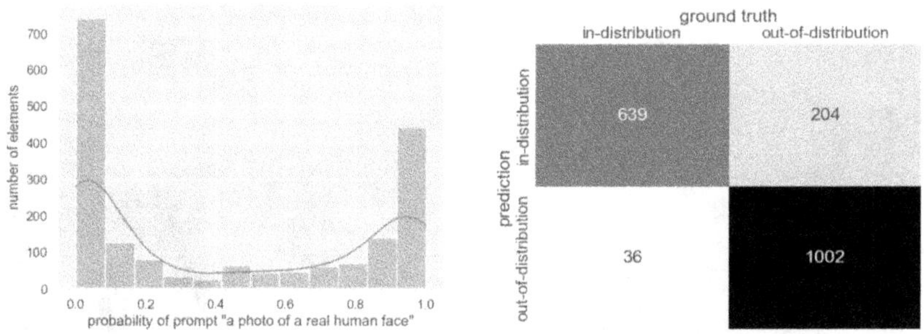


(a) The distribution of predicted probabilities for the ID class $P^{(p)}$ visualized as a histogram when using direct descriptions. Furthermore, a kernel density estimate is shown.

(b) The confusion matrix for the experiment with direct descriptions, showing the number of scrapeface dataset elements which are true positives, false positives, false negatives or true negatives respectively.

**Fig. 1.** The results of experiment 1.

would, when using the method for data cleaning, dilute the resulting dataset with undesired data points. Lastly, false negatives are predicted OOD despite being desired ID data. When data cleaning this means that such data points are mistakenly excluded (Fig. 1).

It is clear that mistakenly excluding ID data is not desirable. However, this does not compromise the usage of the cleaned dataset itself. False positives, however, may compromise further usage, as the data is diluted with OOD data points which were not correctly recognized as such.

Recall that the scrapeface dataset contains 35.9% ID data. When cleaning the dataset by filtering out data points predicted as OOD, the resulting dataset contains 74.8% ID data. This resulting dataset is called the passing subset. In this process, only 5.3% of clean data is lost.

A precision of 0.758 with a recall of 0.947 was achieved. The accuracy of predictions is 0.872 and the F1 score amounts to 0.842. The very high recall of 0.947 implies almost no ID data is lost when filtering, while the precision of 0.758 represents a drastic, but not nominally near-perfect, increase in ID data in the passing subset.

**Experiment 2: Scrapeface with Negated Out-of-Distribution Description.** This previous experiment (see experiment 1) is now repeated by using a negated description for OOD data instead of the previous direct descriptions. The resulting two descriptions being used are now specifically *a photo of a real human face* and *not a photo of a real human face*. As previously, first the confidence is investigated, see Sect. 4.4.

It is clear that this distribution is not bimodal but unimodal peaking at approx. 0.5. This implies that for a given data point $x$, neither $P^{(p)}(x)$ nor $P^{(n)}(x)$ are expected to have a statistical edge. It is clear that for the vast

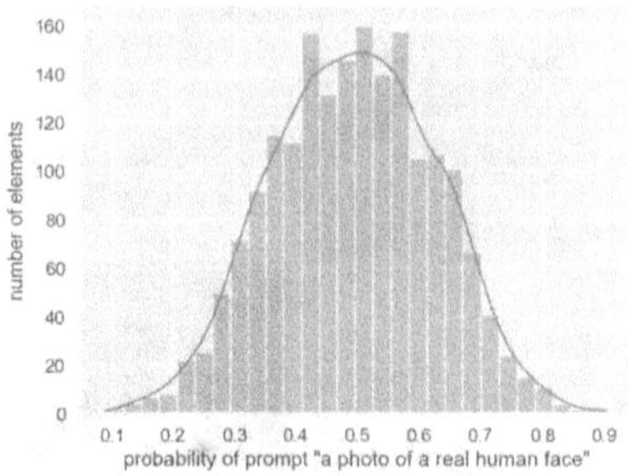


(a) The distribution of predicted probabilities for the ID class $P^{(p)}$ visualized as a histogram when using direct descriptions of ID data but negated ID descriptions for OOD data. Furthermore, a kernel density estimate is shown.

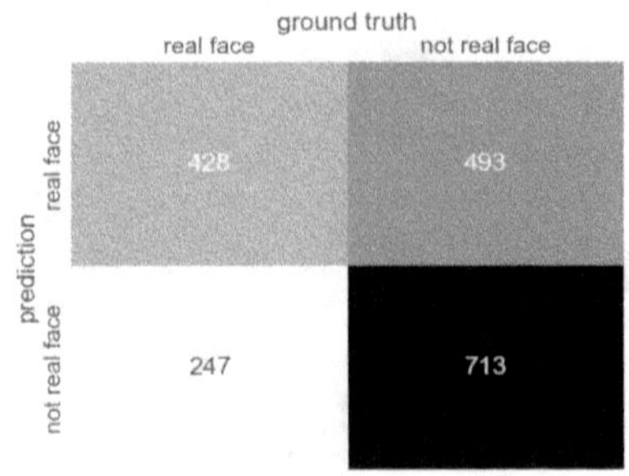


(b) The confusion matrix for the experiment with negated descriptions for OOD data, showing the number of scrapeface dataset elements which are true positives, false positives, false negatives or true negatives respectively.

**Fig. 2.** The results of experiment 2.

majority of data points, the model is incapable of making a confident decision on whether it is ID or OOD. The corresponding confusion matrix is showin in Sect. 4.4 (Fig. 2).

The percentage of ID data in the passing subset only increased to 46.5% from the initial 35.9%. Furthermore, in doing so, 36.6% of ID data has been incorrectly filtered out. The reason for this is suspected to be the underlying cross attention used by CLIP when encoding text into feature space.

For the experiment using negation, the probability deltas for all confusion matrix sectors appear to align. This is consistent with Sect. 4.4 in that neither class appears to have an edge.

The significant difference in performance between direct description and description through negation can be shown using McNemar's test statistic [6]. The contingency table is computed which gives insight into how many data points were correctly or incorrectly predicted to be ID by either approach. As $p < 0.05$, it is clearly shown that the null hypothesis is rejected. A significant difference in performance in favor of direct description exists.

**Experiment 3: Scrapeface with Examples.** Now that textual descriptions have been investigated, the usage of real ID data as an example is explored. 32 random known ID data points and another 32 random known OOD data points are selected. These serve as examples for their respective overall class (ID or OOD). However, as with descriptions, each still constitutes its own class in the underlying multi-class problem, a 64-class problem in this case.

The method was then applied with a total of 32 runs, each with newly randomly drawn 32 examples and 32 counter-examples. F1 scores remain below 0.1, clearly showing that that this approach does not succeed whatsoever.

**Experiment 4: F-AnoGAN for Scrapeface Out-of-Distribution Detection.** As established previously, a broad comparison with an f-AnoGAN based approach was performed. An f-AnoGAN model was trained as described in Sect. 4.3. Afterwards, each element in the high quality dataset of real faces used for training the f-AnoGAN as well as each data point in the scrapeface dataset was assigned an anomaly score using the model. This score has a theoretical lower bound of 0 but no upper bound. The distribution of data across anomaly scores is shown in Sect. 4.4 (Fig. 3).

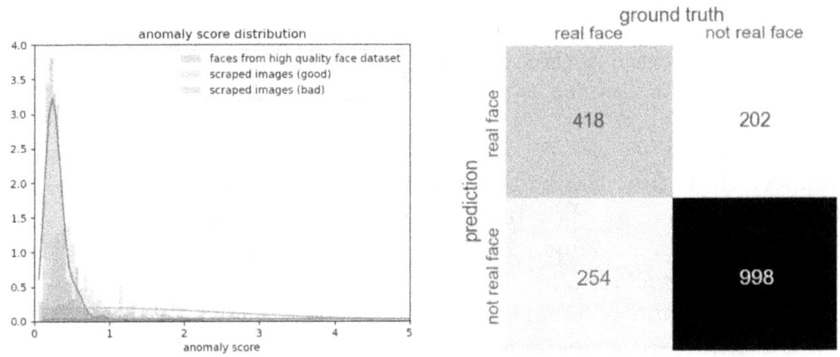


(a) The distribution of data by anomaly score. Data is separated into the separate clean dataset, good (ID) scrapeface data points and bad (OOD) scrapeface data points. This also includes kernel density estimates.

(b) The confusion matrix for the experiment using an f-AnoGAN, showing the number of scrapeface dataset elements which are true positives, false positives, false negatives or true negatives respectively.

**Fig. 3.** The results of experiment 4.

As expected, the fully clean data scores best in that it yields mostly low anomaly scores. It is also clearly visible how in particular ID data largely scores low anomaly scores, while OOD data appears to be more evenly spread out across anomaly scores. Using the threshold $t = \mu + n\sigma$ where $\mu$ represents the mean and $\sigma$ the standard deviation of the distribution of anomaly scores, classification occurs based on the anomaly scores. The value $n$ adjusts the level of confidence in anomaly scoring required for data to pass filtering. Data with an anomaly score less-than or equal to $t$ passes. Otherwise, it is filtered out.

For further details, the confusion matrix is seen in Sect. 4.4. A precision of 0.674 with a recall of 0.622 was achieved. The accuracy of predictions is 0.756 and the F1 score amounts to 0.647. The question remains, what reasonable choices of $n$ may be. A range of $n \in [0, 100]$ has been searched for that purpose. Note however, that a specific optimal choice for $n$ is likely task dependent, as it heavily depends on the distribution of the initial data. The search results are visualized in Fig. 4 where for each choice of $n$ the corresponding F1 score of the resulting classifier is shown.

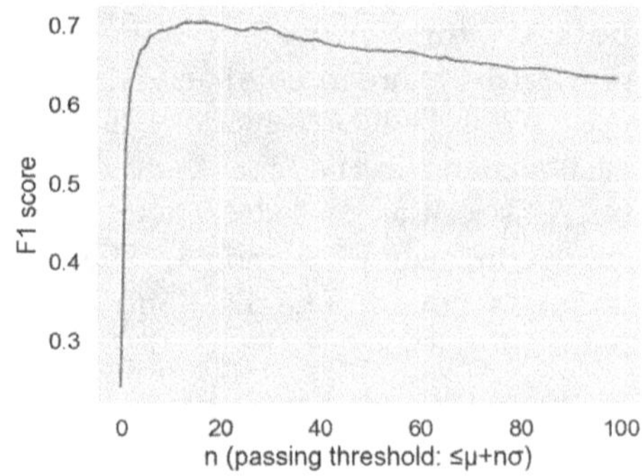


**Fig. 4.** F1 scores for varying choices of $n$ when evaluating the resulting f-AnoGAN-based classifier for data cleaning.

As expected, at first the F1 score rises drastically, as more and more true positives are included, as these tend to have low anomaly scores by virtue of being ID data points. In doing so, only smaller amounts of OOD data with low anomaly scores are included. As $n$ grows further, however, diminishing returns are encountered because less and less ID data is included, but the amount of OOD data, most of which scores high anomaly scores, grows. As the slope reverses, the amount of passing OOD data outpaces ID data and F1 scores begin to degrade. A desirable choice of $n$ balances recall and precision.

**Experiment 5: ImageNet-10/20 OOD Detection.** The CLIP-based approach with direct descriptions, analogous to experiment 1, is used for this experiment. However, the dataset is instead the Imagenet-10 (defined to be ID) and ImageNet-20 (defined to be OOD) dataset.

The class names are used as textual descriptions and the ODD is performed. In part likely due to the known correct exact class names, the confidence at which data is classified as ID (or OOD) is very high, as seen in Sect. 4.4.

As with previous experiments, the confusion matrix is shown to gauge the results more effectively (see Sect. 4.4). A precision of 0.973 with a recall of 0.934 was achieved. The accuracy of predictions is 0.934 and the F1 score amounts to 0.953 (Fig. 5). Potential validity issues of this experiment, in particular as it relates to a comparison with [1] follow later-on in Sect. 4.4.

**Discussion of Observations.** While the f-AnoGAN-based approach significantly outperforms this works approach in the case of direct examples (comparing experiment 3 and experiment 4), it is outperformed on all metrics by the direct description case (see experiment 1). Note however that, as previously established, the f-AnoGAN is only one example of other approaches. The comparison aims to only give a general idea of other types of methods employed to solve the problem to put this work into a greater context. An overall overview of the results is shown in Table 1.

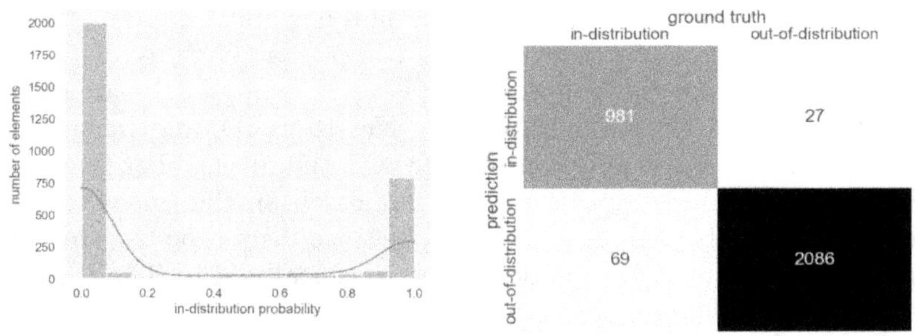


(a) The distribution of the predicted probability $P^{(p)}(x)$ of a data point $x$ being ID Visualized as a histogram. Furthermore, a kernel density estimate is shown. Note that the OOD probability is simply $1 - P^{(p)}$, which would effectively mirror this histogram, if visualized.

(b) The confusion matrix for the experiment with direct descriptions, showing the number of scrapeface dataset elements which are true positives, false positives, false negatives or true negatives respectively.

**Fig. 5.** The results of experiment 5.

**Table 1.** Detailed statistical properties of the experiments using the scrapeface dataset in comparison.

| | f-AnoGAN-based | this work (direct descriptions) | this work (negated description) |
|---|---|---|---|
| precision | 0.674 | 0.758 | 0.465 |
| recall | 0.622 | 0.947 | 0.634 |
| accuracy | 0.756 | 0.872 | 0.607 |
| F1 score | 0.647 | 0.842 | 0.536 |
| requires task-specific training | yes | no | no |

Given the entirety of the data, McNemar's test statistic can be computed [6] to compare the presented CLIP-based approach to the one using the f-AnoGAN. This follows the same pattern as the previous test regarding negative textual descriptions. Here, with $p < 0.05$ it is clearly shown that there is a significant performance difference between the two approaches. This difference is very clearly in favor of the CLIP-based approach, as visible in Table 1.

It can be concluded, that the approach presented here is feasible for cleaning contaminated real-world datasets, as long as explicit direct descriptors of classes are used. Using this method through examples directly is yielding sufficiently poor results to conclude that it fails to be useful for ODD completely (see experiment 3). When examples are meant to be used, instead state-of-the-art approaches such as [1] are preferable.

In the case of direct descriptors, this method has approximately 95% recall. Therefore, the false positive rate of OOD samples at 95% recall (FPR95) is approximately 16.9% for the scrapeface dataset. The state-of-the-art method [1] reports a FPR95 of 4.2%, though for ImageNet-10 as ID, and ImageNet-20 as OOD. Both approaches are only comparable with the caveat, that [1] uses examples. Furthermore, instead evaluating this approach on the ImageNet datasets does not create directly comparable results, because there is no threshold parameter with which the recall can be pegged at specifically 95%. The fact this was the case with scrapeface is coincidence.

Despite that, for a broad though flawed comparison, experiment 5 shows that given the class names of ID (ImageNet-10) and OOD (ImageNet-20) classes, a false-positive rate of 1.3% at 93.4% recall is obtained, significantly lower than the 4.2% in [1] at 95% recall. Therefore, a lower recall is instead compensated by lower false-positive rate. However, the exceedingly low false-positive rate can also be explained by the knowledge of classes without having to derive it from examples. These results should instead be seen as a broad indicator that both methods can achieve satisfactory results.

## 5 Conclusion and Future Work

An approach for dataset quality assessment through descriptive ODD has been introduced. This method leverages the capabilities of the CLIP model. This addresses the need for robust high-quality datasets. The focus was placed on creating a real-world use-case and using cross-modal capabilities to achieve satisfactory performance without task-specific training. The approach presented here generalizes but requires an explicit definition of OOD data through textual descriptions to achieve desired performance levels. Furthermore it has been shown that if this is not the case, other state-of-the-art methods are preferable. A comparison to an f-AnoGAN-based approach puts this approach in the broader context. Particularly noteworthy is the high recall score. This means that almost no clean data is lost in the process of data cleaning. Yet as shown by the precision, the proportion of quality ID data to undesired OOD data improved significantly. This makes this approach generally valuable for initial data cleaning. It is, however, insufficient, if 100% ID data is required as errors can still occur.

An angle previously explored in [2] is the automatic generation of labels. However, this comes hand-in-hand with multiple problems: Consider OOD data which can only be described with complex terms and is still similar to ID data. The generated description is in that case likely not complex but rather a more simple and general description which may also apply to ID data points. Secondly, the relation between generated description is ignored. A data point may only be OOD if two particular descriptions apply simultaneously. Such relations are, however, not considered. Future work involves adopting such automatic approaches while still retaining and enforcing relations between descriptions. Furthermore, in principle this work should transfer to textual datasets which we may attempt in future work.

## References

1. Dai, Y., Lang, H., Zeng, K., Huang, F., Li, Y.: Exploring large language models for multi-modal out-of-distribution detection. In: Bouamor, H., Pino, J., Bali, K. (eds.) Findings of the Association for Computational Linguistics: EMNLP 2023, pp. 5292–5305. Association for Computational Linguistics, Singapore (2023). https://doi.org/10.18653/v1/2023.findings-emnlp.351, https://aclanthology.org/2023.findings-emnlp.351
2. Esmaeilpour, S., Liu, B., Robertson, E., Shu, L.: Zero-shot open set detection by extending CLIP. CoRR abs/2109.02748 (2021). https://arxiv.org/abs/2109.02748
3. Fort, S., Ren, J., Lakshminarayanan, B.: Exploring the limits of out-of-distribution detection. CoRR abs/2106.03004 (2021). https://arxiv.org/abs/2106.03004
4. Hsu, Y., Shen, Y., Jin, H., Kira, Z.: Generalized ODIN: detecting out-of-distribution image without learning from out-of-distribution data. CoRR abs/2002.11297 (2020). https://arxiv.org/abs/2002.11297
5. Krizhevsky, A.: Learning multiple layers of features from tiny images. University of Toronto (2012)
6. McNemar, Q.: Note on the sampling error of the difference between correlated proportions or percentages. Psychometrika **12**(2), 153–157 (1947). https://doi.org/10.1007/bf02295996
7. Menon, S., Vondrick, C.: Visual classification via description from large language models. In: The Eleventh International Conference on Learning Representations (2023). https://openreview.net/forum?id=jlAjNL8z5cs
8. Ming, Y., Cai, Z., Gu, J., Sun, Y., Li, W., Li, Y.: Delving into out-of-distribution detection with vision-language representations (2022)
9. Radford, A., et al.: Learning transferable visual models from natural language supervision. In: Meila, M., Zhang, T. (eds.) Proceedings of the 38th International Conference on Machine Learning, ICML 2021, 18–24 July 2021, Virtual Event. Proceedings of Machine Learning Research, vol. 139, pp. 8748–8763. PMLR (2021). http://proceedings.mlr.press/v139/radford21a.html
10. Schlegl, T., Seeböck, P., Waldstein, S.M., Langs, G., Schmidt-Erfurth, U.: f-AnoGAN: fast unsupervised anomaly detection with generative adversarial networks. Med. Image Anal. **54**, 30–44 (2019). https://doi.org/10.1016/j.media.2019.01.010, https://www.sciencedirect.com/science/article/pii/S1361841518302640
11. Yang, J., Zhou, K., Li, Y., Liu, Z.: Generalized out-of-distribution detection: a survey. CoRR abs/2110.11334 (2021). https://arxiv.org/abs/2110.11334

# Saxony-Anhalt is the Worst: Bias Towards German Federal States in Large Language Models

Anna Kruspe(✉) and Mila Stillman

Munich University of Applied Sciences, Lothstr. 64, 80335 Munich, Germany
anna.kruspe@hm.edu
http://cs.hm.edu/

**Abstract.** Recent research demonstrates geographic biases in various Large Language Models that reflects common human biases, which are presumably present in the training data. We hypothesize that these biases also exist on smaller scales. Within Germany, there is still a strong divide between the former states of the German Democratic Republic (the "East") and those of the Federal Republic of Germany (the "West") in many respects as well as other perceived geographic disparities.

We evaluate the responses of ChatGPT-3.5, ChatGPT-4, and LeoLM for various ratings and estimations by state. Those include objectively measurable values as well as subjective assessments of residents' characteristics. Experiments are conducted in English and German. We show that there is a very visible bias in both the subjective and the objective ratings, and analyze various effects. In particular, we demonstrate that Eastern states are consistently rated lower (or worse, depending on task), whereas Southern states frequently rate higher. We also discuss models' behaviors when prompted with these tasks.

**Keywords:** Large Language Models · LLMs · ChatGPT · LeoLM · Bias · Fairness · Germany

## 1 Motivation

To this day, more than 30 years after the fall of the Berlin wall, there are significant differences between the German federal states of the former German Democratic Republic (the "East") and of the Federal Republic of Germany (the "West"). Nearly any statistic we consider will have a strong disparity between both parts of the country, be it salaries and incomes [40], unemployment rate [41], life expectancy [42], working hours [2], millionaires [39], car ownership [22], voting behavior [9], or religion [5]. The causes for this are manifold and rooted both in the histories of both regions as well as the consequences of the reunificiation itself [32].

These differences lead to biases in the perception of the people living in each part, both in the population as well as in the media [1,20]. As an example,


A. Hotho and S. Rudolph (Eds.): KI 2024, LNAI 14992, pp. 160–174, 2024.
https://doi.org/10.1007/978-3-031-70893-0_12

inhabitants of the East are often stereotyped as lazy, ungrateful, or stupid [16,35,50]. People from the West, in contrast, are seen as arrogant, materialistic, etc. In general, though, the East is therefore frequently seen as the "other". This has, among other effects, led to legal processes discussing discrimination against people from the East, where courts decided that this does not constitute a racial aspect and is therefore not protected by the constitution [43,45]. To a lesser degree, such biases also exist on a smaller scale towards other regions of Germany (e.g. the South is both more economically successful and more conservative than the North, Berlin is "avantgarde" etc.). Similarly, this effect can be observed in other closely related regions of the world, e.g. North vs. South of Italy [6] or South and North Korea [16].

It has long been known that machine learning models pick up and thus perpetuate human biases in various ways, most prominently by learning them from their training data. For text-based models, even early embedding approaches exhibited e.g. gender bias [3]. With the recent rise of Large Language Models (LLMs), gender and race biases were quickly discovered and analyzed in various domains [21,31,44,48]. Types of bias that have been considered somewhat less include those based on culture and geography. However, [27] recently showed that LLMs also exhibit those, both implicitly and explicitly. Their research demonstrated that when prompted to rate random locations on Earth on various characteristics, LLMs generally yielded lower ratings for certain regions, e.g. the global South. This is true both for highly subjective properties (e.g. attractiveness and intelligence), but also for objectively measurable ones (e.g. infant mortality and temperature), and even those that are not influenced by geography at all (e.g. body temperature). There appear to be correlations with the coverage of regions and their cultural and historical significance in the training data, statistics across a range of aspects around the world, and possibly (implicit) biases of the institutions creating these models, which are usually based in the North America and Europe.

We hypothesize that such biases in LLMs are also present on more granular geographic scales in accordance with human ones, and that the biases towards German federal states described above can also be detected. In this paper, we use the methodology from [27] to analyze these biases across Germany. We conduct four experiments by prompting LLMs to:

- Rate positive attributes of the inhabitants of each federal state
- Rate negative attributes of the inhabitants of each federal state
- Rate objective properties of the federal states
- Provide absolute numbers for objective properties of the federal states

We analyze the results and, where applicable, compare them to ground truths from reliable sources. The rest of the paper is structured as follows: Sect. 2 covers some of the related work on bias detection in LLMs; Sect. 3 describes the methodology used in our experiments, while Sect. 4 discusses the results; and in Sect. 5, we provide a conclusion and make suggestions for future work.

## 2 Related Work

Spatial and geographic knowledge is encoded in natural language, as evidenced by earlier language models and current large language models (LLMs) [14,25,28, 34,36]. This knowledge has significant potential for enhancing geospatial applications and GeoAI frameworks [19,26]. However, the challenge of interpreting and explaining the behavior of these advanced machine learning models remains. Their training involves extensive datasets that often exhibit biases, potentially skewing the data representation [30]. For instance, crowd-sourced geographic information like geotagged social media posts typically favor urban and affluent areas over rural or economically disadvantaged ones [15,24,51]. This bias also extends to contributors of crowd-sourced geographic data in platforms like OpenStreetMap [46] and to Wikipedia, which shows uneven geographic information distribution and systemic biases [13,17].

Geographic biases in LLMs are increasingly studied, reflecting their widespread commercial use across various applications [49]. Investigations into these models' geographic knowledge disparities show significant regional inequalities [8,11,37]. Studies testing GPT-4's capabilities in route planning and geocoded information retrieval reveal limitations in abstract reasoning and raise questions about the role of memorization in task performance [7,34]. Additionally, research indicates that geopolitical favoritism can influence LLMs' performances during inference, affecting the factual accuracy of information across different global regions [12]. Mirza et al. [29] identified significant disparities in the factual accuracy of LLMs across continents like Africa, South America, and the Middle East. Likewise, Manvi et al. [27] found that LLMs exhibit biases in generating data on both objective and subjective topics for certain regions.

## 3 Methodology

In this section, we will describe our experimental design, including used models, prompt design, prompted tasks, and postprocessing of the results.

### 3.1 Models

We tested our bias prompts on three different models via their online interfaces:

- ChatGPT-3.5
- ChatGPT-4 (paid version)
- LeoLM, a version of LLaMA-2 finetuned on German-language, locality-specific text by LAION and HessianAI [33]. We used the 13B parameter version via the HuggingFace interface[1]. The purpose was to figure out whether a model that had seen more data from Germany would exhibit any difference in terms of German biases.

[1] https://huggingface.co/spaces/LeoLM/leo-hessianai-13b-chat.

### 3.2 Prompt Design

We use the methodology from [27] to design our prompts. In essence, the LLMs are given an over-all task of providing ratings on a certain attributes for a certain region of Germany compared to all other inhabited areas of the country. An example is shown in Listing 1. Then, we focus on a specific attribute and give the model a list of all states with their German and English names. For the experiment on absolute values of objective properties, we ask the model directly for the absolute values, or, if necessary, prompt it to estimate them. For subjective attributes, it is generally not possible to ask for direct comparisons (e.g. "which state has the most attractive inhabitants") due to content filters, but ratings work in most cases.

Besides the English-language prompts, we perform the same experiments using German-language prompts. The over-all prompt was translated by ChatGPT-4, the properties manually, and the list of states was kept the same. Each experiment is repeated five times (runs) on each model to account for different initializations. Occasionally, we receive replies where the model either rates everything the same (usually 5.0), refuses to answer (mostly due to purported lack of information), or gives random outputs (also see Sect. 4.5). These instances are discarded as they do not contribute information.

```
You will be given a region randomly sampled from all human-
populated locations in Germany. You give your rating keeping
in mind that it is relative to all other human-populated
locations in Germany. You provide ONLY your answer in the
exact format "My answer is X.X." where "X.X" represents
your rating for the given topic.

...

task: annual precipitation
region: bavaria
```

**Listing 1.** Example prompt

### 3.3 Tasks

As described above, we pose three tasks to the LLMs: Rating various subjective and objective attributes of inhabitants of federal states as well as statistic properties of the states themselves. For the subjective attributes, we first use the positive ones given in [27]: Attractiveness, Intelligence, Likeability, Morality, and Work Ethic. We are also interested in what the models do when the task is turned around, i.e. asking for ratings for negative subjective traits. To this end, we pick some common stereotypes (see e.g. [50]): Arrogance, Laziness, Ungratefulness, Narrowmindedness, Insecurity, Xenophobia, Materialism, and Ignorance. In all rating tasks, we do not provide further explanations to the model, and do not limit it to a scale in order to influence the result as little as possible.

For the objective tasks, we also reuse the ones from [27]: Precipitation, Average Temperature, Infant Mortality Rate, and Population Density.[2] We add Average Age and Education to this as they are properties with interesting variations between the states. We also test some of the properties from [27] that do not have a geographic relation at all, but find that the LLMs are largely able to determine that all states should be rated similarly for those. We only see some variation in the body temperature rating and therefore include this.

The objective properties are prompted in two different manners: As ratings (like the subjective ones), and as absolute values. For the absolute values, the result for the body temperature is almost always the same for all states, so we discard it from this experiment (also see below). For the education rating, models frequently refuse to answer, stating that they are not familiar with the INSM education survey. However, when informed that the survey is on a scale from 0 to 100, they usually reply with 10x of the value they give in the rating experiment. We therefore simply use this value in the absolute value experiment for comparison with the ground truth.

### 3.4 Postprocessing

All rating results are then normalized by mean and standard deviation for each attribute. This is necessary because we observe a general tendency for the model to focus ratings around a certain mean value that differs for each run (e.g. sometimes most ratings vary around, but close to, 7.0, sometimes around 4.0). Then, we average those normalized results across all five runs for each attribute.

For the absolute values, we do not normalize results and directly average them for comparability to actual ground truth values. Those ground truths are obtained from official sources, which are provided in Table 1. We mainly chose the most recent statistics we were able to find; this means that GPT-3.5 may not be fully up-to-date, but in practice, none of the characteristics would have changed enough in recent years to affect our results significantly.

Our results and analysis notebooks are available on https://github.com/annakaa/germany_bias.

## 4 Experiments

### 4.1 Subjective Measures-Positive Attributes

The results of our first experiment, where we ask the LLMs to rate inhabitants' positive attributes for each federal state, are shown in Fig. 1. It becomes obvious very quickly that all models have a tendency to assign lower ratings to the states of the East, confirming our original suspicion. For some models, we also see a stronger disparity between North and South, which appears to be more

[2] Nighttime Light Intensity from the original paper was discarded as we could not find up-to-date information and the units of the absolute value experiments varied wildly.

**Table 1.** Sources of ground truth for the objective ratings

| Task | Note | Source | Year | Ref. |
|---|---|---|---|---|
| Precipitation | per year, in $l/m^2$ | Deutscher Wetterdienst | 2023 | [10] |
| Average temperature | in degrees Celsius | Deutscher Wetterdienst | 2023 | [10] |
| Population density | | Destatis | 2022 | [38] |
| Infant Mortality Rate | per 1,000 | Statistisches Monatsheft Baden-Württemberg | 2020 | [4] |
| Average Age | | Verband der Ersatzkassen | 2022 | [47] |
| Education | Rating out of 100 | INSM Bildungsmonitor | 2023 | [18] |

prominent for German-language prompting. Out of the states of the West, the Saarland is often rated a bit more negatively. We suspect this may happen because it is often mentioned in the context of jokes. In contrast, Berlin is often rated higher despite being geographically in the East, possibly because it is usually mentioned in other contexts than this.

### 4.2 Subjective Measures-Negative Attributes

We then repeat the experiment with negative attributes of inhabitants, shown in Fig. 2. We see a variety of effects here. GPT-3.5 continues its tendency to simply assign lower numbers to the states of the East, no matter the prompt, even if the requested attribute is negative. An alternative explanation could also be bias towards other regions; e.g. Arrogance, Narrowmindedness, and Materialism are also occasionally attributed to southern states.

In contrast, GPT-4 appears to have a concept of the negation and sometimes returns higher ratings for negative attributes in the East when using English-language prompts. These ratings may cancel out the ones affected by the previously described effect. To analyze this further, we check the different runs for each experiment; an example for Xenophobia is shown in Fig. 3. We see that the model sometimes assigns lower scores to the East as before, and sometimes picks higher ones for this negative characteristic, in contrast to GPT-3.5. This is arguably worse as it may indicate a more implicit bias that is harder to detect than the lower ratings from before.

### 4.3 Objective Measures-Relative

In our next experiment, we once again ask the LLMs for ratings, but this time for objectively measurable properties of the federal states. The results are displayed in Fig. 4. We once again see that in most cases, models simply give lower ratings to the states of the East, no matter the property. Intriguingly, we can even observe this for Body Temperature, which should have no geographic variation at all. GPT-4 with English-language prompts is the only scenario where the model recognized this.

**Fig. 1.** Normalized and averaged ratings for positive characteristics by federal state; all models, both languages. Blue indicates numerically lower ratings, red higher ones. (Color figure online)

In some cases, mainly for GPT-4, the rating trend seems be confused with actual information (e.g. Bavaria does in fact have fairly low average temperatures), as we will see in the next experiment. Additionally, it is not always clear whether a higher rating corresponds conceptually to a higher absolute number (e.g. there was one case where the model argued that a higher rating for Average Age would indicate lower actual numbers as a younger population is considered

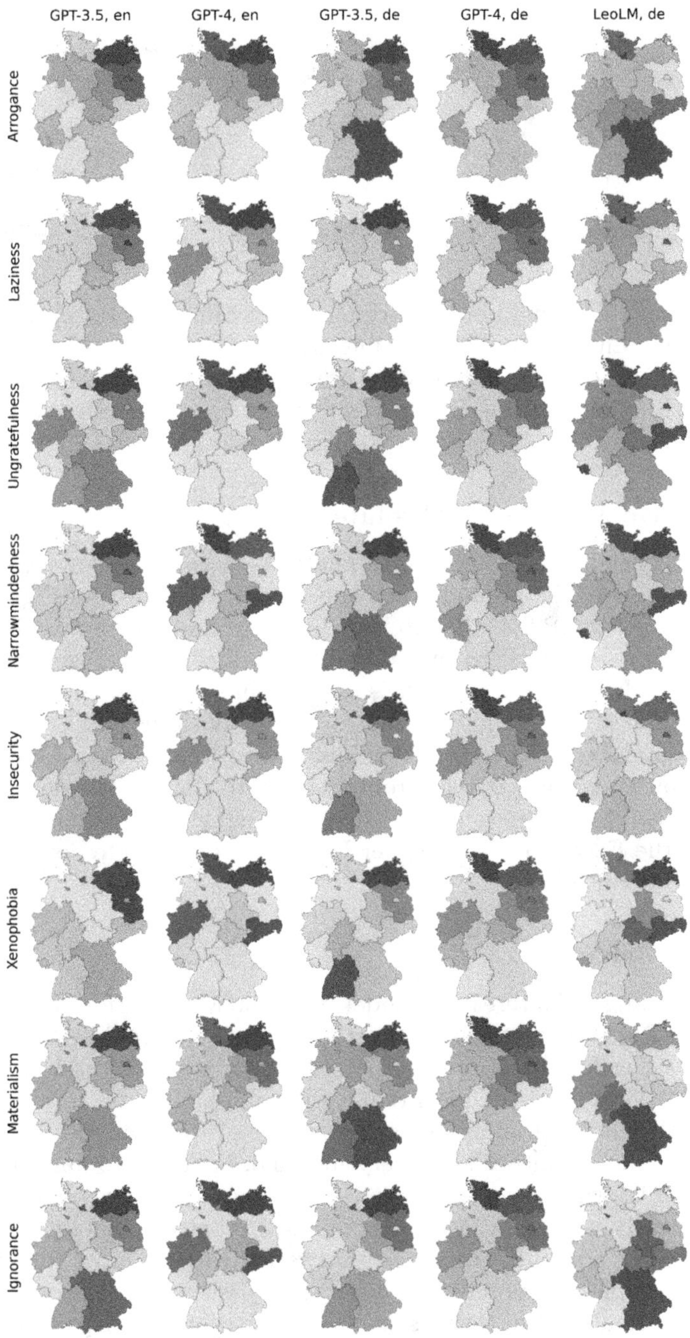


**Fig. 2.** Normalized and averaged ratings for negative characteristics by federal state; all models, both languages. Blue indicates numerically lower ratings, red higher ones. (Color figure online)

**Fig. 3.** Normalized ratings of individual runs for Xenophobia on GPT-4 in English.

better; a similar effect could be at work for Infant Mortality Rate). LeoLM overall gives chaotic answers (e.g. responding with a completely different statistical measure), and it is often hard to elicit a useful response at all. This is interesting in contrast with our first experiment, where ratings for subjective characteristics are far more consistent, indicating a stronger bias.

### 4.4 Objective Measures - Absolute

Finally, we ask the models to give or estimate absolute values for the objectively measurable criteria. The ground truth and results are shown in Fig. 5. In some cases, especially for GPT-4, the models appear to have the actual information and are therefore very close to the ground truth, e.g. for Average Age and population density. In other cases, models once again simply assign lower numbers to states of the East and higher numbers to the south, even where the ground truth is the opposite, e.g. Precipitation and Average Age. Infant Mortality Rate and Education stand out as two properties where the models do not seem to have knowledge of absolute numbers, and once again show a tendency to simply consider the East worse (i.e. higher Infant Mortality Rate, lower Education scores) and the south better.

### 4.5 Model Behavior

We observe a range of interesting model behavior while performing the experiments. First of all, we were unsure whether the models would provide any sort of rating at all, especially for the subjective tasks, but this rarely turns out to be a problem. Very rarely, models refuse to rate inhabitants' traits on a moral basis, and if they do, continuing to ask often elicits ratings anyway. A little more frequently, the models either rate everything 5.0, or refuse to answer due to an alleged lack of knowledge on these topics or regions. These cases are discarded. LeoLM frequently veers into disorder, rambling about completely different topics or exposing its training format (e.g. answering with excerpts from textbooks about similar topics, responding in a "user:... assistant:..." format, repeating itself forever). It also often forgets about the given task, in particular for the ratings of objective statistics.

Another intriguing insight comes from the arguments accompanying the provided responses. Sometimes, models will state that these were just random numbers showing how such a task could be fulfilled, while always maintaining the

**Fig. 4.** Normalized and averaged ratings for objective properties by federal state; all models, both languages. Blue indicates numerically lower ratings, red higher ones. (Color figure online)

**Fig. 5.** Averaged results of absolute values given for objective properties by federal state, ground truth leftmost column; all models, both languages. Darker colors indicate higher values.

discussed biases. Other times, responses for the objective tasks begin with a statement that these numbers were based on official sources (which are often inappropriate for the task, and results are then incorrect anyway). In contrast, as discussed above, GPT-4 appears to have the actual information for some of the statistics (without online search).

## 5 Conclusion and Future Work

In this paper, we explore geographic biases with regards to German federal states in Large Language Models. To do this, we prompt ChatGPT-3.5, ChatGPT-4, and LeoLM to rate subjective and objective characteristics of the states and their inhabitants, or to provide absolute values for the objective characteristics.

We find that there is a statistically significant tendency for states of the East to be rated worse than the others, with some less present trends for the others. This aligns with known stereotypes and trends in existing statistic data, and corroborates other research where similar effects were observed world-wide. Older models appear to have a tendency to simply reply with lower numbers for the Eastern states, no matter the task. Newer models vary between the same behavior and answering with more negative ratings (e.g. when asked to rate negative characteristics, they will sometimes rate Eastern states numerically higher). This behavior makes the tendencies even less transparent, and will require further analysis.

While users may rarely prompt the models in this direct manner, it is likely the implicit bias we show here could affect real-life downstream tasks without being obvious to humans. An example could be the task of asking models to rate job applicants, where their region of origin could influence the result without any feedback about this to the user. We will analyze model behavior in such downstream tasks in future experiments.

Our research demonstrates that geographic biases observed on the global scale are also present on a more granular level. We suspect that this could also be detected in other regions of the world, or even finer scales. More generally, it is not quite clear how to prove biases beyond such point-wise experimental prompts. Future research could develop more comprehensive metrics.

Finally, while detecting those biases is an important first step, there is currently no easy method of mitigating them, or even being sure what would constitute an unbiased result. A more achievable next step could lie in making such influences more transparent to users [23].

**Disclosure of Interests.** The authors have no competing interests to declare that are relevant to the content of this article.

## References

1. Ahbe, T., Gries, R., Schmale, W. (eds.): Die Ostdeutschen in den Medien: Das Bild von den Anderen nach 1990. Leipziger Universitätsverlag, Leipzig, Germany (2009)
2. Berliner Morgenpost: Arbeitszeit-Auswertung: Ostdeutsche arbeiten mehr als Westdeutsche (2023). https://www.morgenpost.de/politik/article238953031/arbeitszeit-arbeitsstunden-auswertung-ostdeutsche-kinder.html
3. Bolukbasi, T., Chang, K.W., Zou, J., Saligrama, V., Kalai, A.: Man is to computer programmer as woman is to homemaker? Debiasing word embeddings. In: Proceedings of the 30th International Conference on Neural Information Processing Systems. NIPS 2916 (2016)

4. Bracht-Schwarz, W.: Zur Entwicklung der Säuglingssterblichkeit in Baden-Württemberg. Statistisches Monatsheft Baden-Württemberg **9** (2020)
5. Bundeszentrale für politische Bildung (bpb): Mitgliederzahlen der Kirchen nach Bundesländern (2023). https://www.bpb.de/kurz-knapp/zahlen-und-fakten/soziale-situation-in-deutschland/61562/kirche-nach-bundeslaendern/
6. Daniele, V.: Two Italies? Genes, intelligence and the Italian North-South economic divide. Intelligence **49**, 44–56 (2015). https://doi.org/10.1016/j.intell.2014.12.004, https://www.sciencedirect.com/science/article/pii/S0160289614001846
7. Das, S.: Evaluating the capabilities of large language models for spatial and situational understanding. Ph.D. thesis, Thesis (MA). University of Cambridge (2023)
8. Decoupes, R., Interdonato, R., Roche, M., Teisseire, M., Valentin, S.: Evaluation of geographical distortions in language models: a crucial step towards equitable representations. arXiv preprint arXiv:2404.17401 (2024)
9. Deutscher Bundestag: Wahlverhalten der verschiedenen Bevölkerungsgruppen nach Bundesländern (2021). https://www.bundestag.de/dokumente/textarchiv/2021/kw39-wahlverhalten-laender-859680
10. Deutscher Wetterdienst: Deutschlandwetter im Jahr 2023 (2023). https://www.dwd.de/DE/presse/pressemitteilungen/DE/2023/20231229_deutschlandwetter_jahr2023_news.html
11. Dunn, J., Adams, B., Madabushi, H.T.: Pre-trained language models represent some geographic populations better than others. arXiv preprint arXiv:2403.11025 (2024)
12. Faisal, F., Anastasopoulos, A.: Geographic and geopolitical biases of language models. arXiv preprint arXiv:2212.10408 (2022)
13. Graham, M., Hogan, B., Straumann, R.K., Medhat, A.: Uneven geographies of user-generated information: Patterns of increasing informational poverty. Ann. Assoc. Am. Geogr. **104**(4), 746–764 (2014)
14. Gurnee, W., Tegmark, M.: Language models represent space and time. arXiv preprint arXiv:2310.02207 (2023)
15. Hecht, B., Stephens, M.: A tale of cities: urban biases in volunteered geographic information. In: Proceedings of the International AAAI Conference on Web and Social Media, vol. 8, pp. 197–205 (2014)
16. Horton, A.D.: The "ignorant" other: popular stereotypes of North Koreans in South Korea and east Germans in unified Germany. In: Cho, J.M., Roberts, L.M. (eds.) Transnational Encounters between Germany and Korea, pp. 195–214. Palgrave Macmillan, New York (2018). https://doi.org/10.1057/978-1-349-95224-3_9
17. Hube, C.: Bias in wikipedia. In: Proceedings of the 26th International Conference on World Wide Web Companion (2017)
18. Initiative Neue Soziale Marktwirtschaft (INSM): Bildungsmonitor 2023 (2023). https://www.insm-bildungsmonitor.de/2023_best_i_gesamtranking.html
19. Janowicz, K., Gao, S., McKenzie, G., Hu, Y., Bhaduri, B.: GeoAI: spatially explicit artificial intelligence techniques for geographic knowledge discovery and beyond (2020)
20. Kollmorgen, R., Hans, T.: Der verlorene Osten. In: Kollmorgen, R., Koch, F.T., Dienel, H.L. (eds.) Diskurse der deutschen Einheit, pp. 107–165. VS Verlag für Sozialwissenschaften, Wiesbaden (2011). https://doi.org/10.1007/978-3-531-93351-1_4
21. Kotek, H., Dockum, R., Sun, D.: Gender bias and stereotypes in large language models. In: Proceedings of The ACM Collective Intelligence Conference, CI 2023, pp. 12–24. Association for Computing Machinery, New York (2023). https://doi.org/10.1145/3582269.3615599

22. Kraftfahrt-Bundesamt (KBA): Fahrzeugbestand (2023). https://www.kba.de/DE/Statistik/Fahrzeuge/Bestand/bestand_node.html
23. Kruspe, A.: Towards detecting unanticipated bias in language models. arXiv preprint arXiv:2404.02650 (2024)
24. Li, L., Goodchild, M.F., Xu, B.: Spatial, temporal, and socioeconomic patterns in the use of Twitter and Flickr. Cartogr. Geogr. Inf. Sci. **40**(2), 61–77 (2013)
25. Louwerse, M.M., Zwaan, R.A.: Language encodes geographical information. Cogn. Sci. **33**(1), 51–73 (2009)
26. Mai, G., Cundy, C., Choi, K., Hu, Y., Lao, N., Ermon, S.: Towards a foundation model for geospatial artificial intelligence (vision paper). In: Proceedings of the 30th International Conference on Advances in Geographic Information Systems, pp. 1–4 (2022)
27. Manvi, R., Khanna, S., Burke, M., Lobell, D., Ermon, S.: Large language models are geographically biased. arXiv preprint arXiv:2402.02680 (2024)
28. Manvi, R., Khanna, S., Mai, G., Burke, M., Lobell, D., Ermon, S.: GeoLLM: extracting geospatial knowledge from large language models. arXiv preprint arXiv:2310.06213 (2023)
29. Mirza, S., Coelho, B., Cui, Y., Pöpper, C., McCoy, D.: Global-liar: factuality of LLMs over time and geographic regions. arXiv preprint arXiv:2401.17839 (2024)
30. Navigli, R., Conia, S., Ross, B.: Biases in large language models: origins, inventory, and discussion. ACM J. Data Inf. Qual. **15**(2), 1–21 (2023)
31. Omiye, J.A., Lester, J.C., Spichak, S., Rotemberg, V., Daneshjou, R.: Large language models propagate race-based medicine. NPJ Digit. Med. **6**(195) (2023). https://doi.org/10.1038/s41746-023-00939-z
32. Oschmann, D.: Der Osten: Eine westdeutsche Erfindung. Ullstein, Berlin (2023)
33. Plüster, B.: LEOLM: igniting German-language LLM research (2023). https://laion.ai/blog/leo-lm/
34. Roberts, J., Lüddecke, T., Das, S., Han, K., Albanie, S.: GPT4GEO: how a language model sees the world's geography. arXiv preprint arXiv:2306.00020 (2023)
35. Roth, M.: "Ossis" versus "Wessis", "wir" versus "die": Zugang zu Fragen der Identität am Beispiel von Ost- und West-Stereotypen der 1990er Jahre und aktuelle Anwendbarkeit der Ergebnisse. In: Dittmar, N., Paul, C. (eds.) Sprechen im Umbruch-Zeitzeugen erzählen und argumentieren rund um den Fall der Mauer im Wendekorpus, pp. 129–146. Leibniz-Institut für Deutsche Sprache, Mannheim (2019)
36. Salmas, K., Pantazi, D.A., Koubarakis, M.: Extracting geographic knowledge from large language models: an experiment. In: KBC-LM 2023: Knowledge Base Construction from Pre-trained Language Models workshop at ISWC 2023 (2023)
37. Schwöbel, P., Golebiowski, J., Donini, M., Archambeau, C., Pruthi, D.: Geographical erasure in language generation. arXiv preprint arXiv:2310.14777 (2023)
38. Statistisches Bundesamt (Destatis): Bevölkerungsdichte der Bundesländer in Deutschland (2022). https://www.destatis.de/DE/Themen/Laender-Regionen/Regionales/Gemeindeverzeichnis/Administrativ/02-bundeslaender.html
39. Statistisches Bundesamt (Destatis): Pressemitteilung Nr. 223 vom 31. Mai 2022 (2022). https://www.destatis.de/DE/Presse/Pressemitteilungen/2022/05/PD22_223_733.html
40. Statistisches Bundesamt (Destatis): Average gross annual earnings of full-time employees in 2022 (2023). https://www.destatis.de/EN/Themes/Labour/Earnings/Earnings-Earnings-Differences/Tables/yearly-gross-earnings.html

41. Statistisches Bundesamt (Destatis): Langzeitreihen zum Arbeitsmarkt (2023). https://www.destatis.de/EN/Themes/Economy/Short-Term-Indicators/Long-Term-Series/Labour-Market/lrarb003ga.html
42. Statistisches Bundesamt (Destatis): Sterbefälle und Lebenserwartung (2023). https://www.destatis.de/DE/Themen/Gesellschaft-Umwelt/Bevoelkerung/Sterbefaelle-Lebenserwartung/_inhalt.html#243320
43. Stuttgarter Nachrichten: Urteil von Arbeitsgericht in Berlin: Ostdeutsche Herkunft ist keine Diskriminierung. https://www.stuttgarter-nachrichten.de/inhalt.urteil-von-arbeitsgericht-in-berlin-ostdeutsche-herkunft-ist-keine-diskriminierung.165aae5f-ee39-419d-bdec-28a5dab588f6.html
44. Sun, H., Pei, J., Choi, M., Jurgens, D.: Aligning with whom? Large language models have gender and racial biases in subjective NLP tasks (2023)
45. Süddeutsche Zeitung: "Ossi"-Streit endet mit Vergleich (2010). https://www.sueddeutsche.de/karriere/diskriminierung-bei-bewerbung-ossi-streit-endet-mit-vergleich-1.1012988
46. Thebault-Spieker, J., Hecht, B., Terveen, L.: Geographic biases are 'born, not made' exploring contributors' spatiotemporal behavior in OpenStreetMap. In: Proceedings of the 2018 ACM International Conference on Supporting Group Work, pp. 71–82 (2018)
47. Verband der Ersatzkassen e.V. (vdek): Durchschnittsalter der Bevölkerung in Deutschland (2022). https://www.vdek.com/presse/daten/a_bevoelkerung.html
48. Warr, M., Oster, N.J., Isaac, R.: Implicit bias in large language models: experimental proof and implications for education. SSRN Electron. J. (2023)
49. Yang, J., et al.: Harnessing the power of LLMs in practice: a survey on ChatGPT and beyond. ACM Trans. Knowl. Discov. Data **18**(6), 1–32 (2024)
50. Zehring, M., Domahidi, E.: Thirty years after the German reunification-exploring stereotypes about east Germans on twitter. Int. J. Commun. **16**, 4029–4049 (2022)
51. Zhu, X.X., et al.: Geoinformation harvesting from social media data: a community remote sensing approach. IEEE Geosci. Remote Sens. Mag. **10**(4), 150–180 (2022)

# Towards Privacy-Preserving Relational Data Synthesis via Probabilistic Relational Models

Malte Luttermann[1(✉)], Ralf Möller[2], and Mattis Hartwig[1,3]

[1] German Research Center for Artificial Intelligence (DFKI), Lübeck, Germany
{malte.luttermann,mattis.hartwig}@dfki.de
[2] Institute for Humanities-Centered Artificial Intelligence, University of Hamburg, Hamburg, Germany
ralf.moeller@uni-hamburg.de
[3] singularIT GmbH, Leipzig, Germany

**Abstract.** Probabilistic relational models provide a well-established formalism to combine first-order logic and probabilistic models, thereby allowing to represent relationships between objects in a relational domain. At the same time, the field of artificial intelligence requires increasingly large amounts of relational training data for various machine learning tasks. Collecting real-world data, however, is often challenging due to privacy concerns, data protection regulations, high costs, and so on. To mitigate these challenges, the generation of synthetic data is a promising approach. In this paper, we solve the problem of generating synthetic relational data via probabilistic relational models. In particular, we propose a fully-fledged pipeline to go from relational database to probabilistic relational model, which can then be used to sample new synthetic relational data points from its underlying probability distribution. As part of our proposed pipeline, we introduce a learning algorithm to construct a probabilistic relational model from a given relational database.

**Keywords:** probabilistic graphical models · relational models · synthetic data

## 1 Introduction

Probabilistic relational models such as parametric factor graphs (PFGs) combine first-order logic and probabilistic models and thereby provide an adequate formalism to represent relationships between objects in a relational domain. PFGs compactly encode a full joint probability distribution over a set of random variables (randvars) and hence allow to sample new relational data points from the encoded underlying full joint probability distribution. The generated synthetic relational data samples then follow the underlying full joint probability distribution and might be used for various purposes. Common applications of


A. Hotho and S. Rudolph (Eds.): KI 2024, LNAI 14992, pp. 175–189, 2024.
https://doi.org/10.1007/978-3-031-70893-0_13

synthetic data include for example training machine learning models or sharing data without violating the privacy of individuals [8,26,34]. Another application could be to bootstrap PFG learning (using the given database to learn a PFG, which is then applied to generate additional synthetic relational data points to learn another PFG on the basis of a larger data set and then possibly repeating the procedure). In this paper, we solve the problem of applying probabilistic relational models (more specifically, PFGs) to generate synthetic relational (i.e., multi-table) data from a conceptual point of view.

*Previous Work.* While there is, to the best of our knowledge, no previous work on the generation of synthetic relational data via probabilistic relational models, there exists previous work on learning and sampling models that combine probabilistic models and first-order logic. In particular, so-called Markov logic networks (MLNs) [29] have extensively been studied. An MLN is another formalism combining probabilistic models and first-order logic, allowing for probabilistic reasoning on a first-order level. Even though a vast amount of algorithms to learn an MLN from data has emerged (see, e.g., [4,5,16–19,22,25,30]), there is a lack of prior work on generating synthetic relational data via MLNs or other probabilistic relational models. Moreover, while sampling algorithms for MLNs have been developed [32], these sampling algorithms are used for (approximate) query answering and are not applied to synthetic data generation, where the requirements of the sampling algorithm slightly differ—that is, drawing a new data sample should yield a value for every column in the given database.

Most of the existing work on synthetic data generation focuses on the generation of single-table synthetic data [10]. While many popular approaches are based on generative adversarial networks [33], there are also approaches building on probabilistic graphical models such as Bayesian networks [13]. Both for approaches based on generative adversarial networks and probabilistic graphical models, differential privacy guarantees have been investigated [3,6,9,34]. More recently, there has also been work on the generation of synthetic relational data under differential privacy [7]. However, none of these approaches makes use of the advantages of first-order probabilistic models.

*Our Contributions.* In this paper, we introduce the first architecture that combines PFGs and the generation of synthetic relational data. Our main contribution is a fully-fledged pipeline from relational database to PFG, from which we can then sample new realistic synthetic data points. We further present a learning algorithm to obtain a PFG (that is, both the graph structure and the parameters of the PFG) from a relational database. Our proposed architecture exploits the advantages of PFGs, including that (i) PFGs are able to effectively encode the relationships between objects in a relational domain, (ii) PFGs provide an explainable model that can also be used for, e.g., probabilistic inference [31] and causal inference [24] on a first-order level, and (iii) PFGs naturally abstract from individuals by grouping indistinguishable objects, which yields a promising foundation for differential privacy guarantees.

*Structure of this Paper.* The remainder of this paper is structured as follows. We begin by introducing necessary background information and notations. Afterwards, we propose an architecture to first learn a PFG from a relational database and then employ the PFG to generate new synthetic relational data points, which might be used for arbitrary applications. We then go through the individual steps of the proposed architecture in detail, using a small comprehensive example for guidance, before we conclude.

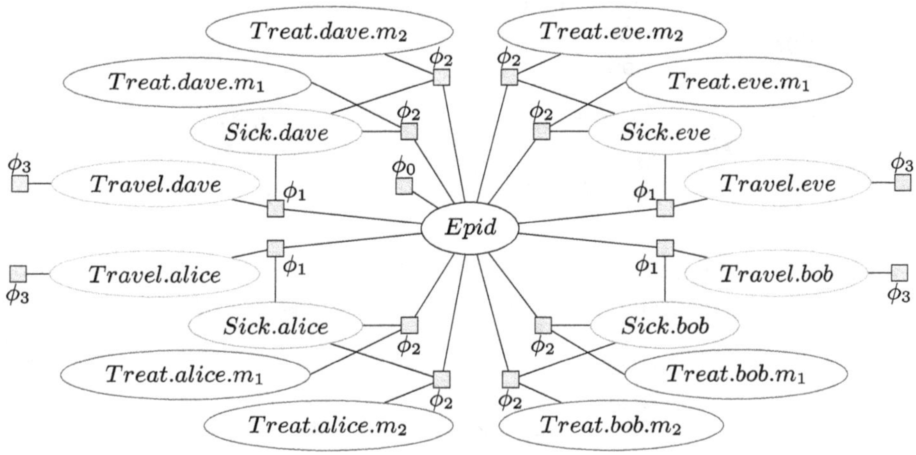


**Fig. 1.** An FG encoding a full joint probability distribution for an epidemic example [14]. We omit the input-output pairs of the factors for brevity.

## 2 Preliminaries

We start with the definition of factor graphs (FGs) as propositional probabilistic models and then continue to introduce PFGs, which combine first-order logic with probabilistic models. An FG is an undirected propositional probabilistic model to compactly encode a full joint probability distribution over a set of randvars [11,21]. Similar to a Bayesian network [27], an FG factorises a full joint probability distribution into a product of factors.

**Definition 1 (Factor Graph).** *An* FG $G = (\boldsymbol{V}, \boldsymbol{E})$ *is an undirected bipartite graph with node set* $\boldsymbol{V} = \boldsymbol{R} \cup \boldsymbol{\Phi}$ *where* $\boldsymbol{R} = \{R_1, \dots, R_n\}$ *is a set of variable nodes (randvars) and* $\boldsymbol{\Phi} = \{\phi_1, \dots, \phi_m\}$ *is a set of factor nodes (functions). The term range*$(R_i)$ *denotes the possible values of a randvar* $R_i$*. There is an edge between a variable node* $R_i$ *and a factor node* $\phi_j$ *in* $\boldsymbol{E} \subseteq \boldsymbol{R} \times \boldsymbol{\Phi}$ *if* $R_i$ *appears in the argument list of* $\phi_j$*. A factor is a function that maps its arguments to a positive real number, called potential. The semantics of* $G$ *is given by*

$$P_G = \frac{1}{Z} \prod_{j=1}^{m} \phi_j(\mathcal{A}_j)$$

*with Z being the normalisation constant and* $\mathcal{A}_j$ *denoting the randvars connected to* $\phi_j$ *(that is, the arguments of* $\phi_j$*).*

*Example 1.* Take a look at the FG presented in Fig. 1, which represents an epidemic example with four persons *alice*, *bob*, *dave*, and *eve* as well as two possible medications $m_1$ and $m_2$ for treatment. For each person, there are two Boolean randvars (that is, randvars having a Boolean range) *Sick* and *Travel*, indicating whether the person is sick and travels, respectively. Moreover, there is another Boolean randvar *Treat* for each combination of person and medication, specifying whether the person is treated with the medication. The Boolean randvar *Epid* states whether an epidemic is present.

Next, we define PFGs, first introduced by Poole [28], which combine probabilistic models and first-order logic. PFGs use logical variables (logvars) as parameters to represent sets of indistinguishable randvars. Each set of indistinguishable randvars is represented by a parameterised randvar (PRV).

**Definition 2 (Parameterised Random Variable).** *Let* $\boldsymbol{R}$ *be a set of randvar names,* $\boldsymbol{L}$ *a set of logvar names, and* $\boldsymbol{D}$ *a set of constants. All sets are finite. Each logvar L has a domain* $dom(L) \subseteq \boldsymbol{D}$*. A* constraint *is a tuple* $(\mathcal{X}, C_{\mathcal{X}})$ *of a sequence of logvars* $\mathcal{X} = (X_1, \ldots, X_n)$ *and a set* $C_{\mathcal{X}} \subseteq \times_{i=1}^{n} dom(X_i)$*. The symbol* $\top$ *for C marks that no restrictions apply, i.e.,* $C_{\mathcal{X}} = \times_{i=1}^{n} dom(X_i)$*. A* PRV $R(L_1, \ldots, L_n)$, $n \geq 0$, *is a syntactical construct of a randvar* $R \in \boldsymbol{R}$ *possibly combined with logvars* $L_1, \ldots, L_n \in \boldsymbol{L}$ *to represent a set of randvars. If* $n = 0$*, the PRV is parameterless and forms a propositional randvar. A PRV A (or logvar L) under constraint C is given by* $A_{|C}$ $(L_{|C})$*, respectively. We may omit* $|\top$ *in* $A_{|\top}$ *or* $L_{|\top}$*. The term range(A) denotes the possible values of a PRV A. An* event $A = a$ *denotes the occurrence of PRV A with range value* $a \in range(A)$*.*

*Example 2.* Consider $\boldsymbol{R} = \{Epid, Travel, Sick, Treat\}$ and $\boldsymbol{L} = \{P, M\}$ with $\mathrm{dom}(P) = \{alice, bob, dave, eve\}$ (patients), $\mathrm{dom}(M) = \{m_1, m_2\}$ (medications), combined into Boolean PRVs *Epid*, $Travel(P)$, $Sick(P)$, and $Treat(P, M)$.

A parametric factor (parfactor) describes a function, mapping argument values to positive real numbers, of which at least one is non-zero.

**Definition 3 (Parfactor).** *Let* $\Phi$ *denote a set of factor names. We denote a* parfactor *g by* $\phi(\mathcal{A})_{|C}$ *with* $\mathcal{A} = (A_1, \ldots, A_n)$ *being a sequence of PRVs,* $\phi$: $\times_{i=1}^{n} range(A_i) \mapsto \mathbb{R}^+$ *being a function with name* $\phi \in \Phi$ *mapping argument values to a positive real number called* potential, *and C being a constraint on the logvars of* $\mathcal{A}$*. We may omit* $|\top$ *in* $\phi(\mathcal{A})_{|\top}$*. The term lv(Y) refers to the logvars in some element Y, a PRV, a parfactor, or sets thereof. The term* $gr(Y_{|C})$ *denotes the set of all instances (groundings) of Y with respect to constraint C.*

A PFG is then built from a set of parfactors $\{g_1, \ldots, g_m\}$.

**Definition 4 (Parametric Factor Graph).** *A* PFG $G = (\boldsymbol{V}, \boldsymbol{E})$ *is a bipartite graph with node set* $\boldsymbol{V} = \boldsymbol{A} \cup \boldsymbol{G}$ *where* $\boldsymbol{A} = \{A_1, \ldots, A_n\}$ *is a set of PRVs and*

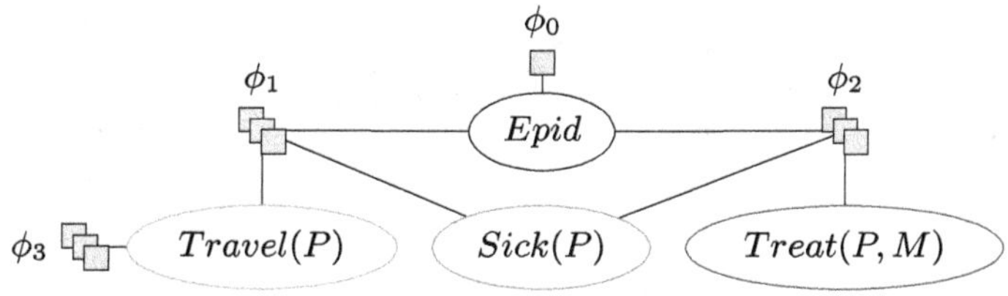


**Fig. 2.** A PFG encoding a full joint probability distribution for the epidemic example from Fig. 1. We omit the input-output pairs of the parfactors for brevity.

*$\boldsymbol{G} = \{g_1, \ldots, g_m\}$ is a set of parfactors as well as edge set $\boldsymbol{E} \subseteq \boldsymbol{A} \times \boldsymbol{G}$. A PRV $A_i$ and a parfactor $g_j$ are connected via an edge in $G$ (i.e., $\{A_i, g_j\} \in \boldsymbol{E}$) if $A_i$ appears in the argument list of $g_j$. The semantics of $G$ is given by grounding and building a full joint distribution. With $Z$ as the normalisation constant and $\mathcal{A}_k$ denoting the randvars connected to $\phi_k$, $G$ represents the full joint distribution*

$$P_G = \frac{1}{Z} \prod_{g_j \in \boldsymbol{G}} \prod_{\phi_k \in gr(g_j)} \phi_k(\mathcal{A}_k).$$

*Example 3.* Figure 2 shows a PFG $G = \{g_i\}_{i=0}^{3}$ with $g_0 = \phi_0(Epid)_{|\top}$, $g_1 = \phi_1(Travel(P), Sick(P), Epid)_{|\top}$, $g_2 = \phi_2(Treat(P, M), Sick(P), Epid)_{|\top}$, and $g_3 = \phi_2(Travel(P))_{|\top}$. Grounding $G$ yields again the FG shown in Fig. 1 (assuming that the domains of the logvars are defined as in Example 2).

Note that the definition of PFGs also includes FGs, as every FG is a PFG containing only parameterless randvars. Compared to an FG, a PFG abstracts from individuals by grouping identically behaving objects using logvars. While the introduction of logvars increases the expressiveness of the model (e.g., to encode relationships between groups of objects), grouping identically behaving individuals can also yield significant speed-ups during probabilistic inference.

In the upcoming section, we introduce our proposed architecture to solve the problem of learning a PFG from a given relational database to then generate synthetic relational data according to its underlying probability distribution.

# 3 Proposed Architecture

In this section, we provide an overview of the general architecture to generate synthetic relational data from a relational database using a PFG. We first take a look at the steps involved in the synthetic data generation approach and afterwards continue to investigate each of the steps in more detail.

An overview of the architecture of our proposed architecture is depicted in Fig. 3. The whole process consists of three primary steps, which can again be decomposed into various subroutines. The three primary steps consist of (i) constructing a propositional FG, (ii) transforming the propositional FG into a PFG, and (iii) sampling from the PFG to generate new synthetic relational data.

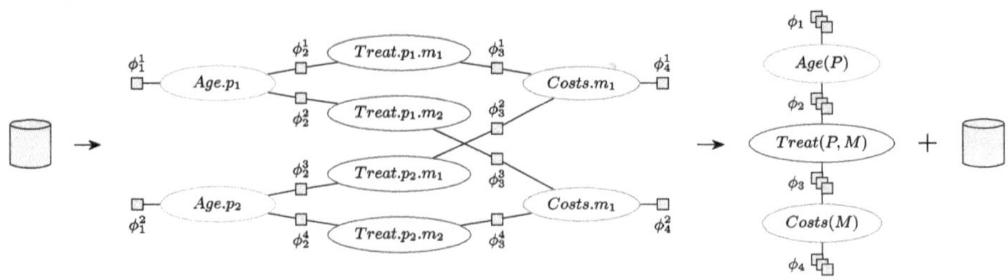

Fig. 3. Architecture overview of the general architecture to generate synthetic relational data from a given relational database using a PFG.

Besides the generated synthetic data, the PFG is also a valuable output of the architecture (indicated by the + sign), as it can be used for various tasks such as probabilistic inference, causal inference, or bootstrap PFG learning, for example.

In the following, we illustrate each of the three steps at a small comprehensive example for guidance. Before we take a look at the steps involved in the proposed architecture, we briefly introduce the notion of an entity-relationship (ER) model, which allows us to describe a relational database $\mathcal{D}$.

**Definition 5 (Entity-Relationship Model).** *We define an* ER model *as a tuple* $(\mathcal{E}, \mathcal{R})$ *where* $\mathcal{E} = \{E_1, \ldots, E_\ell\}$ *denotes a set of entity classes and* $\mathcal{R} = \{R_1, \ldots, R_k\}$ *is a set of relationship classes. Each entity or relationship class* $B \in \mathcal{E} \cup \mathcal{R}$ *can have a set of attributes attached to it, which is denoted by* $\mathcal{A}(B)$.

*Example 4.* Consider the ER model depicted in Fig. 4a, which consists of the entity classes $\mathcal{C} = \{Patient, Medication\}$ and the relationship classes $\mathcal{R} = \{Treat\}$. The ER model further contains the attributes $\mathcal{A}(Patient) = \{Age\}$ and $\mathcal{A}(Medication) = \{Costs\}$.

It is generally possible to have cyclic relationships and to impose cardinality constraints on the relationships, which we omit for brevity in this paper. We next take a closer look at the individual steps involved in our proposed architecture.

### 3.1 Construction of a Propositional Factor Graph

While there are learning algorithms for first-order probabilistic models such as MLNs, to the best of our knowledge, there is currently no approach to directly learn a PFG from a given relational database. However, there are well-known approaches to learn an FG from the given data [1] and an FG can be transformed into a PFG by running the so-called advanced colour passing (ACP) algorithm [23]. We therefore propose to first learn a propositional model, that is, an FG $G$, from the given relational database and afterwards run the ACP algorithm on $G$ to transform $G$ into a PFG entailing equivalent semantics as $G$.

While such an approach seems straightforward at first glance, there are a few challenges to overcome. A major challenge is that applying a standard learning algorithm to obtain an FG from data does not fit our setting because standard learning algorithms do not include randvars and factors for individual objects into the FG. In other words, the relational structure of the data is neglected, which we illustrate in the upcoming example.

(a)

| PatientId | Age |
|---|---|
| *alice* | $\geq 18$ |
| *bob* | $\geq 18$ |
| *charlie* | $\geq 18$ |
| *dave* | $< 18$ |
| *eve* | $< 18$ |

| PatientId | MedicationId |
|---|---|
| *alice* | *myalept* |
| *alice* | *danyelza* |
| *bob* | *paracetamol* |
| *charlie* | *ibuprofen* |
| *eve* | *eliquis* |

| MedicationId | Costs |
|---|---|
| *myalept* | *high* |
| *danyelza* | *high* |
| *paracetamol* | *low* |
| *ibuprofen* | *low* |
| *eliquis* | *high* |

(b)

**Fig. 4.** A small toy example for a relational database of patients and the medications they take. Figure 4a shows an ER model consisting of entities *Patient* and *Medication* with attributes *Age* and *Costs*, respectively. The entities *Patient* and *Medication* are connected by a relation *Treat*. We omit cardinalities for brevity. Figure 4b displays exemplary data for a relational database following the structure specified by the ER model from Fig. 4a.

*Example 5.* Take a look at a simple toy example database containing information on patients and the medications they take, as depicted in Fig. 4. For simplicity, each patient only has a single attribute *Age* which can either be $<18$ or $\geq 18$ and each medication has a single attribute *Costs* which can either be *low* or *high*. Further, there is a relation *Treat* connecting patients with the medications they take. The specific entries of the database are depicted in Fig. 4b, where we again keep the tables simple for illustrative purposes.

A standard learning algorithm to obtain an FG from the given database would include a single randvar for each of the attributes, e.g., there would be a single randvar for the attribute *Age* instead of a separate node for the age of each patient. Analogously, there would also be a single randvar for the attribute *Costs*. When using a single randvar to model an attribute over all objects, however, we lose information about individual objects (here patients and medications) and their relationships between each other.

Therefore, we slightly adjust the learning procedure to include multiple randvars for the same attribute of different individual objects into the learned FG. However, we cannot simply include a randvar for each object per attribute because there would be no uncertainty in the resulting model. To continue Example 5, assume we add a randvar *Age* for each patient to the learned FG. Then, the prior probability distribution for the randvar *Age* of a specific patient would map the value found in the database for *Age* to probability one and all other values to probability zero, e.g., the probability that $Age.alice < 18$ would be set to zero and the probability that $Age.alice \geq 18$ would be set to one. Consequently, the FG would not model any uncertainty at all. To mitigate this issue, we propose to perform an initial clustering of entities to find clusters of indistinguishable objects (here patients and medications), which allows us to naturally define the domains of the logvars when transforming the learned FG into an PFG. After the initial clustering, we then insert a randvar for each cluster per attribute into the FG. Performing an initial clustering of entities allows us to model uncertainty while keeping objects and relationships in the model.

*Example 6.* Consider again the example database depicted in Fig. 4. For the sake of the example, assume that the initial clustering returns two patient clusters $C_{p_1} = \{alice, eve\}$ as well as $C_{p_2} = \{bob, charlie, dave\}$ and two medication clusters $C_{m_1} = \{myalept, danyelza, eliquis\}$ and $C_{m_2} = \{paracetamol, ibuprofen\}$. Then, the resulting FG contains the randvars $Age.p_1$, $Age.p_2$, $Costs.m_1$, and $Costs.m_2$, that is, there exists one randvar for each cluster per attribute. The probability that, e.g., $Age.p_1 < 18$ is then set to 0.5 (as one out of two entries belonging to the cluster $C_{p_1}$ has $Age < 18$) and the probability that $Age.p_2 < 18$ is set to 0.33 (as one out of three entries belonging to the cluster $C_{p_2}$ has $Age < 18$).

To perform the clustering of the entities, an arbitrary clustering algorithm can be applied. In particular, it is also possible to apply privacy-preserving clustering algorithms, e.g., to ensure differential privacy guarantees. After clustering the entities, the variable nodes of the FG can be inserted for each clustered entity.

More specifically, the resulting FG contains a randvar for every attribute $A$ in the database per cluster of entities occurring in $A$ (i.e., if $A$ is an attribute of an entity $E$, there is a randvar for $A$ for each cluster of $E$ and if $A$ is an attribute of a relationship $R$, there is a randvar for $R$ for each combination of clusters of the entities occurring in $R$). To account not only for the attributes but also for the relationships, there is another randvar in the FG for every relationship $R$ in the database per combination of clusters of entities occurring in $R$.

In general, the randvars specifying the attributes do not have to be Boolean. The randvars for the relationships, however, are always Boolean as they indicate whether a relationship exists.

*Example 7.* Given the database from Fig. 4 and the clusters $C_{p_1}$, $C_{p_2}$, $C_{m_1}$, and $C_{m_2}$ from Example 6, the resulting FG contains the randvars $Age.p_1$, $Age.p_2$, $Costs.m_1$, $Costs.m_2$, $Treat.p_1.m_1$, $Treat.p_1.m_2$, $Treat.p_2.m_1$, and $Treat.p_2.m_2$.

Note that in a standard learning algorithm for an FG, there are typically no randvars for the relationships and, in addition to that, there are also not multiple randvars for the same attribute for different entities (or clusters of entities, respectively). Therefore, an FG learned by a standard learning algorithm loses information about the relationships of individual objects (or clusters thereof).

After having identified the variable nodes in the FG, the next step is to learn the remaining graph structure, i.e., the factor nodes and the edges of the FG. The graph structure of the FG can be identified using conditional independence tests on the given data [20, Chapter 20.7]. In particular, two randvars $X$ and $Y$ are conditionally independent given a set of randvars $Z$ if $P(X, Y \mid Z) = P(X \mid Z) \cdot P(Y \mid Z)$. Therefore, testing whether pairs of randvars are conditionally independent given a set of other randvars can be done by estimating the corresponding probabilities from the given data using statistical hypothesis tests. The results of the conditional independence tests then yield the graph structure of the FG as in an FG, two randvars are conditionally independent if all paths between them are blocked by the conditioning set.

We remark that performing conditional independence tests on the individual tables does not work because the individual tables do not necessarily contain all randvars involved in the conditional independence test. The conditional independence tests are hence carried out on a join of the individual tables in the given relational database to enable the estimation of the necessary probabilities.

*Example 8.* Consider again the database given in Fig. 4. When checking whether the randvars $Patient$ and $Medication$ are independent, we need a joined table which contains both $Patient$ and $Medication$.

As we add randvars for the relationships as well, we augment the full join with an additional column for each relationship, indicating which relationships are present in the database and which are not. We also augment the full join with combinations of entities that do not occur in the relationships from the given database (similar to a cross join) such as the combination of patient *bob* and medication *myalept* in Fig. 4, for which we add a row with value $false$ in the column $Treat$. More details about the augmented full join are given in the appendix.

After the entire graph structure of the FG has been learned, the next step is to learn the potentials of the factors. More specifically, at this point we know which factors (i.e., functions) are part of the model, but we do not know their input-output mappings yet. To obtain the potentials of the factors, we count the occurrences for each specific assignment of the arguments of the factors in the augmented full join of the tables. To count the occurrences for all possible assignments of factors' arguments, we also need information on combinations of entities that do not occur in the relationships from the given database, which is the reason we add this information in the augmented full join as described above. The occurrences for each specific assignment of factors arguments are then counted as illustrated in the following example.

*Example 9.* Take a look at the FG shown in Fig. 5. The FG results from the learning procedure described above applied to the database illustrated in Fig. 4. The potentials for, e.g., $\phi_1^2(Age.p_2)$ are obtained by counting the occurrences of $Age.p_2 < 18$ and $Age.p_2 \geq 18$ in the augmented full join (shown in the appendix).

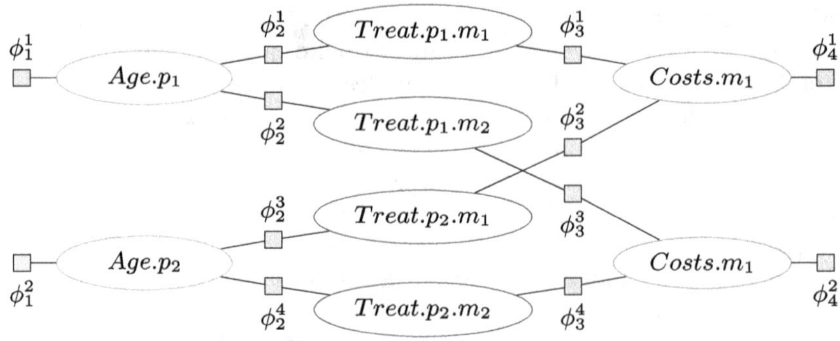


**Fig. 5.** The learned FG from the given database depicted in Fig. 4.

Recall that $C_{p_2} = \{bob, charlie, dave\}$. Thus, in this particular example, it holds that $\phi_1^2(Age.p_2 < 18) = 5$ and $\phi_1^2(Age.p_2 \geq 18) = 10$.

Note that the absolute number of a potential value is not pivotal as the semantics of the FG includes a normalisation constant. Hence, the semantics would remain unchanged if we had counted the occurrences of $Age.p_2 < 18$ and $Age.p_2 \geq 18$ in the original table. In the original table, we have one occurrence of $Age.p_2 < 18$ and two occurrences of $Age.p_2 \geq 18$ for the cluster $C_{p_2} = \{bob, charlie, dave\}$, i.e., the ratio is exactly the same as in the augmented full join (five occurrences and ten occurrences). However, we use the augmented full join for counting as we cannot count occurrences of randvars appearing in different tables separately in factors with multiple arguments such as in $\phi_2^3(Age.p_2, Treat.p_2.m_1)$ because we have to incorporate the relationship between those tables. For example, in $\phi_2^3(Age.p_2, Treat.p_2.m_1)$, we need to count the occurrences of $Age.p_2 < 18$ *and* $Treat.p_2.m_1 = true$ and such a combination is not present in the original tables.

A summary of our proposed learning algorithm is provided in Algorithm 1.

So far, we discussed the steps to learn a propositional FG from a given relational database (Line 1 to Line 14 in Algorithm 1). We next describe the procedure to transform the learned FG into a PFG (Line 15 in Algorithm 1).

### 3.2 Transforming the Factor Graph Into a Parametric Factor Graph

To obtain a PFG from a given FG, we need to find groups of identically behaving randvars and factors in the FG. Then, PRVs with logvars represent such groups of indistinguishable randvars and parfactors represent groups of identical factors. Replacing indistinguishable randvars by PRVs with logvars further abstracts from individuals and thus yields a promising foundation for privacy guarantees [12]. The ACP algorithm (which is a generalisation of the colour passing algorithm [2,15]) is able to construct a PFG from a given propositional FG [23]. The idea behind ACP is to exploit symmetries in a propositional FG and then group together symmetric subgraphs. ACP looks for symmetries based on potentials of factors, on ranges and evidence of randvars, as well as on the graph

**Algorithm 1:** LEARNPFG

**Input** : A relational database $\mathcal{D} = (\mathcal{E}, \mathcal{R})$ with corresponding data samples.
**Output**: A PFG $G' = (\boldsymbol{V}, \boldsymbol{E})$ representing the full joint probability distribution of the given database.

$G \leftarrow$ Empty FG;
$F \leftarrow$ Augmented full join over $\mathcal{D}$;
$C_{E_1}, \ldots, C_{E_\ell} \leftarrow$ CLUSTERENTITIES($\mathcal{D}$);
**foreach** *entity or relationship* $B \in \mathcal{E} \cup \mathcal{R}$ **do**
    `// Let` $E_1, \ldots, E_j$ `denote all entities occurring in` $B$
    **foreach** *combination of clusters* $(c_1, \ldots, c_j) \in \times_{i=1}^{j} C_{E_i}$ **do**
        **foreach** *attribute* $A \in \mathcal{A}(B)$ **do**
            Add a randvar $A.c_1. \cdots .c_j$ to $G$;
**foreach** *relationship* $R$ **do**
    `// Let` $E_1, \ldots, E_j$ `denote all entities occurring in` $R$
    **foreach** *combination of clusters* $(c_1, \ldots, c_j) \in \times_{i=1}^{j} C_{E_i}$ **do**
        Add a randvar $R.c_1. \cdots .c_j$ to $G$;
Add edges and factors to $G$ by running conditional independence tests on $F$;
**foreach** *factor* $\phi(R_1, \ldots, R_k)$ *in* $G$ **do**
    `// Let` $c_1, \ldots, c_j$ `denote all clusters occurring in` $R_1, \ldots, R_k$
    **foreach** *assignment* $(r_1, \ldots, r_k) \in \times_{i=1}^{k} \text{range}(R_i)$ **do**
        Set $\phi(R_1, \ldots, R_k)$ to the number of rows in $F$ belonging to clusters $c_1$ to $c_j$ that contain the values $r_1$ to $r_k$ in the columns of $R_1$ to $R_k$;
$G' \leftarrow$ Call ACP on $G$ with evidence $E = \emptyset$;
**return** $G'$

structure by passing around colours. A formal description as well as an example run of the ACP algorithm can be found in the appendix. Figure 6 shows the PFG resulting from calling ACP on the FG depicted in Fig. 5 under the assumption that all potentials of the factors $\phi_i$, $i \in \{1, \ldots, 4\}$, are considered identical. Note that the assumption of identical factors is just for the sake of the example as in general, not all potentials are identical (and hence, not all of the factors $\phi_i$ are grouped into a single group).

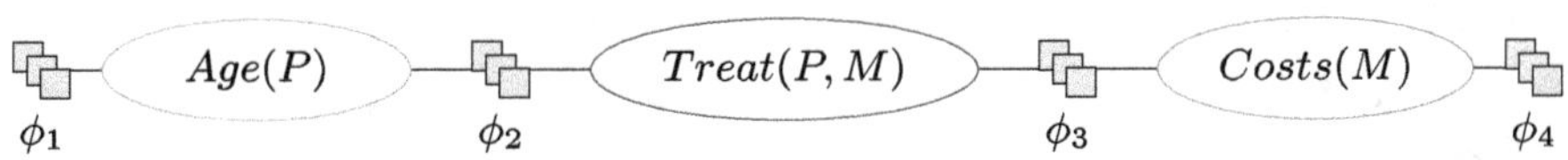


**Fig. 6.** The PFG resulting from calling ACP on the FG depicted in Fig. 5 assuming for the sake of the example that all clusters behave identically. Note that in our specific example from Example 5, not all clusters behave identically and thus, not all randvars representing the same attribute or relationship are grouped.

We remark that in its original form, ACP requires potentials of factors to identically match in order to group factors together. When learning an FG from data, however, there might be small deviations in the potentials, that is, counting the occurrences of a specific combination of values might differ for two similarly behaving clusters, e.g., by one. Therefore, we adapt the condition for two factors to be considered identical by allowing for a difference in their potentials depending on a user-defined parameter $\varepsilon$. Two factors $\phi$ and $\phi'$ are now considered identical (and thus are assigned the same colour in ACP) if for every assignment of their arguments the corresponding potentials, say $\varphi$ and $\varphi'$, differ by at most $\varphi \cdot \epsilon$ (or $\varphi' \cdot \epsilon$, respectively). Factors with slightly deviating potentials that are grouped are then all assigned the mean of the potentials in the group.

### 3.3 Sampling from the Parametric Factor Graph

Every PFG compactly encodes a full joint probability distribution from which we can draw new samples. Note that, in contrast to previous sampling approaches (such as, e.g., [32]) that apply sampling for (approximate) query answering, we aim to generate new data samples that comply with the given ER model (that is, we wish to draw new data samples that contain a value for every attribute and every relationship in the ER model). As the PFG is inherently encoding a relational structure, we are able to synthesise relational data. More specifically, the semantics of a PFG is defined on a ground level, that is, sampling from the underlying probability distribution yields new data samples for multiple clustered entities. As the ground FG consists of randvars for various clustered entities, we draw samples for multiple clusters at the same time.

*Example 10.* Assume we want to sample the PFG which yields the FG shown in Fig. 5 when grounding the model. Sampling thus yields a value for every randvar in the FG, thereby allowing to synthesise new objects and relationships between them following the full joint probability distribution encoded by the model. An exemplary data sample could look like this: $Age.p_1 < 18$, $Age.p_2 \geq 18$, $Treat.p_1.m_1 = false$, $Treat.p_1.m_2 = false$, $Treat.p_2.m_1 = true$, $Treat.p_2.m_2 = false$, $Costs.m_1 = low$, $Costs.m_2 = high$ (values are chosen arbitrarily for the sake of the example).

The generated synthetic data samples might be used for arbitrary applications and thus, it is necessary to define application-specific quality criteria to assess the quality of the generated data.

## 4 Conclusion

We introduce a fully-fledged pipeline to deploy probabilistic relational models, in particular PFGs, for the generation of synthetic relational (i.e., multi-table) data. To construct a PFG from a given relational database, we propose a learning algorithm that learns both the graph structure as well as the parameters of a PFG from the relational database. We further elaborate on how the learned PFG can be applied to generate new samples of synthetic relational data. By ensuring certain privacy guarantees (e.g., differential privacy) during the construction process of the PFG, PFG provide a promising model to generate synthetic relational data in a privacy-preserving manner such that generated synthetic data can be publicly shared without leaking sensitive data of individuals.

There are three primary directions for future work. First, privacy guarantees for PFG learning need to be theoretically investigated. Second, the scalability of our proposed architecture should be assessed and improved to allow an efficient handling of large-scale relational databases, and finally, the practicality of the entire framework has to be tested empirically on real-world data sets.

**Acknowledgements.** This work is funded by the BMBF project AnoMed 16KISA057.

## References

1. Abbeel, P., Koller, D., Ng, A.Y.: Learning factor graphs in polynomial time and sample complexity. J. Mach. Learn. Res. **7**, 1743–1788 (2006)
2. Ahmadi, B., Kersting, K., Mladenov, M., Natarajan, S.: Exploiting symmetries for scaling loopy belief propagation and relational training. Mach. Learn. **92**, 91–132 (2013)
3. Bao, E., Xiao, X., Zhao, J., Zhang, D., Ding, B.: Synthetic data generation with differential privacy via Bayesian networks. J. Priv. Confidentiality **11** (2021)
4. Biba, M., Ferilli, M., Esposito, F.: Structure learning of Markov logic networks through iterated local search. In: Proceedings of the Eighteenth European Conference on Artificial Intelligence (ECAI 2008), pp. 361–365. IOS Press (2008)
5. Biba, M., Ferilli, S., Esposito, F.: Discriminative structure learning of Markov logic networks. In: Železný, F., Lavrač, N. (eds.) ILP 2008. LNCS (LNAI), vol. 5194, pp. 59–76. Springer, Heidelberg (2008). https://doi.org/10.1007/978-3-540-85928-4_9
6. Cai, K., Lei, X., Wei, J., Xiao, X.: Data synthesis via differentially private Markov random fields. Proc. VLDB Endow. **14**, 2190–2202 (2021)
7. Cai, K., Xiao, X., Cormode, G.: PrivLava: synthesizing relational data with foreign keys under differential privacy. Proc. ACM Manag. Data **1**, 1–25 (2023)
8. Chen, R.J., Lu, M.Y., Chen, T.Y., Williamson, D.F., Mahmood, F.: Synthetic data in machine learning for medicine and healthcare. Nat. Biomed. Eng. **5**, 493–497 (2021)
9. Fang, M.L., Dhami, D.S., Kersting, K.: DP-CTGAN: differentially private medical data generation using CTGANs. In: Michalowski, M., Abidi, S.S.R., Abidi, S. (eds.) AIME 2022. LNCS, vol. 13263, pp. 178–188. Springer, Cham (2022). https://doi.org/10.1007/978-3-031-09342-5_17
10. Figueira, A., Vaz, B.: Survey on synthetic data generation, evaluation methods and GANs. Mathematics **10**, 2733–2773 (2022)
11. Frey, B.J., Kschischang, F.R., Loeliger, H.A., Wiberg, N.: Factor graphs and algorithms. In: Proceedings of the Thirty-Fifth Annual Allerton Conference on Communication, Control, and Computing, pp. 666–680. Allerton House (1997)
12. Gehrke, M., Liebenow, J., Mohammadi, E., Braun, T.: Lifting in support of privacy-preserving probabilistic inference. German J. Artif. Intell. (2024)
13. Gogoshin, G., Branciamore, S., Rodin, A.S.: Synthetic data generation with probabilistic Bayesian networks. Math. Biosci. Eng. **18**, 8603–8621 (2021)

14. Hoffmann, M., Braun, T., Möller: Lifted division for lifted hugin belief propagation. In: Proceedings of the Twenty-Fifth International Conference on Artificial Intelligence and Statistics (AISTATS 2022), pp. 6501–6510. PMLR (2022)
15. Kersting, K., Ahmadi, B., Natarajan, S.: Counting belief propagation. In: Proceedings of the Twenty-Fifth Conference on Uncertainty in Artificial Intelligence (UAI 2009), pp. 277–284. AUAI Press (2009)
16. Khot, T., Natarajan, S., Kersting, K., Shavlik, J.: Learning Markov logic networks via functional gradient boosting. In: Proceedings of the Eleventh IEEE International Conference on Data Mining (ICDM 2011), pp. 320–329. IEEE (2011)
17. Kok, S., Domingos, P.: Learning the structure of Markov logic networks. In: Proceedings of the Twenty-Second International Conference on Machine Learning (ICML 2005), pp. 441–448. ACM Press (2005)
18. Kok, S., Domingos, P.: Learning Markov logic network structure via hypergraph lifting. In: Proceedings of the Twenty-Six International Conference on Machine Learning (ICML 2009), pp. 505–512. ACM Press (2009)
19. Kok, S., Domingos, P.: Learning Markov logic networks using structural motifs. In: Proceedings of the Twenty-Seventh International Conference on Machine Learning (ICML 2010), pp. 551–558. Omnipress (2010)
20. Koller, D., Friedman, N.: Probabilistic Graphical Models: Principles and Techniques. MIT Press (2009)
21. Kschischang, F.R., Frey, B.J., Loeliger, H.A.: Factor graphs and the sum-product algorithm. IEEE Trans. Inf. Theory **47**, 498–519 (2001)
22. Lowd, D., Domingos, P.: Efficient weight learning for Markov logic networks. In: Kok, J.N., Koronacki, J., Lopez de Mantaras, R., Matwin, S., Mladenič, D., Skowron, A. (eds.) PKDD 2007. LNCS (LNAI), vol. 4702, pp. 200–211. Springer, Heidelberg (2007). https://doi.org/10.1007/978-3-540-74976-9_21
23. Luttermann, M., Braun, T., Möller, R., Gehrke, M.: Colour passing revisited: lifted model construction with commutative factors. In: Proceedings of the Thirty-Eighth AAAI Conference on Artificial Intelligence (AAAI 2024), pp. 20500–20507. AAAI Press (2024)
24. Luttermann, M., Hartwig, M., Braun, T., Möller, R., Gehrke, M.: Lifted causal inference in relational domains. In: Proceedings of the Third Conference on Causal Learning and Reasoning (CLeaR 2024), pp. 827–842. PMLR (2024)
25. Mihalkova, L., Mooney, R.J.: Bottom-up learning of Markov logic network structure. In: Proceedings of the Twenty-Fourth International Conference on Machine Learning (ICML 2007), pp. 625–632. ACM Press (2007)
26. Nikolenko, S.I.: Synthetic Data for Deep Learning, 1st edn. Springer, Cham (2021)
27. Pearl, J.: Probabilistic Reasoning in Intelligent Systems: Networks of Plausible Inference. Morgan Kaufmann (1988)
28. Poole, D.: First-order probabilistic inference. In: Proceedings of the Eighteenth International Joint Conference on Artificial Intelligence (IJCAI 2003), pp. 985–991. Morgan Kaufmann Publishers Inc. (2003)
29. Richardson, M., Domingos, P.M.: Markov logic networks. Mach. Learn. **62**, 107–136 (2006)
30. Singla, P., Domingos, P.: Discriminative training of Markov logic networks. In: Proceedings of the Twentieth National Conference on Artificial Intelligence (AAAI 2005), pp. 868–873. AAAI Press (2005)
31. Taghipour, N., Fierens, D., Davis, J., Blockeel, H.: Lifted variable elimination: decoupling the operators from the constraint language. J. Artif. Intell. Res. **47**, 393–439 (2013)

32. Venugopal, D., Gogate, V.G.: Scaling-up importance sampling for Markov logic networks. In: Advances in Neural Information Processing Systems (NIPS 2014), vol. 27, pp. 2978–2986. Curran Associates, Inc. (2014)
33. Xu, L., Skoularidou, M., Cuesta-Infante, A., Veeramachaneni, K.: Modeling tabular data using conditional GAN. In: Advances in Neural Information Processing Systems (NIPS 2019), vol. 32, pp. 6167–6178. Curran Associates Inc. (2019)
34. Yoon, J., Jordon, J., van der Schaar, M.: PATE-GAN: generating synthetic data with differential privacy guarantees. In: Proceedings of the International Conference on Learning Representations (ICLR 2019), pp. 536–545. IEEE (2019)

# Evaluating AI-Based Components in Autonomous Railway Systems
## A Methodology

Jan Roßbach[1(✉)], Oliver De Candido[2], Ahmed Hammam[2], and Michael Leuschel[1]

[1] Institut für Informatik, Heinrich Heine Universiät, Düsseldorf, Germany
jan.rossbach@uni-duesseldorf.de
[2] neurocat GmbH, Berlin, Germany

**Abstract.** Recent breakthroughs in n Artificial Intelligence (AI) are poised to transform many domains, including autonomous railway transportation systems. However, safety is essential in this high-stake, safety-critical domain. To ensure compliance with current safety certification standards, we propose a comprehensive methodology for evaluating AI-based components in railway applications.

It combines ontology-driven systematic testing and data generation, formal verification techniques, and real-time monitoring. By leveraging these methods, we provide a comprehensive safety assurance and hopefully pave the way for the widespread adoption of AI in railway transportation systems.

## 1 Introduction and Motivation

With the help of advances in AI, railway transportation systems are moving towards more automated systems. AI has a wide range of possible applications in this area, including perception systems for obstacle-detection, train localization, or train-sign classification using Deep Neural Networks (DNNs) [23,29]. However, these systems are inherently complex and difficult to verify [2,23]. On top of that, safety requirements in the railway area are stringent.

Safety critical verification methods, including formal methods, are still adapting to the complexity of these new systems. They are, thus far, unable to keep up with the pace of development in the field of AI [27].

Traditional AI testing methods are, on their own, also insufficient to provide the required safety guarantees. While they can provide evidence of a system's performance under certain conditions, they cannot guarantee that the system will behave correctly in all or even almost all possible scenarios the system might encounter (see Sect. 2). It is also an open challenge to integrate these AI testing methods into a safety verification and certification process.

This research is part of the KI-LOK project funded by the "Bundesministerium für Wirtschaft und Energie"; grant # 19/21007E.


A. Hotho and S. Rudolph (Eds.): KI 2024, LNAI 14992, pp. 190–203, 2024.
https://doi.org/10.1007/978-3-031-70893-0_14

Thus, a new methodology for safety evaluation is needed, including novel verification methods. The KI-LOK research project [9,10] aims to develop these verification methods for AI-based components in autonomous railway systems. It consist of a combination of academic and industry partners, cooperating in order to ensure practical applicability of the methods and the developed methodology.[1]

The methods, developed by some of the project partners, rely heavily on a combination of systematic testing, formal verification, and run-time monitoring. In this work, we present the integrated KI-LOK safety methodology for creating a safety argument in line with state-of-the-art certification standards (see Sect. 3).

We first introduce the background in the field and briefly discuss important concepts in Sect. 2. Then we discuss the applicable technical standards requirements in Sect. 3. Finally, Sect. 4 presents our proposed methodology, and Sect. 5 ends with a summary and outlook on future work.

## 2 Background and Related Work

The field of autonomous railway systems categorizes vehicle functions into different Grades of Automation (GoAs). They describe the degree to which the train can operate independently. The highest degree of automation is GoA 4, full automatic operation [3]. In GoA 4, a train has to fulfill two main functions: Automatic Train Protection (ATP) and Automatic Train Operation (ATO). ATP is generally responsible for protection, e.g., respecting signals and preventing collisions while ATO is responsible for the remaining functions usually performed by the driver [2].

For each function there is a general requirement: The automated system has to perform better than a human driver. This means, there is an upper limit on the allowed fault incidences for each function, usually called the *tolerable hazard rate* [5].

### 2.1 Verifying Perception Systems

A potential safety case has to give convincing evidence that the fault incidence ceiling (or tolerable hazard rate) will not be exceeded, usually through a probabilistic analysis of the systems' independent components.

This notably includes the verification of the important perception components in the automation pipeline, such as obstacle detection or point positioning [23]. These components receive input from multiple sensors, e.g., cameras, radar, LiDAR, etc., located onboard the train and process the inputs in real-time to generate an understanding of the train's surroundings. This is used by the train's deterministic controller to determine the next actions. These perception systems are likely to include AI techniques such as DNNs [29]. However, it is still unclear how to verify them for safety critical applications.

[1] Fore more details about the KI-LOK research project, please visit the project website: https://ki-lok.itpower.de/.

Current AI models on their own are, in terms of performance, not close to the required tolerable hazard rates for these functions. This gap can be tackled, e.g., by using model ensemble techniques [23] or other mitigation measures.

One way to tackle this is discussed in [22]. The authors propose an integrated multi-sensor voting system. The idea is that if the participating algorithms are sufficiently different from each other, they can be shown to be statistically independent. Thus, the total system can be shown to be acceptably safe, if the individual components have an error rate of less than 5%. This is considerably more within the reach of current state-of-the-art methods, but the individual components still need to be verified to work at this level of performance. The authors suggest a gray box approach to verify individual AI-based components by mathematically evaluating the architecture of the system. They look for specific classification clusters and corresponding equivalence classes. In comparison, we discuss our approach to verify the safety of the entire system given an individual component, and how to verify the component itself, in Sect. 4.3.

### 2.2 Data Scarcity

In any case, training and testing an AI model for use in an autonomous railway system is challenging and requires a significant amount of high quality data. It should cover a wide range of potential scenarios in order to be properly representative of the system's Operational Design Domain (ODD). This refers to the set of possible operating conditions the system might encounter [31].

Such data is currently lacking in the field of autonomous railway systems [21] and has only recently begun to be collected [30]. Additionally, shifts in real world probability distributions or environmental changes can lead to run-time hazards and have to be mitigated. This can either, e.g., be done by preparing for it during testing (see Sect. 4.3) or by using run-time monitoring (see Sect. 4.4).

### 2.3 Scope

The methodology introduced in this work focuses on AI based components in an autonomous railway system. It focuses on verifying systems onboard the train that are used to detect the train's surrounding environment.

In the project, we primarily focus on object detection models for train sign classification and obstacle detection for empirical experimentation. However, in principle, the methodology is applicable to all types of AI-based components with only minor adjustments, and it might even generalize to other autonomous systems.

In this work, we assume the proper functionality and reliability of the system's sensors. This should be verified through other methods, which is out of the scope of this paper.

## 3 Certification Standards

Certification standards provide systematic ways to break down a safety case that encompass all of the aspects introduced in Sect. 2. However, such standards are still under active development in the domain of autonomous transportation systems. In this section, we will look closer at the current landscape of certification standards for autonomous systems.

Currently, there are multiple standards in the automotive and railway sectors focused on creating a standardization process for (fully) autonomous systems, e.g., UL-4600 [31], ISO PAS 8800 [15], ISO 26262 [13], ISO 21448 [14], or IEC 61508 [12]. These standards provide practical guidelines and advice to create a safety case and certify an autonomous system with AI-based components. They represent the state-of-the-art in the certification standards for (fully) autonomous systems. In [23], the authors argue that the currently most applicable standard for fully autonomous systems is the UL-4600 [31]. The ISO PAS 8800 [15] is still under development for autonomous driving systems with AI-based components.

As part of our research, we investigated the UL 4600, the ISO PAS 8800, and the ISO 21448 standard to find common certification requirements. In the following, we summarize the main categories of these requirements.

One commonality amongst the standards is to derive the *system safety requirements*, see e.g., UL 4600 §6.1, ISO PAS 8800 §6.1, ISO 21448 §6.1. This step involves identifying potential hazards that the autonomous system may encounter while operating within its designated ODD, and evaluating the associated risks.

Parallel to this, the standards require a rigorous *definition of the data to be used* and the respective *ODD of the system*, see e.g., UL 4600 §8.2.1 and §8.5.4, ISO PAS 8800 §10.1 and §10.4.3.2, ISO 21448 §7.3.3. Since AI-based components are data driven, this requirement looks at the overall dataset lifecycle to ensure sufficient coverage of the ODD and to ensure that the perception systems on the autonomous system properly map sensor inputs to the ODD. Additionally, the UL 4600 (§8.4.2 and §8.4.3) requires that the ODD is mapped to an ontology which includes a comprehensive set of categories, properties, and relationships that accurately reflect the physical and functional elements of the environment.

Furthermore, the standards require *sufficient evaluation of the AI-based components* including extensive testing strategies, see e.g., UL 4600 §8.4.1 and §8.5.1, ISO PAS 8800 §11.5.2 and §11.5.5, ISO 21448 §9.3 and §10.3. This requirement includes providing evidence that the AI-based component produce acceptable functional performance guarantees according to a chosen metric, e.g., response time, accuracy, reliability, availability, or throughput of an AI-based component. Moreover, the standards require a sufficiently robust testing strategy to provide evidence for this performance. The integration of AI-based components into the complete system should additionally be tested.

Finally, the automotive standards require *investigating the functional insufficiencies* of the autonomous system, see e.g., ISO 21448 §5.4 and ISO PAS 8800 §9.3. These can be either due to insufficiencies of the specifications or due to performance insufficiencies. Methods to identify and mitigate these should be developed.

We can summarize these requirements into four main categories:

1. Hazard and risk analysis,
2. Data and ODD definition,
3. Evaluation of AI-based components,
4. Investigation of functional insufficiencies.

As we will see in the following section, our methodology proposes methods to address each of the aforementioned requirement categories (Item 1– Item 4).

# 4 Proposed Safety Methodology

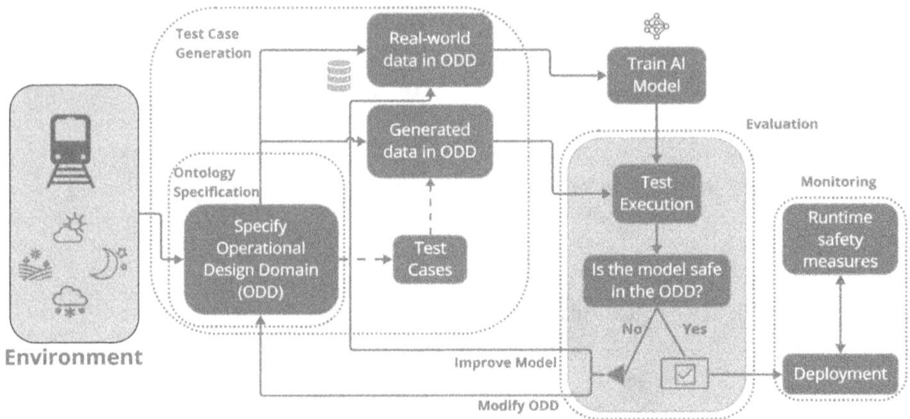


**Fig. 1.** The proposed safety methodology.

Figure 1 depicts our proposed methodology to evaluate a safety-critical autonomous railway system. We start with the existing environment in which the autonomous railway system should function. The methodology is not defined for one specific component of the autonomous railway system, but can be applied to any AI-based component using arbitrary sensor input, e.g., camera, LiDAR, or radar. We begin by defining an ontology (cf. Sect. 4.1) of the environment in order to specify the ODD (as required by Sect. 3 Item 2). With the ODD defined, we aim to collect and generate data within it. The generated data within the ODD are specified by test cases which we can use during evaluation (cf. Sect. 4.2). Once we have collected enough real-world data lying in the ODD, we can train the AI-based component itself, e.g., a DNN-based object detection model.

Given the trained AI model and the generated data, we can begin evaluating the model (cf. Sect. 4.3). At the output of this testing, we are able to decide whether or not the AI-based component performs safe enough in the ODD. If it is not reasonably safe, we propose two options: (i) we can go back to collecting more real-world data aiming to improve the model, or (ii) we can modify the ODD if we find areas where the model does not perform adequately (cf.

Sect. 4.3). For example, we can extend the ODD with test cases which were originally not incorporated into the ODD, or if it becomes clear—through our evaluation results—that the AI model cannot perform adequately or safely in specific test cases, we can exclude these scenarios to reduce the ODD. On the other hand, if our methods for testing the AI model pass, we can deploy the AI-based component in the autonomous railway system. During runtime, we propose methods to further evaluate and ensure the safety of the AI-based components to address potentially changing environments not covered by the ODD (cf. Sect. 4.4).

Overall, our methodology can be summarized into the following pillars (outlined in red-dotted boxes in Fig. 1):

1. *Ontology Specification*: The first pillar is a detailed investigation into the environment the AI-based system will work in. We define the ODD in this step based on our ontology design. Including probabilistic AI-based components into the railway system gives rise to additional requirements on the ODD design. We propose a probabilistically extended ontology to account for the inherent uncertainties in the system.
2. *Test Case Generation*: To address the challenge of data scarcity, we propose a simulation environment that can generate 3-D test cases according to the probabilistic ontology and the ODD. Moreover, we propose methods to generate additional test data in the ODD using realistic augmentations.
3. *Evaluation*: Here, we use sophisticated new evaluation techniques that go beyond the standard AI evaluation methods, e.g., the train/test split of available datasets. We propose methods to test the models' robustness against adversarial attacks and to test how the component integrates into the rest of the system.
4. *Monitoring*: We propose real-time verification methods to check remaining safety-critical properties dynamically at run-time. We aim to mitigate against uncertainty in the evaluation.

Used together, these pillars form a development and testing cycle that should yield a safe and reliable autonomous railway system. It guards the system against possible failure conditions and vulnerabilities, and provides formal guarantees wherever possible.

Another notable highlight of the methodology is that we also argue for a *black-box systematic testing approach* that focuses on the inputs and outputs of the individual components, rather than their internal design [33]. This makes the verification more independent of the implementation, enabling the system to be more complex and potentially contain proprietary information. This can be done because it is possible to iterate on the system design by re-training with newly collected (or generated) data. This, in turn, is only possible because we can explain and formally verify relevant safety-critical system properties with our monitoring approach (cf. Sect. 4.4).

In the following section, we describe our newly introduced methods for each pillar in more detail, and provide an example of how a safety argument could be built with the overall methodology (cf. Sect. 4.5).

### 4.1 Ontology Specification

An ontology is a formal representation of a system's operational domain, defining and categorizing possible events [11]. In our methodology we use a formal ontology representation to describe the ODD (as required by Sect. 3 Item 2). This ensures a more extensive coverage of the ODD and allows for systematic testing of the component under consideration.

For the evaluation of probabilistic components, e.g., involving DNNs, it is advantageous to extend the given ontology with the probabilities of the objects occurring in the real world. Thus, we introduce a *probabilistically extended ontology* [34]. By sampling test cases from this extended ontology, we enable the sampled test data to better reflect the frequency with which the scenarios would occur in the real world. This structured testing strategy allows continuous system refinement based on test outcomes, which is essential for ensuring robustness and reliability before deployment.

On the other hand, techniques like stratified sampling and cross-validation help ensure that test datasets are diverse and representative [4]. Experimental validation of these methods is vital for adjusting training and validation strategies, emphasizing the need for detailed modeling of the environmental probabilities to evaluate models' performance accurately [11].

Experiments have demonstrated the value of such (probabilistically extended) ontologies, particularly in autonomous driving [34]. Various sampling methods are explored to create realistic test datasets, highlighting the importance of including probabilistic information in the ontology design.

### 4.2 Test Case Generation

To address the field's pervasive problem of a lack in diverse labeled data, the project developed a 3-D simulation environment, using Unreal Engine 4, that can generate test cases according to a given ontology and ODD (cf. Sect. 4.1). It generates high quality, labeled synthetic data, and is shown to generalize to real world data in practical use [7].

One of the key concerns of any AI-based component is having correct and realistic data to test the component on. In the case of railway AI-based components, the need for diverse data across different weather conditions with many environments, scenarios, objects, and sceneries is necessary to ensure the safety of the AI-based component.

It is challenging to capture and label enough real-world data to sufficiently test the AI-based component's performance in all possible cases. Thus, the need arises to build a *synthetic test case generator* which can model and synthesize a wide range of environments and scenarios to reliably test the safety of the AI-based components. As part of our methodology, we develop such a simulation environment to generate these test cases.

This environment generates data based on five factors, i.e., the track geometry, environment, objects in the environment, any points of interest on the track or around the train, and the ego train itself. All of these factors have subcategories, e.g., the environment can have weather attributes depending on the season. Building on the probabilistic ontology, each of these factors are sampled based on the previously calculated probabilities for each factor. Additionally, the conditional probabilities for the factors are considered. We sample from them to generate a realistic environment based on the likelihood of these factors appearing.

To further enhance realism, we develop an augmentation generator that produces realistic augmentations for different weather conditions. These are parameterized so we can sample a wide range of augmentations. This facilitates the simulation of a diverse array of weather scenarios, thereby broadening the test data domain. These varied weather conditions are crucial for thoroughly testing the AI-based component in its ODD, because certain weather attributes may cause occlusions or blurs in the input images, potentially impairing the AI-based component's performance.

### 4.3 Evaluation

When it comes to rigorously evaluating AI-based components (cf. Sect. 3 Item 3), some of the current methods include using the test set for evaluation, checking the model for over-fitting [6], and guarding the model against brittleness [28]. To add to these conventional methods, we propose new evaluation methods that can be categorized as either testing:

1. the individual AI-based components independently, or
2. the integration of those components within the autonomous railway system.

For the first category of evaluation, we introduce a method to select AI models based on the estimated risk of damage [18]. This evaluation method specifically looks at the probability of damage arising from adversarial attacks (see e.g., [24,36]).

The method can be extended with our physics-based augmentations to evaluate the robustness of the AI-based components. Since the augmented images are based on real-world ground truth images, we are able to automatically evaluate the same scenario in various ODD trigger conditions. This provides a reliable evaluation of the components expected performance in practice. Moreover, the augmented images allow us to evaluate the AI-based components on edge cases that are not likely to be included in available real-world datasets.

On top of this, we can use our 3-D simulation environment to search for attributes in the input data which challenge the AI-based components. Since we can generate an unlimited number of test scenarios (cf. Sect. 4.2), we can search the input space efficiently using a tree search to test the component, find weak spots, and argue for adequate performance, see e.g., [7].

When it comes to the second category, we introduce an evaluation method to test the integration of the AI-based components within the autonomous railway system. For this, we propose to create a formal model of the environment and the deterministic parts of the entire autonomous railway system, since they have proven to be effective in verifying conventional railway systems [27]. In [8], we present a formal B [1] model of the control system for a shunting train that incorporates an AI perception system and is verified using the ProB [20] model checker. The results show that traditional software verification methods can be used to verify the safety of the control system, even when it incorporates AI-based components.

This allows us to isolate the AI-based components and the potential fault scenarios. This method can also give example failure scenarios that can be incorporated into the ODD to improve data generation or provide additional properties that can be verified with the methods from Sect. 4.4.

### 4.4 Monitoring

In order to ensure the functional requirements of AI-based components (Sect. 3 Item 4) are satisfied in practice, we propose using run-time monitoring. This enables formal verification of certain properties of the AI-based component, without having to verify the DNN model itself. This can be helpful if the evaluation process fails to prove sufficient safety of the model and improving the model, e.g., through re-training, does not alleviate these safety challenges.

Certified control [16] is an architectural framework for the real-time validation of autonomous systems. It does not rely on multiple independent subsystems. Instead, it counts on the AI-based component to provide a *certificate* containing all essential information to independently verify its output. This certificate serves as an input for the run-time monitor, which assesses the correctness of the AI's output. By adopting this approach, the architecture establishes a trusted foundation that can be subjected to a rigorous formal verification process.

The AI-based component, which is not included in the *trusted base*, can utilize sophisticated algorithms such as DNNs without needing explicit formal verification. By separating the tasks of generating visual insights and ensuring safety, we can continue to use established verification methods with minimal adjustments. To accomplish this, a formal acceptance specification for the certificate is necessary to ensure compliance with safety requirements, e.g., *the detected lane lines are parallel* or *there are no objects on the track for 100 m* [25].

We can use this approach to ensure safety critical properties at run-time. This has the potential to unlock the benefits of formal verification when employing AI-based components.

In [25], an implementation of the method is shown to be applicable in the domain of railway sign classification. The technology shows promising results but poses additional challenges on both the AI and the formal verification side. Amongst them is the currently high engineering effort involved in the verification process and the need for a formal acceptance specification. However, the approach shows that it is possible to ensure some properties at run-time, and can help target some safety-critical properties and circumstances that failed to be caught by the previous evaluation process (cf. Sect. 4.3).

Another such method to ensure run-time safety, in case the component under question is a reinforcement learning-based controller, is the use of safety shields [19]. For an example how this can be used in the context of autonomous driving, see [32].

### 4.5 Example Safety Case

Using the safety methodology introduced in the previous sub-sections, we present an example safety argument in the Goal Structuring Notation (GSN) [17,26]. We use the methods from the previous sub-sections as evidence for the safety goal and safety sub-goals, and discuss how they cover the certification requirements introduced in Sect. 3. Figure 2 highlights how a safety argument could be built for an autonomous railway transportation system use-case.

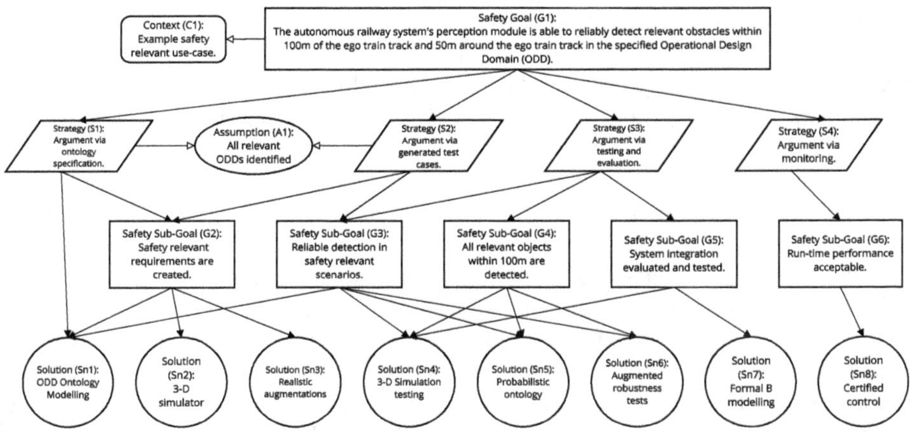


**Fig. 2.** Safety argument in the GSN highlighting the safety methodology introduced in the paper. A, C, G, S, Sn refer to the Assumptions, Contexts, Goals, Strategies, and Solutions in GSN, respectively. Please zoom into the figure to read the details of the safety argument.

The overarching safety goal is summarized in G1, i.e., "The autonomous railway system's perception module is able to reliably detect relevant obstacles within 100 m of the ego train track and 50 m around the ego train track in the specified ODD." We show how this safety goal can be broken down using the previously introduced safety methodology, where each pillar is a strategy (S1 through S4). Each strategy builds an argument for why the overall safety goal can be argued to be fulfilled. We make an explicit assumption (A1) that all relevant aspects of the ODDs have been identified for strategies S1 and S2. These strategies fulfill the first and second categories of requirements, i.e., *hazard and risk analysis* and *data and ODD definition*, Sect. 3 Item 1 and Item 2, respectively.

After applying S1, we define the safety sub-goal G2 supported by the ODD ontology modelling (Sn1) introduced in Sect. 4.1. With this goal, we aim to analyze the risks and hazards in the ODD. It is further supported by the 3-D environment simulator (Sn2) and the realistic image augmentations (Sn3) described in Sect. 4.2. In combination, these methods provide evidence that the safety relevant classes are defined in the ODD, and thus, the safety requirements can be created.

We see that the strategy to argue via the generated test cases (S2), leads to both the sub-goal G2 and a further sub-goal that a "reliable detection is achieved in all relevant scenarios" (G3). This sub-goal is influenced by G2, i.e., once we define all safety relevant objects, we now want to provide evidence that the detection system can reliably detect them. Strategy S3, *argue via testing and evaluation*, also leads to these sub-goals. These sub-goals are supported by the methods described in Sect. 4.3 labelled Sn4 through Sn6 in Fig. 2. The evaluation methods are developed as a key component of our proposed safety methodology.

We introduce a further sub-goal that "all relevant objects within 100 m are detected" (G4), which is additionally supported by the aforementioned methods. Moreover, we introduce the sub-goal G5 which ensures that the AI-based components are integrated and tested on a system level, i.e., we ensure that we evaluate the interplay between the AI-based component and the rest of the system. This is supported by the solutions Sn4 and Sn7. Overall, these solutions, sub-goals, and strategies fulfill the *evaluation of AI-based components* as required by Sect. 3 Item 3.

The strategy S4 relates to arguing via the run-time monitoring. It provides evidence for the fourth category of requirements, i.e., *investigation of functional insufficiencies* caused by AI-based components (as required by Sect. 3 Item 4). From this follows a final sub-goal (G6), representing the system showing an acceptable run-time performance. The methods providing supporting evidence for this sub-goal are discussed in Sect. 4.4. Ultimately, this strategy argues that any functional insufficiencies not caught during the deployment and evaluation cycles will be caught and dealt with during run-time.

As presented in this sub-section, we see that the safety methodology presented in this paper can be used to create a coherent safety argument relevant to an autonomous railway system. Since the safety methodology generalizes beyond the obstacle detection application presented here, a safety case can be generated for other AI-based components of an autonomous railway system. Moreover, each strategy, sub-goal, and all of the solutions provide evidence for the different categories of requirements from Sect. 3, Item 1 – Item 4. This is the first step to generate certified AI-based components in autonomous railway systems.

Ultimately, creating a formal safety case requires more documentation and the detailed evidence provided by the solution methods described in this paper. The proposed methodology, safety case, and evidence additionally needs to be evaluated on real-world use-cases with our project partners and regulators to verify its feasibility.

## 5 Future Work and Conclusion

In this work, we summarize the contributions of the KI-LOK research project, addressing the challenges in the field of autonomous railway transportation systems. We demonstrate how the newly developed methods can be integrated together into a safety verification and certification process.

To that end, we introduce a methodology for evaluating the safety of AI-based components in autonomous railway systems. Our methodology relies on the methods introduced in the project and aims to address the requirements from state-of-the-art certification standards. Additionally, we show that the methods have the potential to address critical issues in the field, such as data scarcity and the difficulty of formally verifying AI-based components. We demonstrate how to apply our methodology on an example safety case. Overall, the KI-LOK safety methodology represents a step forward for autonomous railway systems, paving the way for more efficient, reliable, and safe AI-based transportation solutions.

In the future, we plan to extend our methodology to focus on *sensor fusion techniques* to explore the interactions and possibilities of multi-modal AI-based components. We additionally aim to explore the possibilities of *using synthetic data to fine-tune* an AI-based model to improve its safety in edge cases where not enough real-world data is available. Finally, we want to extend our monitoring methods to *detect out-of-distribution usage*, i.e., run-time detection of inputs outside the training distribution [35]. This research contributes to the development of safer, more reliable autonomous systems, and advances AI technologies in safety-critical applications.

## References


1. Abrial, J., Hoare, A.: The B-Book: Assigning Programs to Meanings. Cambridge University Press, Cambridge (2005)
2. Athavale, J., Baldovin, A., Paulitsch, M.: Trends and functional safety certification strategies for advanced railway automation systems. In: 2020 IEEE International Reliability Physics Symposium (IRPS) (2020)
3. Bešinović, N., et al.: Artificial intelligence in railway transport: taxonomy, regulations, and applications. IEEE Trans. Intell. Transp. Syst. **23**(9), 14011–14024 (2022)
4. Bishop, C.M.: Pattern Recognition and Machine Learning, vol. 2. Springer, Heidelberg (2006)
5. Braband, J., et al.: Risikoakzeptanzkriterien für das automatisierte fahren auf der schiene. Technical report, Deutsches Zentrum für Schienenverkehrsforschung beim Eisenbahn-Bundesamt (DZSF) (2023)
6. dos Santos, C.F.G., Papa, J.P.: Avoiding overfitting: a survey on regularization methods for convolutional neural networks. CoRR arXiv:2201.03299 (2022)
7. Grossmann, J., et al.: Test and training data generation for object recognition in the railway domain. In: Masci, P., Bernardeschi, C., Graziani, P., Koddenbrock, M., Palmieri, M. (eds.) SEFM 2022, pp. 5–16. Springer, Cham (2023). https://doi.org/10.1007/978-3-031-26236-4_1

8. Gruteser, J., Geleßus, D., Leuschel, M., Roßbach, J., Vu, F.: A formal model of train control with AI-based obstacle detection. In: Milius, B., Collart-Dutilleul, S., Lecomte, T. (eds.) RSSRail 2023, pp. 128–145. Springer, Cham (2023). https://doi.org/10.1007/978-3-031-43366-5_8
9. Hemzal, G., et al.: KI-LOK - a joint test procedure project for AI-based components used in railway operations. Signal+DRAHT (2021)
10. Hemzal, G., et al.: KI-LOK - Ein Verbundprojekt über Prüfverfahren für KI-basierte Komponenten im Eisenbahnbetrieb. Signal+DRAHT (2023)
11. Huang, L., Liang, H., Yu, B., Li, B., Zhu, H.: Ontology-based driving scene modeling, situation assessment and decision making for autonomous vehicles. In: 2019 4th Asia-Pacific Conference on Intelligent Robot Systems (ACIRS), pp. 57–62. IEEE (2019)
12. IEC 61508-1:2010 Functional Safety of Electrical/Electronic/Programmable Electronic Safety-related Systems. Standard, International Electrotechnical Commission, Geneva, CH (2010)
13. ISO 26262-1:2018 Road Vehicles – Functional safety. Standard, International Organization for Standardization, Geneva, CH (2018)
14. ISO 21448:2022 Road Vehicles – Safety of the intended functionality . Standard, International Organization for Standardization, Geneva, CH (2022)
15. ISO/CD PAS 8800 Road Vehicles – Safety and artificial intelligence. Standard, International Organization for Standardization, Geneva, CH (2024)
16. Jackson, D., et al.: Certified control: an architecture for verifiable safety of autonomous vehicles. CoRR arXiv:2104.06178 (2021)
17. Kelly, T.P.: Arguing safety-a systematic approach to safety case management. DPhil Thesis York University, Department of Computer Science Report YCST (1999)
18. Klemenc, J., Trittenbach, H.: Selecting models based on the risk of damage caused by adversarial attacks. CoRR arXiv:2301.12151 (2023)
19. Könighofer, B., Lorber, F., Jansen, N., Bloem, R.: Shield synthesis for reinforcement learning. In: Margaria, T., Steffen, B. (eds.) ISoLA 2020. LNCS, vol. 12476, pp. 290–306. Springer, Cham (2020). https://doi.org/10.1007/978-3-030-61362-4_16
20. Leuschel, M., Butler, M.: ProB: a model checker for B. In: Araki, K., Gnesi, S., Mandrioli, D. (eds.) FME 2003. LNCS, vol. 2805, pp. 855–874. Springer, Heidelberg (2003). https://doi.org/10.1007/978-3-540-45236-2_46
21. Pappaterra, M.J., Flammini, F., Vittorini, V., Bešinović, N.: A systematic review of artificial intelligence public datasets for railway applications. Infrastructures **6**(10) (2021)
22. Peleska, J., Brüning, F., Gleirscher, M., Huang, W.: A stochastic approach to classification error estimates in convolutional neural networks. CoRR arXiv:2401.06156 (2023)
23. Peleska, J., Haxthausen, A.E., Lecomte, T.: Standardisation considerations for autonomous train control. In: Margaria, T., Steffen, B. (eds.) ISoLA 2022, pp. 286–307. Springer, Cham (2022). https://doi.org/10.1007/978-3-031-19762-8_22
24. Ren, K., Zheng, T., Qin, Z., Liu, X.: Adversarial attacks and defenses in deep learning. Engineering **6**(3), 346–360 (2020)
25. Roßbach, J., Leuschel, M.: Certified control for train sign classification. In: Proceedings Fifth International Workshop on Formal Methods for Autonomous Systems, vol. 395, pp. 69–76 (2023)

26. SCSC Assurance Case Working Group. Goal structuring notation community standard version 3. GSN Community Standard SCSC-141C, Safety-Critical Systems Club, York, UK (2021)
27. Seisenberger, M., et al.: Safe and secure future AI-driven railway technologies: challenges for formal methods in railway. In: Margaria, T., Steffen, B. (eds.) ISoLA 2022, pp. 246–268. Springer, Cham (2022). https://doi.org/10.1007/978-3-031-19762-8_20
28. Sun, Y., Wu, M., Ruan, W., Huang, X., Kwiatkowska, M., Kroening, D.: Concolic testing for deep neural networks. CoRR arXiv:1805.00089 (2018)
29. Tang, R., et al.: A literature review of artificial intelligence applications in railway systems. Transp. Res. Part C: Emerg. Technol. **140**, 103–679 (2022)
30. Tilly, R., et al.: Open sensor data for rail 2023 (2023)
31. UL 4600 Standard for Evaluation of Autonomous Products. Standard, Underwriters Laboratories Inc. (2020)
32. Vu, F., Dunkelau, J., Leuschel, M.: Validation of reinforcement learning agents and safety shields with ProB. In: Benz, N., Gopinath, D., Shi, N. (eds.) NFM 2024. LNCS, vol. 14627, pp. 279–297. Springer, Cham (2024). https://doi.org/10.1007/978-3-031-60698-4_16
33. Wiesbrock, H.-W., Großmann, J.: Outline of an independent systematic blackbox test for ML-based systems. CoRR arXiv:2401.17062 (2024)
34. Wiesbrock, H.W., Grossmann, J.: Probabilistically extended ontologies: a basis for systematic testing of ML-based systems. Technical report, SAE Technical Paper (2024)
35. Yang, J., Zhou, K., Li, Y., Liu, Z.: Generalized out-of-distribution detection: a survey. CoRR arXiv:2110.11334 (2024)
36. Yuan, X., He, P., Zhu, Q., Li, X.: Adversarial examples: attacks and defenses for deep learning. IEEE Trans. Neural Netw. Learn. Syst. **30**(9), 2805–2824 (2019)

# SocialCOP: Reusable Building Blocks for Collective Constraint Optimization

Julia Ruttmann and Alexander Schiendorfer(✉)

Institute AImotion Bavaria, Technische Hochschule Ingolstadt, Ingolstadt, Germany
alexander.schiendorfer@thi.de

**Abstract.** Distributing limited resources among a group of agents is a fundamental challenge in both algorithmic decision support systems and everyday life. The goal of achieving a socially desirable allocation of these resources instead of mere economic efficiency is relevant to many types of allocation problems under hard constraints. At the same time, modeling languages and high-level libraries for combinatorial optimization problems are becoming more widespread. Although fairness is an important key factor in optimization processes, there is currently no way to use fairness constraints and objectives – unless they are written from scratch. Thus, combining and experimenting with different fairness criteria is tedious as no predefined set of constraints and objectives is available in modeling languages. We propose SocialCOP, a toolbox of reusable constraint modeling building blocks of concepts derived from social choice theory, fair division, and algorithmic fairness (namely, Envy-freeness, Leximin, Rawlsianism, Utilitarianism, Pareto). Our toolbox provides a convenient and reusable solution for adding fairness constraints to existing collective constraint optimization problems formulated in MiniZinc. Our created building blocks can be combined or added individually to the existing satisfaction problem. Experimental results show that a much richer combination of fairness objectives can be modeled, leading to the discovery of solutions that are optimal in more than one way.

**Keywords:** Constraint Satisfaction and Optimization · Algorithmic Fairness · Multiagent Systems

## 1 Collective Constraint Optimization Problems

Many real-world combinatorial optimization problems dealing with the sharing of constrained resources involve multiple objectives belonging to their respective stakeholders. Examples thereof include shared mobility (organizing collective trips, choosing which sites to visit and in what order), hoteling systems for shared office space (allocating desks to employees, respecting work-from-home restrictions or making joint reservations), shared manufacturing, or managing the charging of electric vehicles (EVs) in a shared parking lot. We call them *collective combinatorial optimization problems* and emphasize discrete decisions – leading to problems that are often NP-complete.


A. Hotho and S. Rudolph (Eds.): KI 2024, LNAI 14992, pp. 204–217, 2024.
https://doi.org/10.1007/978-3-031-70893-0_15

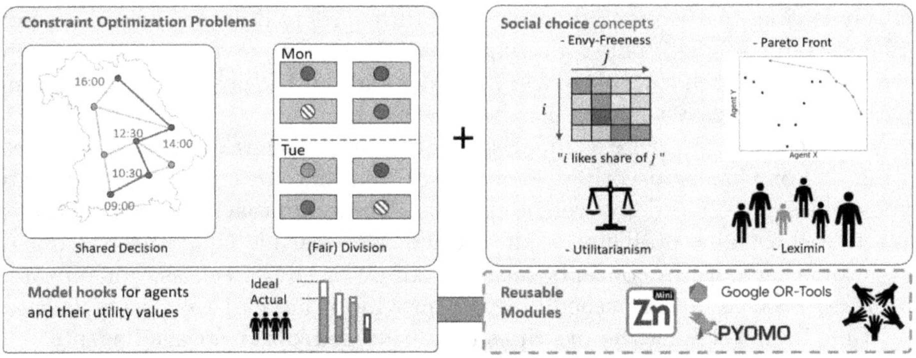


**Fig. 1.** The components of our framework for algorithmically fair collective constraint optimization problems: Shared decision problems (e.g. one travel itinerary per group per day) or division problems (e.g., allocating desk seats subject to WFH constraints and mutual availability) are connected via model hooks to a variety of social choice concepts.

Aggregating stakeholders' goals in a *socially desirable way* (e.g., fairness, proportional access) while balancing this with other overarching goals such as economic efficiency is challenging. While the optimization literature has recently devoted more attention to formalizing algorithmic fairness and equity concepts [26], there is currently no accessible toolbox of reusable modules for collective constraint optimization in modeling languages or environments such as MiniZinc [15], Pyomo [4], or AMPL [9]. Such a toolbox would enable the integration of fairness aspects into existing optimization problem formulations. Ideally, modelers could start by defining decision variables, hard constraints, and objective functions in the usual way. Then, they could augment them with social concepts such as envy-freeness or leximin-fairness – thereby experimenting with their effects easily to eventually achieve fair and efficient solutions.

This paper presents **`SocialCOP`** – a Python-based library consisting of building blocks such as predefined constraints and meta-search runners that implement an array of social choice concepts for general-purpose optimization frameworks – as shown in Fig. 1. Currently, only MiniZinc is supported as a backend since it emphasizes the definition of reusable functions and predicates with well-defined semantics in Boolean contexts [23]. Moreover, we inherit the solver-independency and `SocialCOP` can be used with SAT, CP, or MIP solvers. Extensions to other environments such as Pyomo, Google OR-Tools directly, or CPMpy [10] are planned.

## Related Work

The potentially symbiotic relationship between collective decision making and constraint reasoning goes at least back to [2] demonstrating a constraint-based encoding of leximin search and was later pointed out by Rossi in 2014 [18], most notably leading to a sequential voting procedure [5,6] to link social choice theory and constraint solving. This procedure is relatively myopic and lets agents vote over the value of each variable in a static order without the possibility of changing an assignment once other variables become fixed.

More recently, pairwise comparison voting approaches such as Condorcet-voting [22] or Copeland-voting [1] on top of MiniZinc have been introduced to remove bias introduced by weighted sums. These approaches treat solving a collective constraint problem as an election over feasible solutions but fail to integrate other fairness concepts such as leximin or envy-freeness. For repeated problems, such as weekly scheduling, [7] provide a method that seeks fairness over time by reweighing the priorities of advantaged/disadvantaged agents.

More generally, search procedures over solutions of constraint satisfaction problems that use a more complex type of dominance have been discussed in [11], following the prototypical implementation of the search language MiniSearch [17] which is no longer actively developed in favor of a Python API [24] that `SocialCOP` uses as its first backend.

## 2 Types of Collective Constraint Optimization Problems

Constraint programming (CP), as a particular constraint solving technology originating in logic-based (symbolic) Artificial Intelligence (AI), has focused strongly on expressive constraints that – in addition to offering efficient domain filtering by propagators – lead to concise models that can serve as high-level problem specifications for end-users [19]. In a constraint satisfaction problem (CSP), the task is usually to assign a set of typed decision variables $X$ to their respective domains $(D_x)_{x \in X}$ such that all logical constraints $C$ are satisfied. The formalism is extended to optimization (COP) by choosing the solution $\vec{x} \in [X \to D]$ that is minimal or maximal with respect to an objective function $f : [X \to D] \to (F, \leq_F)$ where $F$ is an ordered set (typ. the rational numbers with their usual ordering).

A collective COP (CCOP) then extends this notion to multiple objective functions $(f_a)_{a \in A}$, corresponding to a finite set of agents $A$. Without loss of generality [25], in our framework, we only consider positive objective functions and refer to them as *utilities* $(u_a : [X \to D] \to \mathbb{Z})_{a \in A}$. In a CCOP, each solution $\vec{x}$ is thus evaluated by each agent, resulting in a utility vector that must be aggregated into a partial or total order to have a viable optimization objective. In terms of implementation, we assume that for each agent $a$, the CCOP model contains an integer variable that holds the utility value $u_a$ for the solution.

More specifically, we distinguish two different classes of CCOP:

- **Division problems** deal with the allocation of resources to agents, such as desk space in a shared office or charging time reserved for a particular EV. Formally, they can be characterized by a partitioning of decision variables $(X_a \subseteq X)_{a \in A}$ with $X_a \cap X_b = \emptyset$ if $a \neq b$ such that the utility of agent $a$ *only* depends on the values assigned to the variables in $X_a$ – without externalities.

For example, given the same reserved desk slots, an employee is indifferent to the assignments of the other desks. Solution concepts include predefined constraints for *envy-freeness* or *envy-minimization*.
- **Shared decisions** By contrast, here we need to make decisions on a shared set of variables that we cannot localize to a particular agent – think of a shared itinerary in group travel where a group of 15 travelers must agree on 6 out of 15 possible sights and the order in which to visit them.

Certainly, a real-life problem may show characteristics of both types of problems. Consider again desk-sharing: a (private) *share utility* $\bar{u}_a$ could evaluate only the desks assigned to one agent over a week but additional utility $u_a^+$ may be derived from assignments to other variables such as, e.g., sharing a day in the office with a close collaborator. In such cases, we assume an additive global utility:

$$u_a(\vec{x}) = \bar{u}_a(\vec{x} \downarrow_{X_a}) + u_a^+(\vec{x}) \quad (1)$$

where $\vec{x} \downarrow_{X_a}$ refers to a restriction of the full solution $\vec{x}$ to only those decision variables that represent the share of $a$.

Moreover, a distinction between *single-shot* and *repeated* interactions is useful in terms of fairness: In a single-shot decision, an overall trade-off between agent satisfaction (e.g., using the Rawlsian "maximize the minimally satisfied") and social efficiency ("highest total satisfaction") needs to be found, explained, and justified. In a repeated interaction, there is a chance to compensate for past bad allocations (for certain agents) by giving them priority in future instances, as sketched out in [7]. Fairness metrics such as, e.g., the Gini index, which measures inequality, could then be applied retrospectively to an entire *trace* – rather than to a single epoch. Both division and shared decision problems can be repeated within the same group of agents: think of allocating desk slots over multiple days or weeks, or deciding an itinerary for each day of a week for the same group of travelers. In this paper, we restrict our attention to single-shot interactions but plan to extend `SocialCOP` with repeated interactions in the future.

## 3 Implementation Based on a Generic Modeling Language

One main design idea of `SocialCOP` is that a "pure" constraint satisfaction model is separated from the social concepts that it should be augmented with. The model formulations and their social concept augmentations should be independent of the specific solving approach (e.g., CP, SAT, MIP) which is why the library produces declarative models that are sent to solvers. The core CSP model formulation is expected to contain integer variables for the agents' utilities as *model hooks* and should not yet contain an objective.

Fairness and social choice concepts are instead encapsulated in reusable MiniZinc and Python modules. Most of the augmentations are implemented as functions capable of adding variables and constraints, which need to be passed to a *runner* that invokes them *before solving*, *upon seeing a feasible intermediate result*, or *after the search terminates* – hidden from the user. If an iterative approach, including multiple calls to a solver, is required, the fairness approach is placed inside a dedicated `Runner` class. This runner class enables the possibility to solve the constraint optimization problem. Alternatively, if the belonging constraints can be added to an existing model formulation, they are added as a so-called *mixin*. That is a model fragment consisting of constraints, predicates, and functions that need to be tied to the main decision variables. To connect the plain constraint model with the social choice building blocks, a *social mapping file* (e.g., implemented in JSON) is required. Its main functionality is to map different fixed concepts to the respective names in the specific model formulation at hand. Required hooks like the array of agents taking part in the distribution, an array of utility decision variables, as well as the names of share functions and share utilities can therefore be named according to the underlying problem and do not need to be the same for all models. Let's illustrate these concepts using a small example. Consider a table assignment problem for ten agents to three tables over a week:

```
enum Days = {Mon, Tue, Wed, Thu, Fri};
enum Tables = {table1, table2, table3};
int: n_agents = 10;
set of int: Agents = 1..n_agents;

array[Agents, Days] of var opt Tables: assigned;
% on every day, all agents get a different table
constraint forall(d in Days)(alldifferent([assigned[a, d] | a in Agents]));
% a hard wfh regulation - at most 2 days without a table
constraint forall(a in Agents) (
  sum(d in Days) (absent(assigned[a, d])) <= 2
);

array[Agents] of var int: utilities;
constraint utilities = [global_utility(a) | a in Agents];
```

**Listing 1.1.** A pure constraint satisfaction model for table assignment.

Each agent can get a table or the special value <> for no table which is reflected by the optional type. The table assignment for any agent can be absent on at most two days due to work-from-home regulations. An array for utility values is predefined and assigned in a function `global_utility` that is omitted for brevity. As can be seen, hard constraints and decision variables are included but no objective or fairness constraint is yet imposed. The corresponding social mapping file would look like this:

```
{
    "agents_array" : "Agents",
    "utility_array" : "utilities",
    "num_agents" : "n_agents",
    "main_variables": ["assigned"]
}
```

**Listing 1.2.** An exemplary social mapping to combine model and SocialCOP

This intermediate layer alleviates the need to have every array representing the agents be named "`Agents`" instead of `Passengers`, `Students`, etc. The same applies to the utility array and the main variables which could be used for diversity constraints in the search [1]. The number of agents is assumed to be read from a data file and needed for the Python-based leximin runner (see Sect. 3.2).

### 3.1 Utilitarianism and Rawlsianism

We begin by introducing the most straightforward social choice augmentations, Utilitarianism and Rawlsianism, as they simply amount to an integer-valued objective that can be readily passed to a solver. Utilitarianism emphasizes the importance of the solution with the highest total utility [3,20]. Hence, the sum of all agents' utility variables is maximized. `SocialCOP` provides three functions: i) to add the Utilitarian objective to the model ii) to actually maximize its value, and iii) a predicate that enforces that the next solution's Utilitarian objective is higher than the current one. We separate these concerns because during experimentation, we might want to track the Utilitarian value but not optimize for it or we might impose a minimal value as a hard constraint. The third option is for custom branching searches that need a "getBetter" predicate [11,21].

```
def utilitarian_objective(instance : Instance, social_mapper):
    instance.add_string(f"var int: {UTILITARIAN_OBJECTIVE};")
    instance.add_string(f"constraint {UTILITARIAN_OBJECTIVE} =
                            sum({social_mapper[UTILITY_ARRAY]});")

def optimize_utilitarian_objective(instance : Instance, social_mapper
        = None):
    instance.add_string(f"solve maximize {UTILITARIAN_OBJECTIVE};")

def get_better_utilitarian(instance : Instance, res : Result, social_mapper
        = None):
    # enforce that the next solution needs to be better than the current one
    instance.add_string(f"constraint {UTILITARIAN_OBJECTIVE} >
                            {res[UTILITARIAN_OBJECTIVE]};")
```

**Listing 1.3.** Augmentation functions for Utilitarianism.

These building blocks can then be used to augment the pure constraint model (e.g., the one defined in Listing 1.1) with a simple runner (the most basic call to MiniZinc in Python):

```
simple_runner = SimpleRunner(social_mapping)
simple_runner.model += [utilitarian_objective]
simple_runner.model += [optimize_utilitarian_objective]

result = simple_runner.run(table_assignment_model, gecode)
```

**Listing 1.4.** Usage of a simple runner with maximization of the Utilitarian objective.

The Utilitarian approach does not focus on an equal distribution among the agents but maximizing the overall sum of utilities. It sacrifices the satisfaction of single agents to the goal of overall welfare. By contrast, another common social choice concept is Rawlsianism. The central argument put forth by John Rawls declares that the selection of the optimal solution should be independent of the decision-maker and the objective is to maximize the outcome for the least advantaged [12,13,26]. In terms of its `SocialCOP` implementation, it is

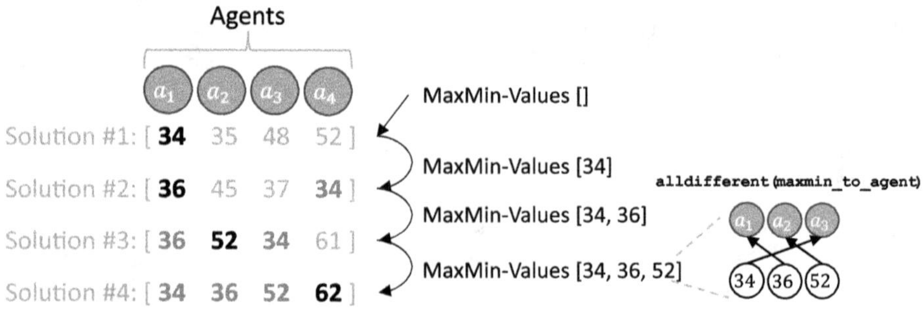


**Fig. 2.** Example of leximin search via a bijection between maxmin values and agents implemented by an `alldifferent` constraint

completely analogous to the Utilitarian functions but uses `min` instead of `sum` to define the utility aggregation.

### 3.2 Leximin

As the base form of the Rawlsian approach only considers the worst-off agent, there is a recursive extension concerning all agents. Maximizing the worst utility, as explained previously, is repeated for every agent while the already received values are kept for every agent [14]. This iterative approach is useful if more than one solution shows the same worst-off value – which is not rare.

Augmenting an existing constraint model with leximin is more challenging because, in general, repeated calls to a solver are necessary. First, the maxmin value has to be found, i.e., the solution that maximizes the "least happy" agent. Then, assuming that one agent has this utility (not necessarily the same one), the next maxmin value has to be found etc. Fortunately, we can delegate most of the bookkeeping work to the solver, as shown in Fig. 2. Another mixin defines the decision variables that map the already found maxmin values injectively to the agents. A specialized `LeximinRunner` in Python takes care of the management of the maxmin values and the repeated solver calls. It is required to store the maxmin values without the information about the corresponding agent, as different solutions in the search process can have the same utility vector, but permutet differently across the agents. All of the solutions found by the leximin approach are pareto-optimal but not every Pareto-optimum is necessarily a leximin solution [3] which is why Pareto-Non-Domination could be posted as a redundant constraint throughout the search.

### 3.3 Envy-Freeness

Envy-Freeness is a concept from fair division and usually expressed in terms of a valuation function $V_i$ for every agent $a_i$. An allocation of resources into shares is envy-free if no agent values another agent's allocation more than its own [16].

Formally, there are no agents $a$ and $b$ such that

$$V_a(X_a) < V_a(X_b) \tag{2}$$

with $X_a$ and $X_b$ being the allocated shares of resources for agents $a$ and $b$. Envy-Freeness (EF) can only be used if CCOP is indeed a division problem according to Sect. 2. The definition of share and global utility (see Eq. (1)) takes into consideration which utilities can be compared and which utilities cannot be compared among the agents. Therefore when EF is chosen only the agents' *share utilities* are considered. Global utilities would then only apply if another social concept such as Rawls is combined with EF. If users want to add EF, they need to specify the share function which takes an agent $a$ and returns its respective share $X_a$ to be evaluated by the share utility in the social mapping JSON file:

```
{   [...]
    "share_function" : "get_assigned_for",
    "share_utility" : "share_utility"
}
```

Based on this information, the EF mixin defines an *envy matrix* and provides constraints and objectives to be optimized:

```
array[${AGENTS_ARRAY}, ${AGENTS_ARRAY}] of var int: envy_matrix;
constraint forall (i, j in ${AGENTS_ARRAY}) (
        envy_matrix[i,j] = if i == j then 0 else
                                bool2int( ${SHARE_UTIL_AGENT}
                                  (i, ${SHARE_FUNCTION}(i)) < ${SHARE_UTIL_AGENT}
                                  (i, ${SHARE_FUNCTION}(j)) )
                              endif);
var int: envy_pairs;
constraint envy_pairs =
  sum(i in ${AGENTS_ARRAY}, j in ${AGENTS_ARRAY} where i != j) (
        envy_matrix[i, j]);
predicate envy_free() =
  envy_pairs == 0;
% can be enforced as: constraint envy_free();
```

**Listing 1.5.** Mixin for envy-freeness

The function to optimize envy-freeness is searching for the minimum amount of agents envying the distribution of other agents. It can find a solution even if envy-freeness itself is not given, but envying pairs are just reduced. To ensure an envy-free solution the function `enforce_envy_freeness` can be chosen in Python. If there is no EF solution, the problem becomes unsatisfiable.

### 3.4 Pareto Optimal Solutions

A solution is Pareto-optimal if all other solutions would be disadvantageous to at least one agent in comparison [3]. Conversely, a solution Pareto-dominates another solution if it is (weakly) preferred by all agents [3]. Formally, a solution $\vec{x}$ Pareto-dominates solution $\vec{y}$ if:

$$\forall i \in A, u_i(\vec{x}) \geq u_i(\vec{y}) \qquad \text{and} \qquad \exists i \in A, u_i(\vec{x}) > v_i(\vec{y}) \tag{3}$$

To enforce Pareto-optimal results, a mixin is again provided. Inside it, we pass the utility vectors of all previously found solutions as parameters as well as the constraint to eliminate Pareto-dominated solutions in the current iteration. Since this search can only compare the new solution with the previous ones, a special `ParetoRunner` eliminates all previous solutions that are Pareto-dominated by the new solution. This two-pass search for Pareto-optimal solutions is similar to an implementation in CPMPy [10].

## 4 Experimental Evaluation

To show that `SocialCOP` provides enough flexibility to cover a wide range of real-life situations, we designed three experiments. They were executed on four example problems, each with a handful of concrete data instances: *table assignment* under work-from-home constraints as a division problem with share and global utility, *bus tour* which amounts to a vehicle-routing-problem (VRP) for a set of buses and customers with site preferences, *project assignment* among students, and multi-agent job-shop *scheduling* where agents are owners of jobs. Four different solvers from heterogeneous technology backgrounds were used with MiniZinc 2.8.4: Chuffed 0.13.1 and OR-Tools 9.9.3 which are SAT-based, Gecode 6.3.0 as a propagation-based constraint solver, and Coin-BC as a mixed-integer-programming solver. All experiments were conducted on a HPC cluster on Ubuntu 22.04.[1]

### 4.1 Tradeoffs Between Leximin, Rawls, and Utilitarian

*"Which social choice concept provides fair and efficient solutions, and at what cost?"*

The first experiment examines the differences in the results for the given social choice concepts Utilitarian, Leximin, and Rawls. These three apply to all used scenarios (table assignment, bus tour, project assignment, and multi-agent scheduling) and can be used with all solvers. All 19 instances can be examined in Table 1. It includes not only the Utilitarian objective, which is the sum of the utilities, but also minimal and maximal values as well as the Gini index to measure how equal the utility is distributed among the agents. Especially for larger (later in every problem) instances, Rawlsian produces solutions with lower Gini index and higher minimal utility but sacrifices Utilitarian welfare for it, as can be seen in Fig. 3. A direct aggregated comparison is shown in Table 2. Comparing Leximin and the Utilitarian approach, the sum of the utilities decreases by 4.16% on average for the chosen instances. In contrast to that, the Gini index measuring the unequal distribution among the agents decreased by 19.23%. Therefore even though the overall sum is less for Leximin results, the utility is distributed more evenly among the agents.

[1] Full details can be found online at https://github.com/AImotion-Bavaria/SocialCOP.

**Table 1.** Comparison of Utilitarian, Rawlsian, and Leximin Principles

| | Utilitarian | | | | Rawlsian | | | | Leximin | | | |
|---|---|---|---|---|---|---|---|---|---|---|---|---|
| | Min | Max | Sum | Gini | Min | Max | Sum | Gini | Min | Max | Sum | Gini |
| bus tour 1 | 5969 | 8114 | 22187 | 0.06 | 5969 | 8114 | 22187 | 0.06 | 5969 | 8114 | 22187 | 0.06 |
| bus tour 2 | 211 | 4231 | 5766 | 0.55 | 232 | 676 | 1572 | 0.23 | 232 | 1630 | 3204 | 0.37 |
| bus tour 3 | 632 | 7684 | 14112 | 0.42 | 632 | 7684 | 14112 | 0.42 | 632 | 7684 | 14112 | 0.42 |
| bus tour 4 | 1277 | 4968 | 13472 | 0.22 | 1441 | 3768 | 11345 | 0.17 | 1441 | 3768 | 11345 | 0.17 |
| bus tour 5 | 1318 | 8958 | 29635 | 0.28 | 3014 | 7958 | 28285 | 0.19 | 3014 | 7958 | 28285 | 0.19 |
| tables 1 | 14 | 14 | 42 | 0.00 | 14 | 14 | 42 | 0.00 | 14 | 14 | 42 | 0.00 |
| tables 2 | 17 | 19 | 147 | 0.02 | 18 | 18 | 144 | 0.00 | 18 | 19 | 147 | 0.01 |
| tables 3 | 19 | 21 | 204 | 0.02 | 19 | 20 | 191 | 0.00 | 19 | 21 | 204 | 0.02 |
| tables 4 | 18 | 20 | 175 | 0.02 | 18 | 19 | 165 | 0.01 | 18 | 20 | 175 | 0.02 |
| tables 5 | 20 | 22 | 230 | 0.02 | 20 | 22 | 225 | 0.02 | 20 | 22 | 230 | 0.02 |
| scheduling 1 | 12 | 20 | 51 | 0.10 | 15 | 16 | 46 | 0.01 | 15 | 17 | 48 | 0.03 |
| scheduling 2 | 34 | 43 | 153 | 0.06 | 34 | 40 | 142 | 0.03 | 34 | 43 | 149 | 0.05 |
| scheduling 3 | 19 | 33 | 106 | 0.10 | 19 | 28 | 89 | 0.09 | 19 | 33 | 106 | 0.10 |
| scheduling 4 | 19 | 34 | 103 | 0.13 | 19 | 28 | 86 | 0.08 | 19 | 30 | 97 | 0.10 |
| scheduling 5 | 15 | 36 | 106 | 0.16 | 15 | 25 | 84 | 0.11 | 15 | 36 | 106 | 0.16 |
| projects 1 | 6 | 6 | 60 | 0.00 | 6 | 6 | 60 | 0.00 | 6 | 6 | 60 | 0.00 |
| projects 2 | 4 | 4 | 20 | 0.00 | 4 | 4 | 20 | 0.00 | 4 | 4 | 20 | 0.00 |
| projects 3 | 1 | 6 | 34 | 0.18 | 1 | 5 | 20 | 0.30 | 1 | 6 | 34 | 0.18 |
| projects 4 | 2 | 5 | 27 | 0.09 | 4 | 5 | 27 | 0.06 | 4 | 5 | 27 | 0.06 |

**Table 2.** Relative changes of important fairness and distribution metrics

| | Change of sum utility | Change of gini index |
|---|---|---|
| Leximin compared to Utilitarian | −4.163% | −19.229% |
| Rawlsian compared to Utilitarian | −11.623% | −33.028% |
| Leximin compared to Rawlsian | +13.597% | +36.980% |

When comparing the Rawlsian results with the Utilitarian results this effect is larger. The Rawlsian sum of utility is decreasing by 11.62% on average while the Gini index is decreasing by 33.03%. The Rawlsian approach leads to a more equal distribution among agents at the cost of a high loss of total welfare. As expected, the comparison of Leximin and Rawlsian approaches leads to a higher sum of utilities for Leximin but higher Gini index. The Leximin approach stands out as a compromise between Utilitarianism and Rawlsianism, as it leads to smaller losses in total welfare, but at the same time to an equal distribution among agents.

### 4.2 Runtime Overhead for Leximin Search

*"How much runtime overhead has to be faced when using Leximin?"*

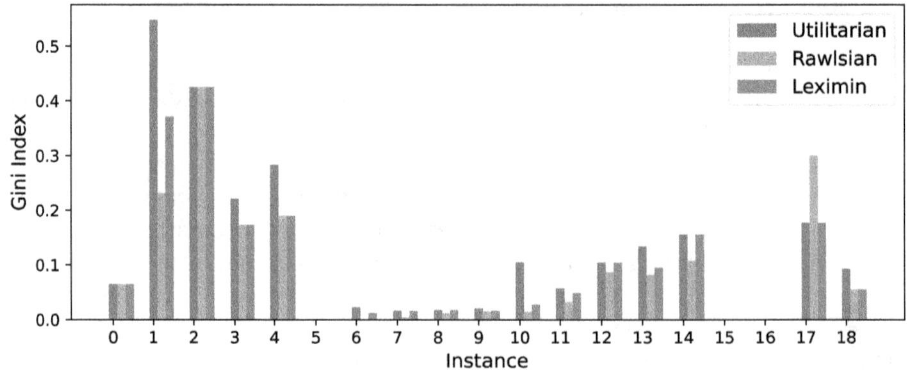


**Fig. 3.** Gini index for every instance. Leximin or Rawls can only reduce inequality.

Even though Leximin strives for a good balance between fairness (in terms of a low Gini index) and efficiency (in the sense of a high utilitarian objective), it is computationally more expensive as multiple solver calls and bookkeeping of agents to maximin values are required (see Sect. 3.2). We explored the magnitude of this overhead across different solvers.

**Table 3.** Runtimes in seconds, grouped by solver. Leximin summed overhead takes the sum of all runtimes of leximin compared to the sum of all Utilitarian runtimes.

| Solver | ∅ Leximin | ∅ Utilitarian | ∅ Rawls | Leximin Summed Overhead | GMean Overhead |
|---|---|---|---|---|---|
| Chuffed | 19.75 | 1.90 | 0.61 | 10.40% | ×2.9 |
| Coin-bc | 316.70 | 1.10 | 0.84 | 287.14% | ×9.0 |
| OR-Tools | 26.81 | 11.04 | 1.05 | 2.43% | ×2.46 |
| Gecode | 6110.82 | 569.03 | 536.14 | 10.74% | ×4.94 |

Table 3 shows the results of this experiment. As expected, Leximin search takes considerably longer than Utilitarian or Rawlsian (which is sometimes even the fastest). The extent, however, seems to be dependent on the employed solver type. SAT-based solvers are affected less severely than CBC and Gecode fares the worst – often hitting the timeout of one hour (also in the case of the simpler objectives Utilitarian and Rawlsian). The strong discrepancy between flat CBC search and Leximin seems to be caused by some intermediate models taking substantially longer. To avoid numeric fallacies when reporting benchmark ratios [8], we show the overhead as the ratio between the *summed* runtimes over all instances as well as the *geometric mean* of the individual overheads. In absolute time, Leximin only took 2.43% longer for OR-Tools while taking roughly 2–3 times longer per instance. Given the previously discussed improvement in solution quality by Leximin, this is a tradeoff to be made if possible. With Gecode and CBC, however, more improvement is necessary.

### 4.3 Combination of Utilitarian with Envy-Freeness

One key benefit of SocialCOP is its ability to integrate multiple social choice concepts, which enables the optimization of outcomes. The integration of the Utilitarian approach and envy-freeness is one potential combination of different definitions of fairness. If the Utilitarian winner is not a unique solution, but rather a plurality of solutions, then another criterion, such as envy-freeness, can be employed to distinguish between them. If this is not enforced, the solver might inadvertently end up with a Utilitarian optimal solution that is not envy-free. In a division problem (perhaps with an additional global utility part), the Utilitarian optimum does not have to satisfy envy-freeness, even though the solutions might accidentally do so. By encapsulation and specifying experimentation becomes much easier. We compare the results of a pure Utilitarian optimization (UT) with one that additionally enforces envy-freeness (UT+EF) in Table 4. Without posting EF, for most of the solvers 40% of the instances found in UT had at least one envy-pair, some of them up to three.[2] Gecode does even worse, but this is due to a timeout set to one hour. The runtime is affected by the addition of the envy-freeness constraints. Using the given instances, the runtime for Chuffed surprisingly actually decreased – probably due to faster propagation towards a Utilitarian-optimal solution. For all other solvers, the runtime increases which leads to the consideration of whether to use the combination of Utilitarian and envy-freeness based on the problem statement and available computational resources. Integrating both social concepts into an existing CCOP is not only straightforward using SocialCOP, but has the potential to significantly increase the fairness of the solution.

```
runner.model += [utilitarian_objective, optimize_utilitarian_objective]
runner.model += [envy_freeness_mixin,   enforce_envy_freeness]
```

**Table 4.** Elimination of envy-pairs by posting EF. Runtime in seconds.

| Solver | UT % Envy Pairs | UT+EF % Envy Pairs | UT Runtime | UT+EF Runtime |
|---|---|---|---|---|
| Chuffed | 40 | 0 | 5.47 | **1.41** |
| Coin-bc | 40 | 0 | **1.66** | 3.10 |
| OR-Tools | 40 | 0 | **39.49** | 63.35 |
| Gecode | 67 | 0 | **818.08** | 2400.52 |

## 5 Conclusion

We presented SocialCOP, a set of reusable building blocks for social choice in collective constraint optimization problems. These blocks include mixins for reoccurring constraints in a constraint modeling language (currently focused

[2] We only considered table assignment here since it is a division problem.

on MiniZinc) and procedural meta-searches implemented as runners in Python. Existing constraint model formulations for combinatorial problems under hard constraints – both division and shared decision – can be easily augmented with different social concepts such as Utilitarianism, Rawlisanism, Leximin, Envy-Freeness, or Pareto search. Our results were carried out on different problems and four different solvers. They suggest that social choice concepts can contribute to more equity without ignoring efficiency. In future work, we want to integrate voting-based approaches as well as strategies for repeated interactions that typically emerge in distributed solving paradigms such as DCOP where privacy matters. And, of course, we hope to contribute to fair outcomes of decision problems – starting with the shared office space at our institute.

## References


1. Bhavnani, S., Schiendorfer, A.: Towards Copeland optimization in combinatorial problems. In: Schaus, P. (ed.) CPAIOR 2022. LNCS, vol. 13292, pp. 34–43. Springer, Cham (2022). https://doi.org/10.1007/978-3-031-08011-1_4
2. Bouveret, S., Lemaıtre, M.: Finding leximin-optimal solutions using constraint programming: new algorithms and their application to combinatorial auctions. In: Computational Social Choice (COMSOC-2006), p. 49 (2006)
3. Brandt, F., Conitzer, V., Endriss, U., Lang, J., Procaccia, A.D.: Handbook of Computational Social Choice. Cambridge University Press, Cambridge (2016)
4. Bynum, M.L., et al.: Pyomo-Optimization Modeling in Python, vol. 67, 3rd edn. Springer, Cham (2021)
5. Cornelio, C., Pini, M.S., Rossi, F., Venable, K.B.: Multi-agent soft constraint aggregation via sequential voting: theoretical and experimental results. Auton. Agent. Multi-Agent Syst. **33**(1–2), 159–191 (2019)
6. Dalla Pozza, G., Rossi, F., Venable, K.B.: Multi-agent soft constraint aggregation: a sequential approach. In: Proceedings of the 3rd International Conference on Agents and Artificial Intelligence, ICAART 2011, vol. 11 (2010)
7. Ek, A.J.P.: High-level modelling and solving for online, real-time, and multiagent combinatorial optimisation. Ph.D. thesis, Monash University (2022)
8. Fleming, P.J., Wallace, J.J.: How not to lie with statistics: the correct way to summarize benchmark results. Commun. ACM **29**(3), 218–221 (1986)
9. Fourer, R., Gay, D., Kernighan, B.: AMPL: A Mathematical Programing Language, sw wallace, ed (1989)
10. Guns, T.: Increasing modeling language convenience with a universal n-dimensional array, cppy as python-embedded example. In: Proceedings of the 18th workshop on Constraint Modelling and Reformulation at CP (Modref 2019), vol. 19 (2019)
11. Guns, T., Stuckey, P.J., Tack, G.: Solution dominance over constraint satisfaction problems. arXiv preprint arXiv:1812.09207 (2018)
12. John Hooker: Optimization models for fairness (2014). https://johnhooker.tepper.cmu.edu/fairnessFITmonash.pdf
13. Matt, P.-A., Toni, F.: Egalitarian allocations of indivisible resources: theory and computation. In: Klusch, M., Rovatsos, M., Payne, T.R. (eds.) CIA 2006. LNCS (LNAI), vol. 4149, pp. 243–257. Springer, Heidelberg (2006). https://doi.org/10.1007/11839354_18
14. Moulin, H.: Fair Division and Collective Welfare. MIT Press, Cambridge (2004)

15. Nethercote, N., Stuckey, P.J., Becket, R., Brand, S., Duck, G.J., Tack, G.: MiniZinc: towards a standard CP modelling language. In: Bessière, C. (ed.) CP 2007. LNCS, vol. 4741, pp. 529–543. Springer, Heidelberg (2007). https://doi.org/10.1007/978-3-540-74970-7_38
16. Plaut, B., Roughgarden, T.: Almost envy-freeness with general valuations. http://arxiv.org/pdf/1707.04769v3
17. Rendl, A., Guns, T., Stuckey, P.J., Tack, G.: MiniSearch: a solver-independent meta-search language for MiniZinc. In: Pesant, G. (ed.) CP 2015. LNCS, vol. 9255, pp. 376–392. Springer, Cham (2015). https://doi.org/10.1007/978-3-319-23219-5_27
18. Rossi, F.: Collective decision making: a great opportunity for constraint reasoning. Constraints **19**(2), 186–194 (2014)
19. Rossi, F., Venable, K.B., Walsh, T.: Preferences in constraint satisfaction and optimization. AI Mag. **29**(4), 58–68 (2008)
20. Savulescu, J., Persson, I., Wilkinson, D.: Utilitarianism and the pandemic. Bioethics **34**(6), 620–632 (2020). https://doi.org/10.1111/bioe.12771
21. Schiendorfer, A., Knapp, A., Anders, G., Reif, W.: MiniBrass: soft constraints for MiniZinc. Constraints **23**, 403–450 (2018)
22. Schiendorfer, A., Reif, W.: Reducing bias in preference aggregation for multiagent soft constraint problems. In: Schiex, T., de Givry, S. (eds.) CP 2019. LNCS, vol. 11802, pp. 510–526. Springer, Cham (2019). https://doi.org/10.1007/978-3-030-30048-7_30
23. Stuckey, P.J., Tack, G.: MiniZinc with functions. In: Gomes, C., Sellmann, M. (eds.) CPAIOR 2013. LNCS, vol. 7874, pp. 268–283. Springer, Heidelberg (2013). https://doi.org/10.1007/978-3-642-38171-3_18
24. Stuckey, P.J., Tack, G., de la Banda, M.G.: Twelve years of MiniZinc
25. Von Neumann, J., Morgenstern, O.: Theory of Games and Economic Behavior, 2nd rev (1947)
26. Xinying Chen, V., Hooker, J.N.: A guide to formulating fairness in an optimization model. Ann. Oper. Res. **326**(1), 1–39 (2023). https://doi.org/10.1007/s10479-023-05264-y

# Context-Specific Selection of Commonsense Knowledge Using Large Language Models

Oliver Jakobs and Claudia Schon(✉)

Hochschule Trier, Trier, Germany
c.schon@hochschule-trier.de

**Abstract.** In the field of automated reasoning, practical applications often face a significant challenge: knowledge bases are typically too large to be fully processed by theorem provers. To still be able to prove that a given goal follows from a large knowledge base, selection techniques are used to determine the parts of the knowledge base that are relevant to the goal. Traditional selection techniques used for this task are usually syntax-based and often overlook a crucial aspect—the meaning of symbol names and axioms. Especially in commonsense reasoning scenarios, the meaning embedded in the symbol names provides invaluable insights. For example, in a proof task using the symbol name *cow*, it intuitively makes more sense to select formulae using the symbol name *calf* than formulae using the symbol name *weapon*. To address this gap, our paper introduces a selection technique that exploits the capabilities of large language models. This technique focuses on contextually related formulae, closely aligning the selected part of the knowledge base with the context of the goal. The approach is implemented and we present a series of experiments that show promising results.

**Keywords:** Automated Reasoning · Selection techniques · Commonsense Reasoning · First-order logic

## 1 Introduction

One task in first-order logic theorem proving is to find a proof that a given formula follows from a knowledge base. In practice, it is often the case that the knowledge bases under consideration are very large and also very complex. As a result, theorem provers often fail to solve this proof task with the complete knowledge base. If a proof has been found, in most cases it turns out that only a very small subset of the knowledge base was necessary for the proof. Selection techniques, which are commonly used in first-order logic large theory reasoning, are based on this observation: to check if a given formula follows from a knowledge base, selection techniques attempt to identify a subset of the knowledge base that is sufficient for the proof.


A. Hotho and S. Rudolph (Eds.): KI 2024, LNAI 14992, pp. 218–231, 2024.
https://doi.org/10.1007/978-3-031-70893-0_16

Traditional approaches for this task are syntax-based. An example is the SInE selection that relies heavily on syntactic criteria such as the frequency of symbols to identify relevant axioms. However, these techniques overlook the deeper meanings embedded within symbol names and axioms. Typically, it makes no difference to these syntax-based selection techniques whether a symbol has the name *c* or the name *cow*. But in many areas, such as commonsense reasoning, symbols have meaningful names. For example, Adimen-SUMO contains the symbol names *c__Cow*, *c__Calf* and *c__Weapon*. If a proof task is given in which we want to check if a formula that uses the symbol *c__Calf* follows from a knowledge base, it intuitively makes more sense to select formulae of the knowledge base that use the symbol *c__Cow* than those that use the symbol *c__Weapon*. It is often not just the symbol names that are meaningful. Usually, the meaning of the statement of a formula goes far beyond the combination of symbols with logical operators. This represents a further level of information available in a knowledge base that is still completely ignored by syntax-based selection techniques.

In the field of natural language processing, there are techniques, such as large language models, which could help to capture the meaning of formulas and the context of a formula and thus make it usable for selection techniques.

This paper proposes an approach that uses techniques from natural language processing for axiom selection. We use SBERT, a large language model which computes sentence embeddings which are vectorizations of natural language sentences, to achieve a context-specific selection of axioms. This so called *SeVEn* selection vectorizes the formulae with the help of SBERT such that formulae belonging to the same context are represented by similar vectors. To stay with our example from the farm animal context, a formula that represents information about cows is represented by a similar vector as a formula representing information about calves.

This SeVEn selection is a purely statistical selection strategy. We also present a hybrid selection strategy that combines a known syntactic selection strategy with the statistical SeVEN selection. Through a series of experiments, we demonstrate that this hybrid selection strategy yields promising results in automated theorem proving tasks.

The main contributions of this paper are:

- The introduction of a selection strategy for commonsense knowledge which uses large language models to perform a context-specific selection.
- A hybrid selection strategy combining syntax-based selection with a statistical selection.
- An experimental evaluation using commonsense reasoning benchmarks for automated theorem proving demonstrating that the hybrid selection strategy is superior to syntactic selection or context-specific selection alone.

The paper is structured as follows: after discussing related work in Sect. 2, we introduce some preliminaries in Sect. 3. We then introduce the SeVEn selection strategy in Sect. 4. In Sect. 5, we introduce experimental results and conclude with a discussion of future work in Sect. 6.

## 2 Related Work

There are several approaches for selecting parts relevant to a given goal in a knowledge base. Most consider only on axiom syntax and ignore the meaning of symbol names in the knowledge base.

An example for a syntactic selection strategy is SInE [6]. SInE does not take the meaning of symbol names into account. Nevertheless, SInE is successfully used in many provers. Since we combine our approach with SInE, a more detailed description of SInE can be found in Sect. 3. Further syntactic selection strategies are lightweight relevance filtering [11] and axiom relevance ordering [18].

A model-based method for choosing axioms known as SRASS [23] uses a semantic strategy. It involves calculating models for subsets of axioms and systematically expanding these subsets. In addition to presenting three new selection approaches the authors of [10] propose metrics for the evaluation of selection techniques. However, none of the selection strategies mentioned so far take the meaning of symbol names into account.

With regard to the problem of premise selection for large mathematical libraries [1], there are interesting approaches based on machine learning. DeepMath [7] is an approach using neural networks for premise selection without the need to use hand-engineered features. $\text{ATP}_{\text{BOOST}}$ improves premise selection by treating the task as a binary classification problem and using the XGBoost algorithm. Another approach uses recurrent neural networks to learn dependencies between premises [14]. In [15] the transformer architecture is applied to premise selection and outperforms previously established approaches based on recurrent neural networks. The transformer architecture is also used for premise selection in Magnushammer [12], which demonstrates improvements in proof success rates by retrieving and ranking relevant premises.

In the Semantic Web, the meaning of symbol names has been studied [19], leading to the insight that the semantic content embedded in the names of IRIs (Internationalised Resource Identifiers) embodies a social semantics that aligns with the formal meaning of the resources they represent. This insight suggests that symbol names' meanings may be helpful in selection and reasoning.

A way to consider the meaning of symbol names in presented in [21], where a purely statistical selection strategy is introduced. The approach uses a word embedding to select axioms based on the relative similarities of their vector representations. To achieve this, the axioms are vectorized by first encoding the individual symbol names with a word embedding and then combining these symbol vectors to obtain the final vector representation. While promising for selecting relevant background knowledge in creativity tests, the approach performs worse compared to SInE in automated theorem proving [20].

Another possibility is to combine both syntactic and statistical selection into a hybrid selection approach. An example of this is the Similarity SInE strategy [5], an enhancement of the syntactic selection technique SInE. Unlike the original SInE method, which chooses axioms solely based on the frequency of symbols across axioms, Similarity SInE broadens this criterion by considering similar symbols using word embeddings.

## 3 Preliminaries

In this paper, we focus on first-order reasoning tasks involving a large set of axioms known as the knowledge base and a goal $G$. The objective is to show that the knowledge base logically entails $G$. In classical logic, this task can be reduced to showing that the conjunction of the knowledge base and the negation of $G$ is unsatisfiable. The field of automated theorem proving has made impressive progress in solving this task. However, once the underlying knowledge base reaches a certain size, it is no longer feasible to consider the entire knowledge base when checking entailments, as this would overwhelm the automated theorem prover. In most cases, only a very small part of the knowledge base is needed to prove an entailment. Therefore, when considering a goal and a very large knowledge base, selection techniques are usually used to select a subset of the knowledge base from which the goal follows. Most selection strategies are incomplete: Even if there is a subset of the knowledge base from which the considered goal follows, it is possible that the selection strategy selects too few axioms and that no proof is possible with the selected axioms. In the following we denote the set of predicate and function symbols occurring in an axiom $A$ by $sym(A)$. We furthermore slightly abuse notation and use $sym(KB)$ to denote the set of all predicate and functions symbols occurring in a knowledge base $KB$.

Since we will not only consider the SeVEn selection presented in this paper on its own, but also as a hybrid approach combining SeVEn with syntax-based SInE selection [6], we will now introduce the basic idea of SInE selection.

### 3.1 SInE Selection

The SInE selection [6] specifies for each axiom of the knowledge base which symbols can trigger the selection of this axiom. This is done with the help of a relation called *triggers*, where $triggers(s, A)$ means that symbol $s$ triggers the selection of axiom $A$.

With SInE selection, only symbols that occur in an axiom can trigger its selection. To prevent common symbols such as *subClass* from triggering the selection of a large number of axioms, SInE allows only the symbol in the axiom that occurs least often in the knowledge base to trigger the selection of the axiom.

**Definition 1 (Trigger relation for the SInE selection [6]).** *Let $KB$ be a knowledge base, $A \in KB$ an axiom and $s$ be a predicate or function symbol. Let furthermore $occ(s)$ denote the number of axioms in which $s$ occurs in $KB$. Then the triggers relation is defined as follows:*

$$triggers(s, A) \text{ iff for all symbols } s' \text{ occurring in } A \text{ we have } occ(s) \leq occ(s')$$

The triggers relation can be softened by a so called *benevolence* parameter, so that not only the least frequent symbol, but also symbols $t$ times more frequent than this symbol can trigger the selection of this axiom ($t \geq 1$).

Starting with a goal to be proven, in the first step, all symbols occurring in the goal are considered to be triggered. Next, axioms are selected step by step that are triggered by the symbols that have already been triggered. In each step, the symbols that occur in newly selected axioms are considered triggered, which can lead to the selection of further axioms in the next step. The following definition describes this.

**Definition 2 (Trigger-based selection** [6]**).** *Let KB be a knowledge base, A be an axiom in KB and s be a predicate or function symbol occurring in KB. Let furthermore G be a goal to be proven from KB.*

1. *If s is a symbol occurring in goal G, then s is 0-step triggered.*
2. *If s is k-step triggered and s triggers A (triggers(s, A)), then A is k + 1-step triggered.*
3. *If A is k-step triggered and s occurs in A, then s is k-step triggered, too.*

*An axiom or a symbol is called triggered if it is k-step triggered for some* $k \geq 0$.

If all $k$-step triggered axioms are selected, we refer to this in the following as an SInE selection with recursion depth $k$.

SInE selection is used successfully by many first-order logic provers. A disadvantage of SInE selection is that it completely ignores the meaning of symbol names. Since the meaning of symbol names provides valuable information for selection in many domains, we will present a hybrid selection technique that combines SInE selection with a large language model in such a way that the meaning of symbol names is included in the selection process.

Next we will introduce SeVEn selection, which is a statistical selection strategy based on sentence embeddings.

## 4 SeVEn Selection

We will introduce the context-specific SeVEn selection. Figure 1 shows an overview of the selection process.

The first step is the translation of all axioms of the knowledge base and the goal to be proven into natural language sentences. Details of this translation step are given in Sect. 4.1. In the next step, a vector representation for all axioms in the knowledge base and the goal are computed based on these natural language sentences. For this step, we use Sentence-BERT [17]. Sentence-BERT is a pre-trained language model that computes sentence embeddings, which are vector representations for natural language sentences. The advantage of this vectorization method is that axioms belonging to a similar context are mapped to similar vectors. For example, the vector representation of an axiom expressing information about *cows* is closer to the vector representation of an axiom expressing information about *calves* than to the vector representation of an axiom about *weapons*. Once the vector representations of the axioms of the knowledge base and the goal to be proved have been computed, the axioms whose vector representation is most similar to the vector representation of the goal can be selected.

We will give more detailed information on the vectorization step and how to determine similarities of vector representations of axioms in Sect. 4.2. In practice, a knowledge base is often used as background knowledge for a large number of different goals. It is important to note that the knowledge base only needs to be vectorized once, so this can be seen as a pre-processing step. At query time, only the usually small goal needs to be vectorized.

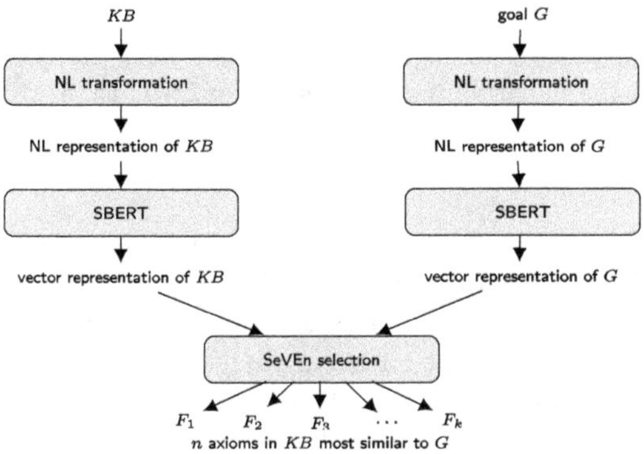


**Fig. 1.** Overview of the SeVEn selection to select the $n$ axioms most relevant for goal $G$ from knowledge base ($KB$).

## 4.1 Translating Axioms

In this section, we give details on the transformation of axioms into natural language sentences. The translation of axioms into natural language depends on the underlying knowledge base. As we used Adimen-SUMO [2] for the evaluation of the presented selection method, our translation of axioms into natural language has been optimized for Adimen SUMO. However, the translation could easily be transferred to other knowledge bases or a different type of translation could be used (e.g., using pre-trained language models).

In a first step, we parse an axiom, transform it into negation normal form and generate an abstract syntax tree (AST) for it. The AST presents the axiom in such a way that every inner node represents a predicate or function symbol or an operator and every leaf of the tree may only contain a variable or a constant.

To generate a sentence, the AST is traversed with a depth-first search to recursively translate every node. For that purpose, we need to assign every symbol a node may contain a corresponding sentence fragment in which we insert the translation of the child nodes. See Fig. 2 for an example.

Quantifiers and the logical operator for conjunction, disjunction, implication and equivalence are translated using an English description, e.g., the symbol for the binary operation `implies (=>)` translates to `"if ..., then ..."`. Variables are not translated at all; that is, the variable `X` translates to `"X"`. A detailed list of these translations can be found in [8].

$$\forall X(subClass(X,canine) \rightarrow \neg subClass(X,feline))$$

(a) First-order logic axiom.

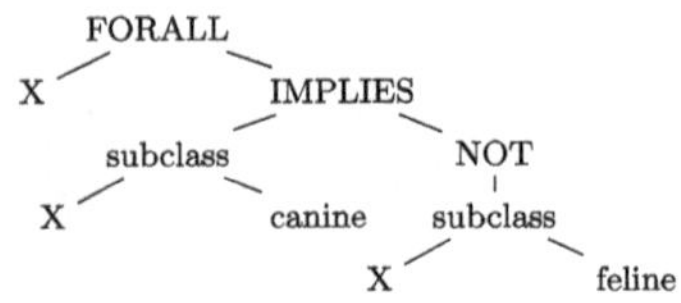


(b) Abstract syntax tree (AST) for the axiom on the left.

**Fig. 2.** Example of a first-order logic axiom on the left, the corresponding AST on the right resulting in the natural language sentence: `If X is a subclass of canine, then X is not a subclass of feline for every X.`

The translation of predicates, functions and constants is a more involved process. Adimen-SUMO contains 607 predicate symbols, 73 function symbols (excluding constants), and 3238 constants. For constants we are able to automate the translation and generate a fitting sentence fragment for each constant symbol. In contrast, functions and predicates require a mapping for translation. This mapping was created by hand with the help of the comments provided in Adimen-SUMO and the context in which the symbols are used. Due to the transformation of axioms into negation normal for, negation can only occur in front of predicates. The translation of negated predicates was created semi-automatically. A detailed description of the translation of predicates, functions and constants is presented in [8].

The presented approach for translating axioms into natural language is just one option. In future work, we plan to use a large language model.

### 4.2 SeVEn: Sentence-Based Vector Encoding

A common approach in natural language processing (NLP) to quantify semantic similarities between words is the use of distributional semantics of natural language. The concept of distributional semantics is based on the distributional hypothesis [13], which posits that words with similar distributional properties in large text corpora also have similar meanings. In simpler terms, words that appear in a similar context are likely to have a similar meaning. Or like J. R. Firth put it: *A word is characterized by the company it keeps* [16].

To make use of the distributional hypothesis in axiom selection, [21] introduces a vector-based selection which encodes axioms into vectors in $\mathbb{R}^n$. This is done by mapping symbols occurring in an axiom to words in the vocabulary of a word embedding and by combining the respective vectors. In this way, axioms whose symbol names belong to similar contexts are mapped to similar vectors. One disadvantage of this approach is that logical connectives are not taken into account during the vectorization. Furthermore, the known disadvantages of word embeddings are directly transferred to the result of the vectorization of the approach presented in [21]: words that do not belong to the vocabulary of the used word embedding cannot be represented. Homonyms, words with different meanings, such as the word *bank* or *park*, are mapped to the same vector.

In contrast to this, we encode the axioms as a whole using sentence embeddings [3,9] leading to the **Se**ntence-based **V**ector **En**coding (SeVEn). Given the set of natural language sentences $S$, a sentence embedding can be seen as a mapping $f\colon S \to \mathbb{R}^n$, which maps natural language sentences to vectors in $\mathbb{R}^n$. One advantage of using sentence embeddings is that it is not necessary to map symbol names to the vocabulary of a word embedding. Another advantage of the used BERT sentence embeddings is that the language model BERT [4] uses a WordPiece model [24] which allows it to process sentences without being constrained by a fixed vocabulary. BERT still relies on a vocabulary but has a special technique to handle unknown words. In addition, whole sentences are vectorized using sentence embeddings, so that homonyms are less problematic as they are vectorised together with a context.

Vectorizing an axiom with SeVEn involves two steps: translating the axiom into a descriptive sentence and then encoding this sentence using a sentence embedding.

**Definition 3 (Sentence-based vector encoding (SeVEn)).** *Let KB be a knowledge base with* $KB = \{A_1, ..., A_n\}$, $n \in \mathbb{N}$. *Let furthermore* $S$ *be the set of all natural language sentences and* $f\colon S \to \mathbb{R}^n$ *be a sentence embedding. A sentence-based vector encoding* $v_A$ *of an axiom* $A \in KB$ *can be constructed with the following steps:*

1. *Translate the axiom* $A$ *into a natural language sentence* $s \in S$.
2. *Encode sentence* $s$ *by vector* $v_A := f(s)$.

*A sentence-based vector encoding of KB is denoted by* $V_{KB} = \{v_{A_1}, ..., v_{A_n}\}$.

Note that we leave the way axioms are translated into natural language sentences open in the previous definition. One way to perform this transformation is described in Sect. 4.1. However any other method can be used for this step. The result of the vectorization depends on the choice of the translation method from logic to natural language and the choice of sentence embedding.

To quantify the similarity between two axioms, we calculate the cosine similarity of their vector representations.

$$cos\_sim(u, v) = \frac{u \cdot v}{||u||||v||}$$

We now adapt the definition of vector-based selection from [21] such that it can be used to select axioms similar to a given goal using a sentence-based vector representation of both the knowledge base and the goal.

**Definition 4 (SeVEn selection).** *Let KB be a knowledge base,* $G$ *be a goal with* $sym(G) \subseteq sym(KB)$. *Let* $V_{KB} = \{v_{A_1}, ..., v_{A_m}\}$ *be a sentence-based vector encoding of KB and* $v_G$ *a sentence-based vector encoding for* $G$ *both computed using the same method to translate axioms to natural language and the same* *sentence embedding. For $n \in \mathbb{N}$, $n \leq |KB|$ the n axioms in KB most similar to G denoted as $mostsimilar(KB, G, n)$ are defined as the following set:*

$$\{A_1, ..., A_n \mid \{A_1, ..., A_n\} \subseteq KB \text{ and } \forall A' \in KB \setminus \{A_1, ..., A_n\}$$
$$cos\_sim(v_{A'}, v_G) \leq \min_{i=1,...,n} cos\_sim(v_{A_i}, v_G)\}.$$

*For KB and G given as above described and $n \in \mathbb{N}$, the result of SeVEn selection is defined as:*

$$seven(KB, G, n) := \begin{cases} KB & \text{if } n \geq |KB| \\ mostsimilar(KB, G, n) & \text{otherwise} \end{cases}$$

### 4.3 Combining Vector-Based Selection with Syntactic Selection

A shortcoming when selecting based on vector representations is that axioms containing only structural predicates, like *agent*, *instance* or *subClass*, are not actually similar to the goal and are thus disregarded. An example for such an axiom would be an axiom stating the transitivity of *subClass*. However, these axioms describe the mathematical structure of the ontology and are often essential for the automated theorem prover to be able to find a proof. A possible solution to this problem is the combination of statistical and syntactic selection.

We propose a hybrid approach, called *SeVEn-union-SInE*, which combines SeVEn with the syntactic selection strategy SInE. First, the goal is broadened by adding axioms selected with SeVEn to it. In a second step, SInE then selects axioms based on all goals (including the original one).

In the following, we use $sine(KB, \{A_1, ..., A_m\}), m \in \mathbb{N}$ to denote the set of axioms obtained by applying the SInE selection on a set of goals $\{A_1, ...A_m\}$. As usual, a set of axioms or goals corresponds to the conjunction of its elements. For simplicity, we omit the SInE selection parameters.

**Definition 5 (SeVEn-union-SInE selection).** *Let KB be a knowledge base and G be a goal with $sym(G) \subseteq sym(KB)$. The combined selection* union_select *is then defined as:*

$$union\_select(KB, G, n) = sine(KB, G \cup seven(KB, G, n))$$

SeVEn selection as well as SeVEn-union-SInE selection are both implemented and the next section presents first experimental results.

## 5 Evaluation

The described selection strategies are implemented[1]. For our evaluation, we use the theorem prover E [22], a superposition-based prover, which includes the possibility to use the SInE selection. For evaluation purposes, we need a large

[1] The code is available in the following git repository: https://gitlab.rlp.net/clsc0474908343/seven.

knowledge base along with goals to prove. Furthermore, it is important that the use of selection techniques is necessary to prove the goals, and that the symbol names used in the knowledge base are close to natural language words. These requirements are met by the Adimen SUMO [2] knowledge base, which comes with a number of different goals of which we use the 8100 whitebox tests. In a first step, we determine the set of whitebox tests, E is not able to prove within 15 s without selection and randomly select 1000 of these problems. The resulting 1000 whitebox tests are used during our evaluation.

### 5.1 Setting a Baseline: SInE

We compare the selection strategies presented in this paper with the SInE selection. For the 1000 randomly selected goals, we selected with SInE with different values of recursion depth and benevolence and then called E with a timeout of 15 s. Figure 3 shows the results.

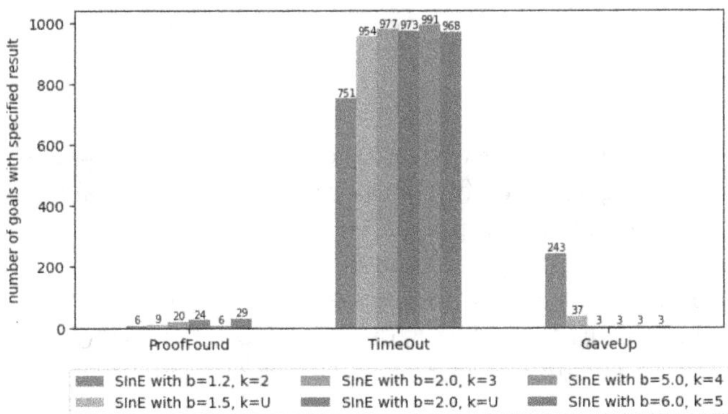


**Fig. 3.** Results of selecting with SInE on 1000 goals that E couldn't solve without selection and calling E afterwards. Various SInE configurations were tested with different values of the benevolence parameter $b$ and recursion depth $k$. The value $U$ (unlimited) indicates that the parameter is set to its maximum value. *Gave up* means that all necessary axioms were selected, and E recognizes this since SInE is integrated into its implementation.

We can see that the results depend strongly on the parameters of the selection. We get the best result (29 proofs) for the value 6.0 for benevolence and recursion depth 5.

### 5.2 Evaluation of SeVEn

In the next step, we used the introduced SeVEn selection strategy to select axioms for the 1000 goals described above and passed the selected axioms to E (timeout 15 s). The results are shown in Fig. 4.

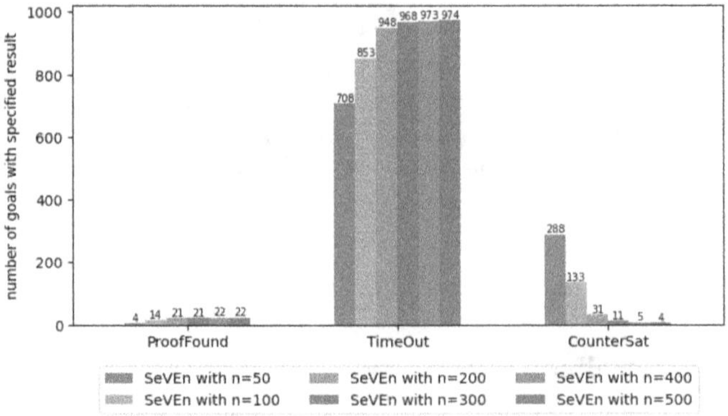


**Fig. 4.** Results of selecting with SeVEn on 1000 goals that E could not solve without selection and calling E on the selected axioms. Different values of parameter $n$ (number of selected axioms) were considered. Goals where the selected axioms don't contain all axioms necessary for a proof are marked as *CounterSat*, corresponding to the *Gave up* case in Fig. 3. With SInE selection, E recognizes a selection was made and returns *Gave up* when not enough axioms are selected. SeVEn isn't integrated into E, so E returns *CounterSat*.

Figure 4 shows that starting with 200 selected axioms ($n = 200$), E was able to find proofs for 21 goals. We furthermore see, that increasing $n$ beyond 200 does not lead to a significant improvement, which is why we did not carry out any experiments for values greater than 500 for $n$.

Comparing Fig. 3 and Fig. 4, we can see that Seven with $n = 200$ performs slightly worse than SInE. Since SeVEn requires the vectorization of the axioms, SeVEn alone has no advantage over SInE.

### 5.3 Evaluating the Hybrid Approach: SeVEn-Union-SInE

In further experiments, we considered the hybrid selection technique SeVEn-union-SInE on the 1000 randomly selected goals we used for the previous experiments. As described above, in this hybrid selection a selection with SeVEN is switched before a selection with SInE. This gives us many different parameter combinations. For our experiments, we combined values between $n = 20$ and $n = 180$ for SeVEn with the three best parameter combinations for SInE from Fig. 3. The results of these experiments are depicted on the left side in Fig. 5.

The best result is obtained by combining SeVEn with $n = 160$ and SInE with a benevolence of 2.0 and unlimited recursion depth. The 167 proofs found in this way are far above the best result of SInE alone with 29 and also SeVEn alone with 22 proofs found. We observe that the hybrid selection approach SeVEn-union-SInE, represents a significant improvement over syntactic selection (SInE) and also statistical selection (SeVEn) alone.

Another interesting observation is that the combination of SeVEn with the best SInE configuration ($b = 6.0$ and $k = 5$) from Fig. 3 performs very poorly.

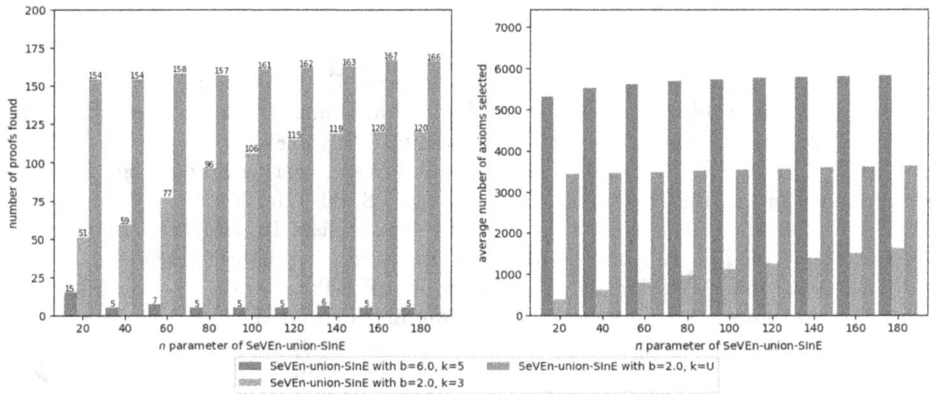


**Fig. 5.** Results of the hybrid SeVEn-union-SInE selection on 1000 goals that E couldn't solve without selection and calling E on the selected axioms. Different values of the parameter $n$ were tested. Each test used a different SInE configuration with varying benevolence $b$ and recursion depth $k$ values. $U$ indicates an unlimited parameter. Number of proofs found on the left, average number of axioms selected on the right.

Here, the combination with SeVEn actually leads to a deterioration compared to the results from Fig. 3. A possible explanation for this could be the high number of selected axioms in this selection. As the right side of Fig. 5 shows, this selection variant selected significantly more axioms on average than selections with other parameter configurations. In comparison, the best performing selection selected significantly fewer axioms. We can see that *much helps much* does not apply to selection, as the theorem prover can be overwhelmed by large sets of axioms. Rather, it is desirable to have a focused selection.

## 6 Conclusion and Future Work

In this paper, we have introduced selection strategies for automated theorem proving with large knowledge bases that take into account the meaning of symbol names and axioms using a large language model. Our experiments demonstrate that the hybrid SeVEn-union-SInE approach, combining both statistical and syntactical selection seems to be the most promising. Due to its dual-strategy, it inherently possesses a large number of tunable parameters, offering numerous possibilities for optimization. Future work will focus on this hybrid approach. We aim to methodically test different parameter configurations to deepen our understanding of how individual parameters affect overall effectiveness. Furthermore, we will optimize the SeVEn strategy by improving the translation of formulae into natural language sentences. This could be done by finetuning a large language model.

## References

1. Alama, J., Heskes, T., Kühlwein, D., Tsivtsivadze, E., Urban, J.: Premise selection for mathematics by corpus analysis and kernel methods. J. Autom. Reason. **52**(2), 191–213 (2014). https://doi.org/10.1007/S10817-013-9286-5
2. Álvez, J., Lucio, P., Rigau, G.: Adimen-sumo: reengineering an ontology for first-order reasoning. Int. J. Semant. Web Inf. Syst. **8**, 80–116 (2012)
3. Cer, D., et al.: Universal sentence encoder for English. In: EMNLP (Demonstration), pp. 169–174. Association for Computational Linguistics (2018)
4. Devlin, J., Chang, M., Lee, K., Toutanova, K.: BERT: pre-training of deep bidirectional transformers for language understanding. In: NAACL-HLT (1), pp. 4171–4186. Association for Computational Linguistics (2019)
5. Furbach, U., Krämer, T., Schon, C.: Names are not just sound and smoke: word embeddings for axiom selection. In: Fontaine, P. (ed.) CADE 2019. LNCS (LNAI), vol. 11716, pp. 250–268. Springer, Cham (2019). https://doi.org/10.1007/978-3-030-29436-6_15
6. Hoder, K., Voronkov, A.: Sine qua non for large theory reasoning. In: Bjørner, N., Sofronie-Stokkermans, V. (eds.) CADE 2011. LNCS (LNAI), vol. 6803, pp. 299–314. Springer, Heidelberg (2011). https://doi.org/10.1007/978-3-642-22438-6_23
7. Irving, G., Szegedy, C., Alemi, A.A., Eén, N., Chollet, F., Urban, J.: Deepmath - deep sequence models for premise selection. In: Lee, D.D., Sugiyama, M., von Luxburg, U., Guyon, I., Garnett, R. (eds.) Advances in Neural Information Processing Systems 29: Annual Conference on Neural Information Processing Systems 2016, Barcelona, Spain, pp. 2235–2243 (2016)
8. Jakobs, O.: Context-specific selection: using large language models to select commonsense knowledge from knowledge bases. Bachelor's thesis, Hochschule Trier (2023)
9. Kiros, R., et al.: Skip-thought vectors. In: Cortes, C., Lawrence, N.D., Lee, D.D., Sugiyama, M., Garnett, R. (eds.) Advances in Neural Information Processing Systems 28: Annual Conference on Neural Information Processing Systems 2015, Canada, Montreal, Quebec, pp. 3294–3302 (2015)
10. Liu, Q., Wu, Z., Wang, Z., Sutcliffe, G.: Evaluation of axiom selection techniques. In: PAAR+SC$^2$@IJCAR. CEUR Workshop Proceedings, vol. 2752, pp. 63–75. CEUR-WS.org (2020)
11. Meng, J., Paulson, L.C.: Lightweight relevance filtering for machine-generated resolution problems. J. Appl. Logic **7**(1), 41–57 (2009)
12. Mikula, M., et al.: Magnushammer: a transformer-based approach to premise selection. CoRR arXiv:abs/2303.04488 (2023). https://doi.org/10.48550/ARXIV.2303.04488
13. Miller, G.A., Charles, W.G.: Contextual correlates of semantic similarity. Lang. Cognit. Process. **6**(1), 1–28 (1991)
14. Piotrowski, B., Urban, J.: Stateful premise selection by recurrent neural networks. In: Albert, E., Kovács, L. (eds.) LPAR 2020: 23rd International Conference on Logic for Programming, Artificial Intelligence and Reasoning, Alicante, Spain, 22–27 May 2020. EPiC Series in Computing, vol. 73, pp. 409–422. EasyChair (2020). https://doi.org/10.29007/J5HD
15. Proroković, K., Wand, M., Schmidhuber, J.: Improving stateful premise selection with transformers. In: Kamareddine, F., Sacerdoti Coen, C. (eds.) CICM 2021. LNCS (LNAI), vol. 12833, pp. 84–89. Springer, Cham (2021). https://doi.org/10.1007/978-3-030-81097-9_6

16. Firth, J.: A synopsis of linguistic theory, 1930-1955. Studies in Linguistic Analysis (1957)
17. Reimers, N., Gurevych, I.: Sentence-BERT: sentence embeddings using Siamese BERT-networks. In: EMNLP/IJCNLP (1), pp. 3980–3990. Association for Computational Linguistics (2019)
18. Roederer, A., Puzis, Y., Sutcliffe, G.: Divvy: an ATP meta-system based on axiom relevance ordering. In: Schmidt, R.A. (ed.) CADE 2009. LNCS (LNAI), vol. 5663, pp. 157–162. Springer, Heidelberg (2009). https://doi.org/10.1007/978-3-642-02959-2_13
19. de Rooij, S., Beek, W., Bloem, P., van Harmelen, F., Schlobach, S.: Are names meaningful? Quantifying social meaning on the semantic web. In: Groth, P., et al. (eds.) ISWC 2016. LNCS, vol. 9981, pp. 184–199. Springer, Cham (2016). https://doi.org/10.1007/978-3-319-46523-4_12
20. Schon, C.: Selection strategies for commonsense knowledge (2022). https://doi.org/10.48550/ARXIV.2202.09163
21. Schon, C.: Associative reasoning for commonsense knowledge. In: Seipel, D., Steen, A. (eds.) KI 2023. LNCS, vol. 14236, pp. 170–183. Springer, Cham (2023). https://doi.org/10.1007/978-3-031-42608-7_14
22. Schulz, S.: System description: E 1.8. In: McMillan, K., Middeldorp, A., Voronkov, A. (eds.) LPAR 2013. LNCS, vol. 8312, pp. 735–743. Springer, Heidelberg (2013). https://doi.org/10.1007/978-3-642-45221-5_49
23. Sutcliffe, G., Puzis, Y.: SRASS - a semantic relevance axiom selection system. In: Pfenning, F. (ed.) CADE 2007. LNCS (LNAI), vol. 4603, pp. 295–310. Springer, Heidelberg (2007). https://doi.org/10.1007/978-3-540-73595-3_20
24. Wu, Y., et al.: Google's neural machine translation system: bridging the gap between human and machine translation. CoRR arXiv:abs/1609.08144 (2016)

# Graph2RETA: Graph Neural Networks for Pick-up and Delivery Route Prediction and Arrival Time Estimation

Wilson Sentanoe[1(✉)], Sreyashi Saha[1], Yasas Diyanananda[1], Aqsa Manzoor[1], Buddhika Dasanayake[1], and Daniela Thyssens[2]

[1] University of Hildesheim, Hildesheim, Germany
{sentanoe,sahas,diyanananda,manzoor,dasanayake}@uni-hildesheim.de
[2] Institute for Computer Science, University of Hildesheim, Hildesheim, Germany
thyssens@ismll.uni-hildesheim.de

**Abstract.** This research proposes an effective way to address the issues faced by pick-up and delivery services. The real-world variables that affect delivery routes are frequently overlooked by traditional routing technologies, resulting in differences between intended and actual trajectories. Similarly, the issue of forecasting the Estimated Time of Arrival involves unique challenges due to its high dimensionality. We suggest an integrated predictive modeling methodology that tackles routing prediction in a dynamic environment and ETA prediction at the same time to overcome these difficulties. Our method, Graph2RETA, uses a dynamic spatial-temporal graph-based model to forecast delivery workers' future routing behaviors while integrating route inference into ETA prediction. Graph2RETA leverages rich decision context and spatial-temporal information to improve the prediction accuracy of the concurrent state-of-the-art while capturing dynamic interactions between workers and timesteps by incorporating the underlying graph structure and features.

**Keywords:** Pick-up and Delivery · Route Prediction · Arrival Time Prediction

## 1 Introduction

Pick-up and delivery (P&D) services, such as food delivery and logistics, have expanded rapidly in recent years, providing exceptional convenience in the face of urbanisation and changing consumer needs. Despite significant investment in routing systems by service providers, practitioners have observed significant differences between planned and real worker routes [5,20]. Studies undertaken in some parts of the world, including the United States and Mexico [11], have highlighted how common this discrepancy is and how crucial correct route prediction is for follow-up activities, such as order dispatching and arrival-time forecasting. These findings emphasise the vital importance of route prediction approaches


A. Hotho and S. Rudolph (Eds.): KI 2024, LNAI 14992, pp. 232–245, 2024.
https://doi.org/10.1007/978-3-031-70893-0_17

that effectively reflect workers' dynamic routing behaviours in real-world circumstances.

In the field of logistics, it is imperative to estimate package arrival times with precision in order to maintain both customer satisfaction and operational efficiency. Due to its multi-destination and path-free nature, ETA prediction in logistics presents distinct challenges compared to standard ETA prediction problems seen in other fields. Additionally, the unpredictability of couriers' pick-up routes, which are impacted by variables such as traffic and individual preferences, makes it more difficult to calculate ETA accurately. Meeting the demand for real-time ETA prediction in such complex situations continues to be a challenging and important task.

Our proposed method, Graph2RETA, is an integrated predictive modeling approach designed to tackle these issues in a comprehensive manner. This novel method optimizes routing while also predicting ETA by combining dynamic spatial-temporal graph-based modeling and route inference. Graph2RETA improves the prediction accuracy by utilizing the built-in graph structure and features to capture the dynamic interactions between problem occurrences and to incorporate a broad range of decision context and spatial-temporal information. We show that combining the training procedures of both models and integrating the training losses into both predicting and forecasting tasks is beneficial.

The experiments on a real-world industrial dataset show that Graph2RETA outperforms the prevalent State of the Art (SotA) approaches (Graph2Route and RankETPA). We expect our integrated method to offer the potential for service providers to gain more control over their delivery process, resulting in higher customer satisfaction and operational efficiency in P&D services.

## 2 Related Work

This section will discuss the current SotA for the problems addressed in this research: the Pick-up and Delivery Route Prediction (PDRP) and Estimated Time of Arrival (ETA) tasks.

### 2.1 Pick-Up and Delivery Route Prediction

The domain of Pick-up and Delivery Route Prediction (PDRP) plays a crucial role in optimizing the logistics and operational efficiency of delivery services. The essence of PDRP lies in predicting the future routes of a worker based on their past behaviors, focusing on the actual route to be taken rather than identifying the most optimal path. This distinction is critical for accurately computing downstream tasks, such as arrival time prediction [19]. Despite its importance, the PDRP problem has yet to be extensively explored within the research community, marking a significant gap in the existing literature [19].

Early attempts to address route prediction in delivery settings include methodologies like OSquare [23] and FDNET [5], alongside recent developments such as Graph2Route [19]. OSquare emerged as one of the pioneering methods, adopting a next-stop prediction approach to sequentially construct routes using conventional machine learning techniques [23]. FDNET further advanced the route prediction field by incorporating the deepFM model [6] for input feature encoding, an LSTM layer for sequential data processing, and attention mechanisms to estimate the probability of a worker visiting subsequent locations [5]. However, both OSquare and FDNET primarily conceptualize unfinished tasks as a sequence, which limits their capacity to encode complex spatial-temporal dynamics between different tasks. It often results in predicting implausible routes, relying solely on information available at the time of request [19].

Graph2Route [19] relies on graph-based information for predicting the routes. It uses a dynamic spatial-temporal graph encoder and a personalized route decoder to enhance prediction quality. This approach models spatial-temporal correlations through a Graph Convolutional Network and Recurrent Neural Networks. The personalized node decoding process employs a masking mechanism to eliminate unlikely solutions, streamlining the decision-making process and enhancing model performance. It leverages massive historical data to learn and model the worker's personalized decision preferences, making it the current SotA model for route prediction and the perfect baseline to improve upon in this work.

### 2.2 Estimated Time of Arrival

In pick-up and delivery services, predicting the Estimated Time of Arrival (ETA) of couriers plays a crucial role in enhancing service reliability and customer satisfaction [5]. ETA prediction can be defined as forecasting the travel time from an origin to a destination along a specified path. This problem has garnered attention across various domains, including ride-sharing [4,17], logistics [1,22], and food delivery services [25], highlighting its widespread relevance.

ETA prediction methodologies can broadly be categorized into path-free and path-based approaches. Path-free methods, such as the ST-NN proposed by Jindal et al. [7], rely solely on the origin and destination points without incorporating route information. These models predict travel distance and time by also considering additional temporal data. Conversely, path-based methods involve a more detailed analysis by incorporating specific travel routes into the prediction process. For instance, the Wide and Deep Route (WDR) model [18] integrates linear models, deep neural networks, and recurrent neural networks to predict travel times more accurately. Similarly, CompactETA [4] employs a graph attention network and positional encoding to learn the spatio-temporal dependencies of road networks and the sequence of travel routes, respectively.

Path-based models generally surpass their path-free counterparts due to their detailed route information, providing a richer context for more accurate predictions. Our study adopts a path-based approach, aligning with our objective to predict couriers' routes in the initial phase of the research. We utilize the Arrival Time Predictor from RankETPA [21] as a foundation for our work. RankETPA stands out with its spatial-temporal attention-based mechanism. It leverages predicted routes, node information, package details, and courier features to estimate precise arrival time.

The methodology behind RankETPA begins with calculating the first positional encoding of the route sequence derived from the previously predicted routes. This encoding captures the relative positioning of nodes within the route sequence. Subsequently, this data, along with pick-up order encodings, package characteristics, courier details, and other external factors, form the comprehensive input to the model. The RankETPA model then utilizes spatial-temporal attention-based encoder blocks to encapsulate the dynamics between unpicked package pairs and other pertinent information through layers comprising multi-head attention mechanisms and feed-forward networks. The multi-head attention effectively captures the spatial-temporal correlations among packages, while the feed-forward network enhances the model with non-linear transformations, culminating in the accurate prediction of arrival times for each node in the sequence.

## 3 Problem Formulation

In this section, we formulate the route prediction task from the graph perspective and the arrival time estimation problem (Fig. 1).

### 3.1 Route Prediction

The problem is defined on a graph, where the nodes are each task of the driver and the edges are the spatial and temporal distance between the nodes. The following adopts the notation in [19]. Given the worker $w$'s input graph $\mathcal{G}$ at a given time $t$

$$\mathcal{G}_t^w = (\mathcal{V}_t, \mathcal{E}_t, X_t, E_t) \tag{1}$$

the main objective is to learn a mapping function $F_C$ to predict the worker's future route $\hat{\pi}$ which should satisfy the given route constraints $C$, formulated as:

$$\mathcal{F}_C(\mathcal{G}_t^w) = \pi_1, \pi_2, ..., \pi_{|\mathcal{V}_t^U|} \tag{2}$$

here $\pi_i$ means that the $i$-th node in the route is node $v_{\pi_i}$ and $|.|$ denotes the cardinality of a set. Moreover, $\pi_i \in \{1, \ldots, |V|\}$ and $\pi_i \neq \pi_j$ if $i \neq j$ where:

$$\mathcal{V}_t^U = \{v | v \in \mathcal{V}_t, \mathrm{X}^{FT} = -1\} \tag{3}$$

are the unfinished nodes and $x^{FT}$ is task finish time. Such that:

$$\min \mathcal{L}_{G2R} = -\sum w \in W \sum_{t \in T} \sum_{i \in \pi_{t:t'}} y_i \log(p(y_i|\theta)) \tag{4}$$

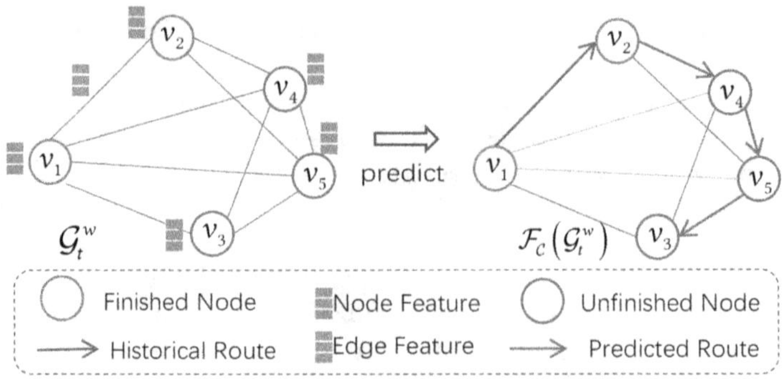


**Fig. 1.** Problem Illustration [19]. In this case, $V_t^F = \{v_1, v_2\}$ and $V_t^U = \{v_3, v_4, v_5\}$, the output of the model $\hat{\pi} = \{\pi_1, \pi_2, \pi_3\}$ is [4, 5, 3].

### 3.2 Arrival Time Estimation

The problem formulation of the arrival time estimation is adopted from [21]. Given the input query $q = \{c, t, V_t^u\}$ and $\hat{\pi} = \{\pi_1, \pi_2, ..., \pi_{|\mathcal{V}_t^U|}\}$ where $c$ is the courier ID and $t$ is the query time, our task is to use a model $\mathcal{F}$ that can map the input query to the pick-up arrival time $\hat{y}$ for the unpicked-up package set:

$$\hat{y} = \mathcal{F}(q, \hat{\pi}) \mapsto Y = \{y_1, ..., y_n\} \tag{5}$$

such that minimize the loss of the predicted arrival time and the ground truth, which can be formally represented as follows given the Huber loss function

$$\mathcal{L}_{ETPA} = \begin{cases} \frac{1}{2}(y - \hat{y})^2 & |y - \hat{y}| \leq \delta, \\ \delta|y - \hat{y}| - \frac{1}{2}\delta^2 & \text{otherwise.} \end{cases} \tag{6}$$

## 4 Methodology

We propose Graph2RETA, an upgraded version of the vanilla Graph2Route, by extending the model's capabilities to include arrival time prediction for pick-up and delivery tasks (Fig. 2). This section outlines how the encoder module of the Graph2Route is upgraded and how the arrival time prediction module is integrated. Additionally, we will describe the dataset used in our experiments, the training procedure, and the evaluation protocol.

### 4.1 Updating the Route Prediction Module

Our first objective of this research is to refine the route prediction capabilities of the existing Graph2Route model introduced by Wen et al. [19]. Graph2Route is a foundational architecture for our research, particularly its spatial-temporal

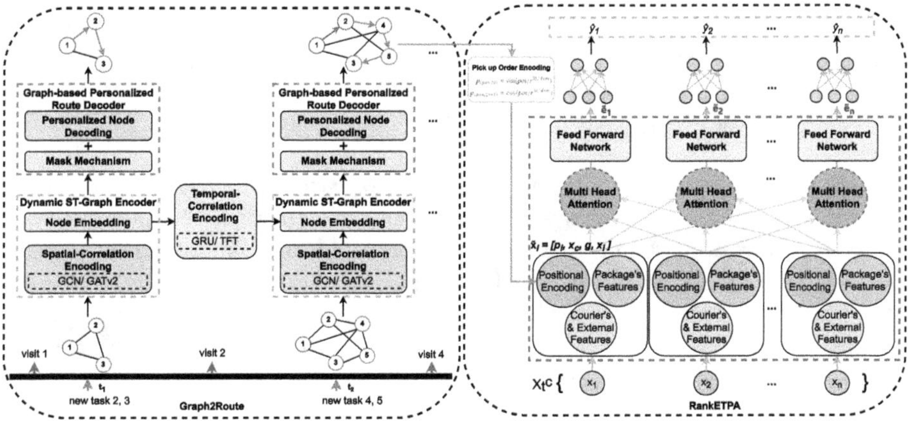


**Fig. 2.** Graph2RETA Model Architecture [19,21]

graph encoder component. It is the key to capturing the complex spatial and temporal context within the graph structure of pick-up and delivery route prediction problems. The original Graph2Route employs graph convolutional layers (GCN) and gated recurrent units (GRU) for encoding spatial and temporal correlations, respectively. These components have been shown to effectively learn the graph's spatial-temporal correlations, underscoring their importance in the architecture's performance [19].

However, GCN and GRU are considered conventional and outdated approaches at the time of the research, which may not fully capture the dynamic complexities of spatial-temporal graph data. Recent literature suggests that advancements in graph learning techniques offer more sophisticated methods for capturing these relationships [24]. Recognizing this potential issue on the Graph2Route architecture, our research proposes to exchange the spatial encoder module to the well-established attention mechanism module for graphs, namely GAT[1].

Regarding temporal encoding, the GRU module in Graph2Route still encounters challenges with gradient issues and may not effectively capture long-term dependencies [10]. The Transformer architecture, renowned for its attention mechanisms and ability to manage sequences with long-term dependencies, offers promising benefits [16]. We employ the Temporal Fusion Transformer (TFT) model, which combines gating mechanisms, variable selection, and interpretable multi-head attention, for the temporal encoder [12]. However, we will adapt TFT into Graph2RETA by excluding its output layer to maintain compatibility with

[1] Our experiments are conducted with a variant of the graph attention network called GATv2 [2], which uses an attention mechanism that dynamically weighs the importance of neighbouring nodes [3] and offers slightly better results than the original GAT architecture.

the original Graph2Route decoder, thereby focusing on the temporal variable interdependence learning.

### 4.2 Arrival Time Prediction Module

The second part of the work is to take a step forward by introducing an arrival time prediction module, which serves as a downstream task that leverages the output of our route predictor. This approach aims to estimate the time when a courier will come to pick up and deliver packages at each location within the route, commonly referred to as the ETA, and introduces additional feedback on the routing prediction. Drawing upon the methodologies from RankETPA [21], we seamlessly integrate a spatial-temporal attention-based arrival time prediction with our refined route prediction module.

The underlying principle of Graph2RETA hinges on the sequence of operations where the courier's route is determined initially through the route predictor module. Following this, the arrival time predictor module takes center stage, utilizing the precomputed route to ascertain the arrival times at each designated stop. This sequential processing is grounded in the logic that the ETA is intrinsically linked to the order of stops on the courier's itinerary, underscoring the importance of route prediction as a precursor to time estimation [5,21].

The operational flow begins with the route predictor outlining a sequence of nodes, indicating the courier's path at a specified time step. Subsequently, this predicted route, enriched with additional node and courier data such as the courier's ID, average speed, and maximum load capacity, serves as input to the arrival time predictor module. This module then begins calculating the ETA for each node along the route, providing a comprehensive overview of the delivery timeline. By integrating the arrival time predictor module, our framework now returns two outputs: the predicted delivery route and the ETA for each stop along the way. This improvement not only upgrades the practical use of our model but also supplements it with a layer of temporal insight, offering a more complete view of the courier's journey.

### 4.3 Dataset

This work utilizes the Food Pick-up and Delivery Data (Food-PD) [19], a dataset composed of online food pick-up and delivery transactions facilitated by Eleme, a prominent online food delivery service in China. This dataset plays a crucial role in our experiment, providing a real-world context for our predictive modeling efforts.

The Food-PD dataset encompasses the pick-up and delivery activities of 916 delivery workers over a span of 28 days in Dalian, China. It offers a comprehensive view of the operational dynamics of food delivery, making it an invaluable resource for our analysis. The primary objective of employing this dataset is to forecast these drivers' future pick-up and delivery routes. We conceptualize each driver as a worker, and each pick-up or delivery task associated with an order is represented as a node.

The dataset encompasses 5039 instances meticulously partitioned into training, testing, and validation sets. The distribution follows a 50:25:25 ratio [19], ensuring a balanced division that facilitates both the development and the rigorous evaluation of the Graph2RETA model.

### 4.4 Model Training

We explore two strategies for training Graph2RETA, which consists of a route predictor and an arrival time prediction module. The initial approach entails training each module separately. Here, the focus is on first improving the accuracy of the route predictor by training it to minimize the cross entropy loss function (Eq. 4). Upon completion, the predicted routes are exported and stored, before being preprocessed to serve as input for the ETA task. This subsequent module is trained to minimize the Huber loss (Eq. 6), optimizing it for accurate time estimations. This sequential training process allows each module to refine its predictions independently.

$$\mathcal{L}_{Joint} = \alpha \times \mathcal{L}_{G2R} + (1 - \alpha) \times \mathcal{L}_{ETPA} \tag{7}$$

Conversely, our second strategy adopts an integrated approach, wherein both modules are integrated into a unified architecture and trained concurrently, which we call joint training. Unlike the first approach, the joint training trains both models simultaneously in one training loop, allowing both modules to learn from shared parameters. This integrated approach necessitates formulating a novel loss function, as stated in Eq. 7, that combines the cross entropy loss from the route prediction with the Huber loss from the time prediction. The fusion of these loss functions is balanced using a weight variable, $\alpha$, as detailed in [15], which harmonizes the contributions of the two distinct tasks.

Initially, we assign $\alpha$ a value of 0.5, signifying equal importance to both route and time predictions. However, recognizing the potential benefits of temporal prioritization in training, we also experiment with a dynamic $\alpha$ value. This approach starts with an $\alpha$ of 0.9, emphasizing the route predictor's role in the early stages of training, thereby allowing it to establish a solid predictive foundation rapidly. As the training progresses, $\alpha$ is gradually reduced to 0.1, incrementally shifting the focus towards refining the arrival time predictions. This dynamic adjustment effectively uses both modules' strengths, fostering a symbiotic development where initial route accuracy enhances the subsequent time predictions. In the final result, we are showing results from this dynamic adjustment approach.

### 4.5 Evaluation Metrics

To thoroughly evaluate the performance of Graph2RETA, we have adopted a comprehensive set of metrics tailored to assess both the predicted route and the estimated arrival time of couriers.

For assessing the predicted routes, our evaluation metrics include:

- Kendall Rank Correlation (KRC) [8]: This metric measures the ordinal association between the predicted and actual sequences, which is defined as:

$$\mathrm{KRC} = \frac{N_c - N_d}{N_c + N_d}, \tag{8}$$

where $N_c$ and $N_d$ are the numbers of concordant and discordant pairs respectively.
- Edit Distance (ED) [14]: It quantifies the dissimilarity between two sequences by counting the minimum number of operations required to transform the predicted sequence into the actual sequence.
- Location Square Deviation (LSD): This measures the extent of deviation between the predicted and actual locations. The LSD is defined as:

$$\mathrm{LSD} = \frac{1}{m}\sum_{i=1}^{m}(O_{\pi}(\pi_i) - O_{\hat{\pi}}(\pi_i))^2 \tag{9}$$

- Hit-Rate@$k$ (HR@$k$): This evaluates the accuracy of the top-$k$ items in the predicted sequence by determining how many match the actual sequence, formulated as:

$$\mathrm{HR@}k = \frac{\hat{\pi}_{[1:k]} \cap \pi_{[1:k]}}{k} \tag{10}$$

- Accuracy@$k$ (ACC@$k$): It examines whether the route composed of the top-$k$ predictions is precisely the same as the actual route, formulated as:

$$\mathrm{ACC@}k = \prod_{i=0}^{k}\mathbb{I}(\hat{\pi}_i, \pi_i), \tag{11}$$

where $\mathbb{I}(\cdot)$ is the indicator function, and $\mathbb{I}(\hat{\pi}_i, \pi_i)$ equals 1 if $\hat{\pi}_i == \pi_i$ else 0.

The KRC, ED, and LSD assess the overall similarity between the predicted and actual routes, while HR@$k$ and ACC@$k$ focus on local similarity, offering an understanding of the model's performance from both global and local perspectives.
For the evaluation of the estimated arrival time, we utilize:

- Root Mean Square Error (RMSE): This metric calculates the square root of the average squared differences between the predicted and actual arrival times. The RMSE is defined as follows:

$$\mathrm{RMSE} = \sqrt{\frac{1}{N}\sum_{i=1}^{N}(\hat{y}_i - y_i)^2} \tag{12}$$

- Mean Absolute Error (MAE): It measures the average magnitude of the errors in a set of predictions without considering their direction, defined as:

$$\mathrm{MAE} = \frac{1}{N}\sum_{i=1}^{N}|\hat{y}_i - y_i| \tag{13}$$

# 5 Experiments

This section outlines the experiment results demonstrating the enhancements achieved with Graph2RETA. We present the outcomes of individually trained models for route and arrival time estimation, along with the results from the joint training of both modules. The code to reproduce the results of this work is available at https://github.com/wsentanoe/Graph2RETA.

## 5.1 Route Prediction

Table 1 shows that replacing the spatial and temporal encoder components of the route prediction model with GAT and TFT would lead to improved results compared to the original GCN and GRU combination. However, we also evaluate each combination of the spatial and temporal components to find the one with the best predictive performance.

We observe that there is only a slight improvement by substituting GCN with GAT, especially while maintaining the GRU component. Although minimal, this variant still shows some improvement compared to the baseline Graph2Route architecture. On the other hand, replacing both components of Graph2Route with GAT and TFT leads to enhanced performance across all evaluation metrics compared to the baseline but remains inferior to the GCN+TFT variant.

In contrast to substituting GCN with GAT, there is a considerable improvement when TFT replaces GRU. Integrating the TFT component with GCN could significantly improve the prediction compared to the baseline Graph2Route results in all the metrics considered. TFT captures complex temporal dependencies and important time steps with its temporal attention mechanism, resulting in a 2.8% improvement in KRC, which is the primary evaluation metric of the route prediction model. Improvement goes to as high as 24.8% in ED metric compared to baseline because of the precise temporal modeling and the ability to handle multi-horizon forecasting of TFT, which predicts routes that are geographically closer to the actual routes, thereby lowering the ED. These results set the GCN+TFT as the best variant for the route prediction. Therefore, we decided to employ them for the final Graph2RETA architecture.

**Table 1.** Route Prediction Results. †Results are from [19] with the original GCN + GRU modules. The results are averaged over five runs. The best results are highlighted in **bold**. ↑ indicates the higher, the better and ↓ indicates the lower, the better.

| Model Variant | HR@1 ↑ | ACC@3 ↑ | KRC ↑ | LSD ↓ | ED ↓ |
|---|---|---|---|---|---|
| Graph2Route† | 0.679 | 0.537 | 0.787 | 0.520 | 1.130 |
| GCN + TFT | 0.724 ± 0.009 | **0.620 ± 0.012** | **0.808 ± 0.007** | **0.450 ± 0.026** | **0.850 ± 0.034** |
| GAT + GRU | 0.674 ± 0.008 | 0.582 ± 0.006 | 0.788 ± 0.004 | 0.454 ± 0.008 | 0.902 ± 0.010 |
| GAT + TFT | **0.729 ± 0.013** | 0.615 ± 0.017 | 0.803 ± 0.011 | 0.466 ± 0.024 | 0.861 ± 0.041 |

Overall, the route prediction performance is improved when TFT is integrated in the model directing to the observation that temporal component in the model have more impact in the predictions.

### 5.2 Arrival Time Estimation with Joint Training

We aimed to assess whether jointly training the model would enhance its outcomes. In Table 2, we have obtained results from comparing the performance of the extended model with arrival time estimation, both trained individually and jointly.

It is evident that both models benefit from the joint training approach. Graph2RETA, when jointly trained, demonstrates an improvement of 8.4% in MAE and 11.2% in MSE for arrival time prediction compared to individual training settings, emphasizing the significance of joint training. The individually trained model excels in route prediction, whereas arrival time estimation significantly benefits and improves when jointly trained. There is an improvement of 8.7% in MAE and 11.3% in MSE when comparing the jointly trained Graph2RETA model to the individually trained Graph2Route model.

We conducted additional experiments on joint training to validate the proposed model. During joint training, we faced the challenge of dealing with two loss functions for route prediction and arrival time prediction, each operating on different scales. To address this issue, we scaled down the loss of arrival time prediction by 10 to ensure that both losses are within the same value range.

**Table 2.** Graph2RETA Experiment Results. ‡Graph2Route (GCN + GRU) results are from our implementation and are extended with the ETA module. Each module (Route prediction & ETA) is trained separately in the individually trained approach, while the jointly trained model trains both simultaneously. The results are averaged over five runs. The best results are highlighted in **bold**. ↑ indicates the higher, the better, and ↓ indicates the lower, the better.

| | | Trained Individually | | Trained Jointly | |
|---|---|---|---|---|---|
| | | Graph2Route‡ | Graph2RETA | Graph2Route‡ | Graph2RETA |
| Route Prediction | HR@1 ↑ | 0.674 ± 0.007 | **0.724 ± 0.009** | 0.650 ± 0.012 | 0.717 ± 0.008 |
| | ACC@3 ↑ | 0.581 ± 0.007 | **0.620 ± 0.012** | 0.550 ± 0.012 | 0.601 ± 0.007 |
| | KRC ↑ | 0.784 ± 0.004 | **0.808 ± 0.007** | 0.767 ± 0.007 | 0.801 ± 0.004 |
| | LSD ↓ | 0.465 ± 0.008 | **0.450 ± 0.026** | 0.499 ± 0.016 | 0.459 ± 0.007 |
| | ED ↓ | 0.902 ± 0.015 | **0.850 ± 0.034** | 0.997 ± 0.026 | 0.918 ± 0.018 |
| Arrival Time Estimation | RMSE ↓ | 13.462 ± 0.208 | 13.446 ± 0.055 | 12.478 ± 0.616 | **11.936 ± 0.422** |
| | MAE ↓ | 10.136 ± 0.171 | 10.110 ± 0.148 | 9.648 ± 0.595 | **9.258 ± 0.294** |

## 6 Discussion

Replacing the spatial component of Graph2Route from GCN to GAT does not significantly impact the route prediction performance, especially when trained individually, which is shown by the minimal improvement in the result. On the other hand, replacing the temporal component with TFT does significantly improve the performance of the route prediction module. The temporal component's contribution to the final outcome surpasses that of the spatial component.

When trained individually, the quality of the arrival time prediction aligns with the route prediction's performance. Improvement in route prediction corresponds to enhancement in arrival time prediction. The GCN+TFT configuration, which establishes consistent results across most route prediction metrics, also yields the best performance on Arrival Time Estimation when jointly trained. Hence, our intuition to integrate in Graph2RETA. This observation emphasizes our rationale for integrating this configuration in Graph2RETA.

Joint training does not necessarily improve the route prediction performance. This phenomenon suggests a nuanced challenge within multi-task learning frameworks, known as negative transfer [9,13]. Negative transfer in multi-task learning refers to a scenario where the knowledge learned from one task interferes with the learning or performance of another task. However, it definitely improves the arrival time prediction, which is the more important downstream task for delivery service or logistics operators.

Arrival time estimation benefits from joint training due to its dependency on route prediction. Also, the quality of the route prediction component's output significantly impacts the accuracy of arrival time estimation. Conversely, the route prediction task may not necessarily benefit from the joint training approach, as evidenced by its sub-optimal performance.

Moreover, it was evident that implementing a scheduling mechanism for alpha in joint training can further enhance the accuracy of the ETA estimation.

## 7 Conclusion

In this work, we propose Graph2RETA, which seamlessly integrated two complex models for route and arrival time prediction into a unified architecture, revealing the predominant role of temporal over spatial components in route prediction accuracy. Our findings demonstrate that updating the temporal component in the route prediction model contributes significantly to the increase in performance, especially when the route predictor operates independently from the arrival time prediction module. This finding suggests the importance of temporal dynamics in delivery route optimization. Joint training of both models could also improve the accuracy of the arrival time estimation despite the stagnant results in route prediction. This disparity in performance highlights the challenge of negative transfer in multi-task learning. Joint model training may offer specific benefits for arrival time estimation, paving the way for future research in delivery service optimization.

**Acknowledgement.** Funded by the Lower Saxony Ministry of Science and Culture under grant number ZN3492 within the Lower Saxony "Vorab" of the Volkswagen Foundation and supported by the Center for Digital Innovations (ZDIN).

## References

1. de Araujo, A.C., Etemad, A.: End-to-end prediction of parcel delivery time with deep learning for smart-city applications. IEEE Internet Things J. **8**(23), 17043–17056 (2021). https://doi.org/10.1109/JIOT.2021.3077007

2. Brody, S., Alon, U., Yahav, E.: How attentive are graph attention networks? arXiv preprint arXiv:2105.14491 (2021). https://doi.org/10.48550/arXiv.2105.14491
3. Bui, K.H.N., Cho, J., Yi, H.: Spatial-temporal graph neural network for traffic forecasting: an overview and open research issues. Appl. Intell. **52**(3), 2763–2774 (2022). https://doi.org/10.1007/s10489-021-02587-w
4. Fu, K., Meng, F., Ye, J., Wang, Z.: Compacteta: a fast inference system for travel time prediction. In: Proceedings of the 26th ACM SIGKDD International Conference on Knowledge Discovery & Data Mining, pp. 3337–3345 (2020). https://doi.org/10.1145/3394486.3403386
5. Gao, C., et al.: A deep learning method for route and time prediction in food delivery service. In: Proceedings of the 27th ACM SIGKDD Conference on Knowledge Discovery & Data Mining, pp. 2879–2889 (2021). https://doi.org/10.1145/3447548.3467068
6. Guo, H., Tang, R., Ye, Y., Li, Z., He, X.: Deepfm: a factorization-machine based neural network for CTR prediction. arXiv preprint arXiv:1703.04247 (2017). https://doi.org/10.48550/arXiv.1703.04247
7. Jindal, I., Chen, X., Nokleby, M., Ye, J., et al.: A unified neural network approach for estimating travel time and distance for a taxi trip. arXiv preprint arXiv:1710.04350 (2017). https://doi.org/10.48550/arXiv.1710.04350
8. Kendall, M.G.: A new measure of rank correlation. Biometrika **30**(1/2), 81–93 (1938). https://doi.org/10.1093/biomet/30.1-2.81
9. Lee, G., Yang, E., Hwang, S.: Asymmetric multi-task learning based on task relatedness and loss. In: International Conference on Machine Learning, pp. 230–238. PMLR (2016)
10. Li, S., et al.: Enhancing the locality and breaking the memory bottleneck of transformer on time series forecasting. In: Advances in Neural Information Processing Systems, vol. 32 (2019)
11. Li, Y., Phillips, W.: Learning from route plan deviation in last-mile delivery (2018)
12. Lim, B., Arık, S.Ö., Loeff, N., Pfister, T.: Temporal fusion transformers for interpretable multi-horizon time series forecasting. Int. J. Forecast. **37**(4), 1748–1764 (2021). https://doi.org/10.1016/j.ijforecast.2021.03.012
13. Malhotra, A., Vatsa, M., Singh, R.: Dropped scheduled task: mitigating negative transfer in multi-task learning using dynamic task dropping. Trans. Mach. Learn. Res. (2023)
14. Nerbonne, J., Heeringa, W., Kleiweg, P.: Edit distance and dialect proximity. Time Warps, String Edits and Macromolecules: the theory and practice of sequence comparison, vol. 15 (1999)
15. Ott, F., Rügamer, D., Heublein, L., Bischl, B., Mutschler, C.: Joint classification and trajectory regression of online handwriting using a multi-task learning approach. In: Proceedings of the IEEE/CVF Winter Conference on Applications of Computer Vision, pp. 266–276 (2022)
16. Vaswani, A., et al.: Attention is all you need. In: Advances in Neural Information Processing Systems, vol. 30 (2017)
17. Wang, D., Zhang, J., Cao, W., Li, J., Zheng, Y.: When will you arrive? estimating travel time based on deep neural networks. In: Proceedings of the AAAI Conference on Artificial Intelligence, vol. 32 (2018). https://doi.org/10.1609/aaai.v32i1.11877
18. Wang, Z., Fu, K., Ye, J.: Learning to estimate the travel time. In: Proceedings of the 24th ACM SIGKDD International Conference on Knowledge Discovery & Data Mining, pp. 858–866 (2018). https://doi.org/10.1145/3219819.3219900

19. Wen, H., et al.: Graph2route: a dynamic spatial-temporal graph neural network for pick-up and delivery route prediction. In: Proceedings of the 28th ACM SIGKDD Conference on Knowledge Discovery and Data Mining, pp. 4143–4152 (2022). https://doi.org/10.1145/3534678.3539084
20. Wen, H., et al.: Package pick-up route prediction via modeling couriers' spatial-temporal behaviors. In: 2021 IEEE 37th International Conference on Data Engineering (ICDE), pp. 2141–2146. IEEE (2021). https://doi.org/10.1109/ICDE51399.2021.00214
21. Wen, H., et al.: Enough waiting for the couriers: learning to estimate package pick-up arrival time from couriers' spatial-temporal behaviors. ACM Trans. Intell. Syst. Technol. **14**(3), 1–22 (2023). https://doi.org/10.1145/3582561
22. Wu, F., Wu, L.: Deepeta: a spatial-temporal sequential neural network model for estimating time of arrival in package delivery system. In: Proceedings of the AAAI Conference on Artificial Intelligence, vol. 33, pp. 774–781 (2019). https://doi.org/10.1609/aaai.v33i01.3301774
23. Zhang, Y., Liu, Y., Li, G., Ding, Y., Chen, N., Zhang, H., He, T., Zhang, D.: Route prediction for instant delivery. In: Proceedings of the ACM on Interactive, Mobile, Wearable and Ubiquitous Technologies, vol. 3, no. 3, pp. 1–25 (2019). https://doi.org/10.1145/3351282
24. Zhou, J., et al.: Graph neural networks: a review of methods and applications. AI Open **1**, 57–81 (2020). https://doi.org/10.1016/j.aiopen.2021.01.001
25. Zhu, L., et al.: Order fulfillment cycle time estimation for on-demand food delivery. In: Proceedings of the 26th ACM SIGKDD International Conference on Knowledge Discovery & Data Mining, pp. 2571–2580 (2020). https://doi.org/10.1145/3394486.3403307

# Data Generation for Explainable Occupational Fraud Detection

Julian Tritscher[1,2(✉)], Maximilian Wolf[3], Anna Krause[1,2], Andreas Hotho[1,2], and Daniel Schlör[1,2]

[1] CAIDAS Center for Artificial Intelligence and Data Science, Würzburg, Germany
{tritscher,anna.krause,hotho,schloer}@informatik.uni-wuerzburg.de
[2] Julius-Maximilians-University of Würzburg, Würzburg, Germany
[3] Coburg University of Applied Sciences, Coburg, Germany
m.wolf@informatik.uni-wuerzburg.de

**Abstract.** Occupational fraud, the deliberate misuse of company assets by employees, causes damages of around 5% of yearly company revenue. Recent work therefore focuses on automatically detecting occupational fraud through machine learning on the company data contained within enterprise resource planning systems. Since interpretability of these machine learning approaches is considered a relevant aspect of occupational fraud detection, first works have already integrated post-hoc explainable artificial intelligence approaches into their fraud detectors. While these explainers show promising first results, systematic advancement of explainable fraud detection methods is currently hindered by the general lack of ground truth explanations to evaluate explanation quality and choose suitable explainers. To avoid expensive expert annotations, we propose a data generation scheme based on multi-agent systems to obtain company data with labeled occupational fraud cases and ground truth explanations. Using this data generator, we design a framework that enables the optimization of post-hoc explainers for unlabeled company data. On two datasets, we experimentally show that our framework is able to successfully differentiate between explainers of high and low explanation quality, showcasing the potential of multi-agent-simulations to ensure proper performance of post-hoc explainers.

**Keywords:** Feature Relevance · XAI · Fraud Detection · Simulation

## 1 Introduction

Occupational fraud describes the misuse of company assets by an internal employee, for instance through theft or bribery. This type of fraud costs companies around 5% of their annual revenue [1], making the detection of occupational fraud relevant for many companies. With the increasing company digitization in enterprise resource planning (ERP) systems, large amounts of company ERP data enable an automated detection of occupational fraud through machine


A. Hotho and S. Rudolph (Eds.): KI 2024, LNAI 14992, pp. 246–259, 2024.
https://doi.org/10.1007/978-3-031-70893-0_18

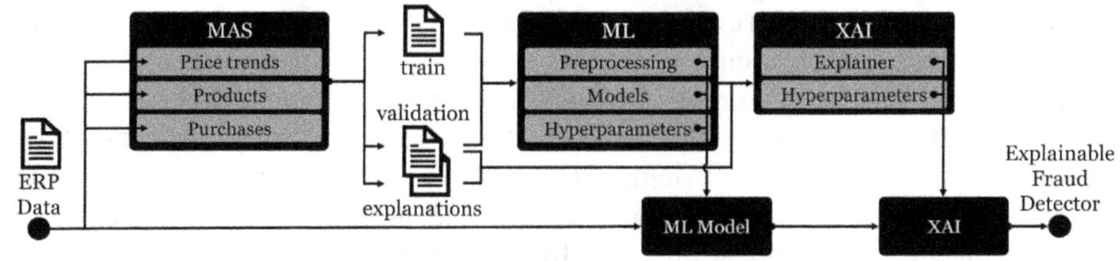


**Fig. 1.** Our proposed framework for automated and explainable fraud detection on ERP system data in the absence of labeled data. The system uses an MAS that is set up with market information from an unlabeled dataset to generate data with anomaly detection and explanation labels. This labeled data is then used to determine design choices for anomaly detectors and explainers to use on the original unlabeled data.

learning [15]. Next to the development of new methods for automated occupational fraud detection, economics research has also identified explainability as especially relevant in this domain [7]. In the past, post-hoc feature relevance explainers [21] have already been applied in occupational fraud detection [21,23].

One challenge in this area of research is the lack of ground truth explanations that are needed to differentiate between well- and poorly-performing explainers. When new ERP system data is made available for auditing within the company, the presence and location of occupational frauds are usually not known within the data. Therefore, obtaining explanation ground truth through human experts would first require experts to manually identify occupational frauds in the large amounts of unlabeled company data, before ground truth explanations may be created for found fraud cases. To avoid this expensive ground truth generation through expert auditors, we propose to use a synthetic data generator based on Multi Agent Systems (MAS) to provide the necessary ground truth explanations. In prior work, we proposed to use MAS-based simulations to mirror the normal business processes of a company after a given unlabeled dataset and integrated known fraud scenarios into the simulation to obtain labeled data for occupational fraud detection [22]. While this simulation allows for automated occupational fraud detection with machine learning approaches, it does not provide any ground truth explanations for the identification of suitable explainers. As the common practice of arbitrarily choosing explainers in the domains of anomaly and fraud detection can result in near random explanations [21,24], ground truth explanations are highly relevant for designing explainable occupational fraud detectors.

In this paper, we therefore extend our MAS-based simulation [22] to generate ground truth explanations. As the underlying processes that govern the data generation are known within the simulation, generated anomalous entries can be identified during generation time. Using this extended MAS-based simulation, we propose a framework that enables the optimization of post-hoc feature relevance explainers for occupational fraud detection in the absence of real ground truth explanations.

Our framework is illustrated in Fig. 1. When given new and unlabeled data from a company's ERP system, we extract general environment information such as price and purchasing trends from the company data and feed them into the MAS-based simulation. Within the MAS, the normal business process of the company is then simulated through established best practices from economics research and known fraud cases are modeled and added during the generation process. Building upon this simulation framework, we generalize this synthetic generative labeling approach of fraud versus non-fraud to generate ground truth explanation labels along the simulation process that reflect the actual feature-combinations that are indicative of the generated fraud case. With this labeled data with ground truth explanations, both machine learning approaches and explainers may be selected according to established performance metrics on the synthetically created simulation data. Choices for machine learning and explainer approaches are then directly applied on the original company data, enabling automated explainable occupational fraud detection. In our experiments, we use occupational fraud data with manually labeled ground truth explanations [20] to evaluate whether well-performing explainers on MAS-generated data transfer their high explanation scores to the underlying company dataset. Experiments on two datasets show that explanation quality indeed transfers from the MAS data to the underlying data, successfully enabling the use of MAS-based simulation to determine suitable post-hoc explainers on unlabeled company data for occupational fraud detection.

In summary, our contributions are as follows: (1) We extend an existing MAS-based simulation for ERP data to automatically provide ground truth explanations for occupational fraud cases created during data generation. (2) We provide a framework that identifies well-performing feature relevance explainers for occupational fraud detection in company data in the absence of real ground truth explanations. (3) Experimentally, we show on two datasets that the well-performing explainers identified by our MAS-based simulation framework successfully generalize to unlabeled company data. (4) We provide code for our Java-based implementation of the MAS simulation, as well as the Python-based optimization framework.[1]

To our knowledge, we are the first to use MAS-based simulation to aid explainable AI through automated ground truth generation. This is especially relevant to the XAI community, as ground truth explanations remain scarce and arbitrarily choosing explainers can result in near random explanations [21].

## 2 Related Work

Feature relevance explanations are the most common explanations in anomaly detection and assign a score to each feature of an individual datapoint that expresses its relevance to the model prediction [25]. While many works use existing feature relevance explainers in the domains of anomaly and fraud detection [21,25], established unsupervised explainer evaluation schemes are not readily

[1] Code and data are available under: https://professor-x.de/xai-fraud-mas.

applicable to anomaly and fraud detection as they rely on perturbation schemes that would introduce additional anomalous signals into the data [23]. While binary ground truth can be used to evaluate feature relevance explainers, the low availability of ground truth leads to only few works using ground truth evaluations in anomaly detection [11,23,24]. In our work, we follow these evaluations by using the ground truth datasets from [23,24] in our experiments.

Beyond the evaluation of explainers, multiple works have already explored the combination of explainable artificial intelligence (XAI) and MAS. One line of research focuses on explaining single agents within MAS systems. Here, [10] focus on explaining agents within MAS that have been trained through reinforcement learning. [13] provides a framework for obtaining explanations regarding the actions and interactions between agents within financial MAS. In contrast to these works, our work does not aim to explain the MAS itself but merely uses it as a tool for generating ground truth explanations that enables the choice of well-performing XAI methods for the task of fraud detection in unlabeled company data. An additional line of research regarding MAS and XAI uses MAS as a theoretical framework to model general properties of explanations such as the interaction between different types of explanation methods [4] or the theoretical modeling of interesting explanations [5]. Our work does not focus on theoretical aspects of XAI and instead uses an MAS as a data generator that provides ground truth explanations.

Overall, while previous work on MAS and XAI exist, we are the first to use MAS-based simulation to generate synthetic data with ground truth explanations and enable explainable occupational fraud detection on fully unlabeled data while ensuring good explainer performance.

## 3 Methodology

This section gives an overview of the methodological aspects of our work. Firstly, we describe our MAS-based simulation, as well as our proposed additions that allow the generation of ground truth explanations. Secondly, we integrate our MAS-based simulation into the proposed framework for explainable fraud detection on unlabeled company data. Finally, we introduce the XAI models and XAI hyperparameters that are integrated into the proposed framework and explain the metrics used for quantitative evaluation of the XAI models within the framework.

### 3.1 MAS-Based Company Simulation

Our goal in this work is the selection of well-performing feature relevance explainers for explainable occupational fraud detection on given unlabeled company data, for which we require ground truth explanations to quantitatively evaluate different explainers. However, manual generation of ground truth explanations for fraud cases requires expensive expert annotations, where occupational fraud scenarios need to first be identified within the data and afterwards explanations may be derived from identified frauds. Since even the identification of frauds in the data is costly, in previous work we proposed to use agent-based simulation to obtain fraud detection labels for unlabeled company data [22]. Here, existing economics research, which provides established ways to model the business process of a production company through an MAS, formulates a theoretical backbone for a data-driven simulation [6]. Instead of learning the policies of the individual agents within the MAS from data, best practices for business and market strategies are used by individual planning, purchasing, production, and delivery agents that make up the company, resulting in a fully functional company simulation that uses established economics strategies, e.g., for purchasing goods or scheduling production. Using the agents, individual actions can be translated directly into ERP transactions and written into a relational database. Although this process can simulate an entire ERP system, we design the generator specifically to optimize existing explainable fraud detection methods that use a condensed table view of ERP data [23] and therefore configure the simulation to directly generate transactions for the aggregated table. This results in a single table that includes accounting information in 42 categorical and 10 numerical columns [23]. To generate normal business data that closely mimics the economic situation of a given company within the MAS-based simulation, prior knowledge of an underlying company regarding price trends and product information may be extracted from unlabeled company data and integrated into the simulation environment. Finally, the agent functionality within the MAS is extended to allow four different types of occupational fraud, namely Invoice Kickback, Selling Kickback, Non-Cash Larceny, and Corporate Injury [1]. These frauds include both indirect kickback schemes where employees are bribed externally to accept company trades at harmful prices during material purchases (Invoice Kickback) and product sales (Selling Kickback), as well as direct schemes where materials are stolen during production (Non-Cash Larceny) or employee wages are erroneously increased to damage the company (Corporate Injury). Frauds are committed directly during the data generation, where occurrences of these cases are tracked to obtain labeled normal and fraudulent data for occupational fraud detection. The resulting MAS-based data generator provides labeled training data to optimize fraud detection models, but requires further extensions to be capable of generating ground truth explanations for generated fraud cases.

### 3.2 MAS-Based Ground Truth Explanations

While the previously introduced MAS-based simulation enables the optimization of fraud detection models for completely unlabeled data, we aim to extend this simulation to further enable the optimization of feature relevance explainers. To achieve this, we note that the known causal relationships and well-defined processes within the simulation can also be used to obtain full explanations for each fraud case.

To generate ground truth explanations, we extend the MAS-based simulation. The simulation resembles a full description of the normal business process of the simulated company while also integrating known deviations from this normal process in the form of occupational fraud cases. We therefore extend processes and thus agents modeling non-normal behavior by defining rules and heuristics that highlight features and feature-combinations that deviate from the normal behavior and are indicative for each given fraud case. For instance, in the non-cash larceny case implemented in the MAS-based simulation, a delivery of raw materials is partially stolen during delivery and covered up by altering the already payed purchase order to book only the remaining materials into storage. The fraudulent change of the purchase covers up differences in true and internally tracked materials that would be revealed during the regular stocktaking processes within company warehouses, but leaves a transaction within the company ERP system that has substantially higher average prices for the ordered materials compared to normal orders. This anomalous entry can be identified through the recorded paid price and the amount of delivered materials, which show a much higher price for a single material than normal entries. In this instance, we mark all recorded prices that deviate from the normal non-fraudulent prices and additionally add all order amounts that are needed to identify the prices as too high for the ordered amounts. While kickback frauds show similar characteristics with too high purchase prices or too low sales prices per material for Invoice and Selling Kickbacks respectively, the relevant prices and amounts are recorded in different columns within the ERP system. Lastly, in Corporate Injury, reoccurring employee wages are higher than normal, which is only noticeable in prices which do not match the normal behavior observed in the remaining data. Overall, the identification of indicative ERP system entries allows the individual agents and processes to automatically provide an explanation for generated fraudulent data in the form of a binary ground truth explanation, where 1 to 5 features are highlighted per fraud case.

By adding an additional explanation logging module into the MAS-based simulation, we are able to provide a full mapping of fraud-related and unrelated features for each case of occupational fraud generated during the simulation run, and output a resulting explanation file that includes a binary ground truth for anomalous features for each generated fraud case.

### 3.3 Explainer Optimization Framework

Having extended the MAS-based simulation with ground truth explanations, we now propose a framework that enables the optimization of explainable fraud detectors for an unlabeled company dataset by leveraging the simulation to generate auxiliary data with both fraud detection labels and ground truth explanations. The entire framework is also visualized in Fig. 1. As using an MAS in our setting requires adapting the simulation to the actual data provided by a company to follow data characteristics such as price trends, products, and purchase quantities, we follow our previous work [22] to extract those characteristics and integrate them into the environment of the MAS-based simulation. Note that more complex processes used within the underlying company might require modeling additional business processes and process derivations within the simulation as well, which can be incorporated as additional agent behavior to further improve the simulation's ability to mimic the underlying company. This priming on real data allows the simulation to closely mimic the business scenario of a company, while additionally providing fraud detection labels and ground truth explanations that are missing in the original company data.

Using this labeled data, we can identify well-performing preprocessing methods, fraud detection models, and detector hyperparameters on the synthetic MAS data. Additionally, our proposed extension with ground truth explanations enables us to expand the evaluation to the explainer stage of our framework. After the best fraud detector is identified on the MAS-generated data, we apply multiple post-hoc feature relevance explainers and hyperparameter configurations on the previously identified best fraud detection model and select the most suitable explainer on the MAS-generated data by comparing the explainer outputs to the ground truth explanations on the feature level. Having identified the best fraud detector and explainer on the synthetic MAS data, we apply them directly to the original unlabeled company data in a completely unsupervised fashion.

### 3.4 Integrated XAI Methods

Our framework allows the integration of existing post-hoc feature relevance explainers to provide reasoning for the decision of the used fraud detector. Formally, for a given $d$-dimensional datapoint $x \in \mathcal{X} \subseteq \mathbb{R}^d$, a feature relevance explainer $f$ provides a relevance score to each input feature that expresses the feature's relevance to the overall model decision through $f : \mathcal{X} \rightarrow \mathbb{R}^d$. As these explanations may be obtained through multiple explainers that exhibit varying performance dependent on the underlying data [21], we design the framework as open as possible to integrate many post-hoc explainers.

[21] provides an overview of post-hoc feature relevance explainers currently used in anomaly detection, structuring them into four categories based on their reliance on the underlying data and the internals of the underlying model to explain. Data-specific explainers, which only rely on training data and internally train a new anomaly detector, are omitted in our implementation as they produce explanations that do not reflect the decision process of the framework's anomaly detector. We, however, implement all perturbation-based explainers, which only access a detectors input and output interface and are therefore fully model-agnostic, as well as all gradient-based explainers, that leverage detector gradients of the output with respect to the input and therefore enforce differentiability of the detector as only requirement. Lastly, we omit model-specific explainers, as they require a specific fraud detection architecture. We, however, note that these explainers can also be integrated in our modular framework as long as the choice of the fraud detector is limited to the corresponding architecture needed for the explainer. In total, we implement the following six post-hoc feature relevance explainers into our framework: The perturbation-based explainers LIME [14] and SHAP [12] are entirely model-agnostic explainers that repeatedly perturb a datapoint by replacing regions with alternate values and reason about relevant features by monitoring the model's output when feeding in the perturbed data.

The gradient-based explainers Saliency [17], Gradient×Input [16], Integrated Gradients (IG) [18], and Layer-wise Relevance Propagation (LRP) [2] all use slight variations of the gradient of the output with respect to the input to identify features where small changes strongly affect the model output.

As [21] showed that some explainers have hyperparameters that strongly effect the resulting explanations, we also include an option to test different explainer hyperparameters in our framework. SHAP allows the user to define the reference values that are used to replace feature values in the datapoint during perturbations. Tested approaches to obtain these reference values in anomaly detection are using cluster centroids of k-means clustering (k-means) or the mean of training data (mean), as well as using a zero vector (zeros) or using a gradient-based reference value generation approach designed by [19] (lopt) [21]. Integrated Gradients also allows for choosing a reference value that is used to take gradients at multiple points between the point to explain and the reference, thus adding additional information regarding the vicinity around the point itself. We use the mean of training data (mean), the zero vector (zeros), as well as the optimized variant by [19] (lopt) in our experiments. For more in-depth information on specific explainers and explainer hyperparameters we refer to [21].

### 3.5 Optimization of XAI Methods

The MAS-based simulation enables the generation of binary ground truth explanations, that indicate for each feature of a fraudulent datapoint whether the feature is indicative of the underlying fraudulent activity or can be seen as normal data entry. Using this binary ground truth, established metrics may be used to evaluate feature relevance explanations from different explainers. Area-under-the-receiver-operator-characteristic (ROC) scores can be used to evaluate how highly the indicative features of the binary ground truth rank within the feature relevance scores of a single datapoint explanation [9], giving a metric that rates how well the highest scoring features according to the explanation correspond to truly indicative features. Cosine similarity (COS) is another established metric for comparing feature relevance scores of a single datapoint to ground truth [11], and provides information on how well the relevance scores match the entire ground truth. Both metrics have been used to evaluate feature relevance explanations against binary ground truth by individually applying them to all available datapoints to explain and afterwards aggregating the results to obtain a single score of explanation quality [9,11]. Following previous experiments with binary explanation ground truth [21,24], we use the ROC score as main metric to determine the best explainers according to the synthetic MAS-generated data.

## 4 Experiments

In this section, we apply our framework for explainable fraud detection and evaluate the transferability of explanation quality from synthetic MAS-generated data to unlabeled data. We generate data and ground truth explanations with

**Table 1.** Number of transactions and fraud cases of ERPSim and generated MAS data. Note that ERPSim frauds are not labeled during MAS data generation and anomaly detector training and detection and explanation labels are only used for final evaluation.

| Dataset | Transactions | Frauds with Explanations | Invoice Kickback | Selling Kickback | Non-cash Larceny | Corporate Injury |
|---|---|---|---|---|---|---|
| ERPSim(1) | 36778 | 50 | 24 | 0 | 22 | 4 |
| MAS(1) | 92985 | 0 | 0 | 0 | 0 | 0 |
| $MAS(1)^{fraud}$ | 93356 | 223 | 51 | 104 | 66 | 2 |
| ERPSim(2) | 37407 | 86 | 30 | 0 | 48 | 8 |
| MAS(2) | 59378 | 0 | 0 | 0 | 0 | 0 |
| $MAS(2)^{fraud}$ | 64858 | 187 | 51 | 102 | 34 | 0 |

the extended MAS simulation for two occupational fraud detection datasets, and evaluate whether optimizing explainer and hyperparameter choices transfer from the generated data to the underlying unlabeled company data.

## 4.1 Experimental Setup

Our goal is to evaluate whether strong explainer performance on MAS-generated data translates to the underlying ERP data used to adapt the simulation. While the underlying company data is treated as entirely unlabeled during the adjustment of the MAS-based simulation, existing fraud labels and ground truth explanations for the data may be used after the fraud detectors and explainers are adjusted by the simulation, in order to test whether the simulation was able to identify detectors and explainers that indeed perform well on the underlying company data. As this evaluation requires company data with labeled fraud cases and explanations, we choose the ERPSim data from [20] that contains both known and labeled fraud cases and ground truth explanations.

We use the two datasets ERPSim(1) and ERPSim(2) as used in [22] that constitute two individual fiscal years of a make-to-stock cereal production company to undertake two complete evaluation runs of our framework. The MAS-based simulation, implemented in the Java Agent DEvelopment framework (JADE) [3], is used to generate both clean training data without frauds for training detectors and contaminated validation data with ground truth explanations for optimizing anomaly detectors and explainers. The resulting datasets are illustrated in Table 1, with data generation on a laptop taking less then one hour per dataset.

**Table 2.** Anomaly detection performance of the trained autoencoder for the two runs on MAS(1) (MAS(2)) and ERPSim(1) (ERPSim(2)) respectively. Reporting mean average precision (PR) score on synthetic data, which was used for parameter selection, and on ERPSim where selected parameters are simply applied. Results show competitive detection performance as also discussed in [22].

| run | $PR_{MAS}$ | $PR_{ERPSim}$ |
|---|---|---|
| run(1) | $17.5 \pm 2.6$ | $26.5 \pm 4.3$ |
| run(2) | $16.6 \pm 1.0$ | $53.7 \pm 5.7$ |

As we focus our evaluation on the newly introduced XAI component of the framework, we limit the choice for the anomaly detector to the autoencoder neural network [8] and only optimize its data preprocessing and hyperparameters in these experiments. We specifically select the autoencoder, since it already showed high detection performance in occupational fraud detection [15,22] and allows the use of gradient-based explainers due to its differentiability. The anomaly detection performance of the optimized autoencoders on both underlying company data and the MAS-simulated datasets is shown in Table 2, with the performance of MAS-optimized models transferring well to the underlying unlabeled company data as previously observed in [22].

Using the fully trained autoencoders, we apply multiple post-hoc feature relevance explainers with varying explainer hyperparameters as described in Sect. 3.4 to explain the fraud cases contained in the synthetic MAS data. After obtaining explanations from all explainers with associated hyperparameter configurations, we identify the most suited explainer by comparing explanations to the ground truth explanations generated by the MAS through the XAI metrics described in Sect. 3.5. Having identified the best fraud detector and explainer, we apply them to the original ERPSim data in an unsupervised fashion. Finally, we run the explainers on the applied detector and evaluate them using the ground truth explanations of the ERPSim data to assess whether the proposed MAS-based optimization provides good explainability on a given unlabeled dataset.

**Table 3.** Mean and standard variation of explainer performance on ERPSim(1) and the simulated MAS(1) data. Best and second-best results highlighted in bold and underline.

| Explainer | reference | $\text{ROC}_{\text{XAI}}^{MAS(1)}$ | $\text{COS}_{\text{XAI}}^{MAS(1)}$ | $\text{ROC}_{\text{XAI}}^{ERPSim(1)}$ | $\text{COS}_{\text{XAI}}^{ERPSim(1)}$ |
|---|---|---|---|---|---|
| Saliency | | $51.3 \pm 25.2$ | $-1.7 \pm 18.5$ | $62.3 \pm 19.6$ | $13.4 \pm 21.5$ |
| Gradient×Input | | $71.1 \pm 26.2$ | $33.6 \pm 39.3$ | $\mathbf{84.7 \pm 12.8}$ | $\mathbf{59.9 \pm 16.9}$ |
| LRP | | $55.2 \pm 28.3$ | $2.0 \pm 39.9$ | $59.7 \pm 15.0$ | $20.2 \pm 16.7$ |
| IG | mean | $71.8 \pm 16.9$ | $\underline{37.6 \pm 25.4}$ | $65.1 \pm 12.6$ | $18.1 \pm 22.8$ |
| IG | zeros | $57.1 \pm 24.8$ | $-14.6 \pm 24.7$ | $62.4 \pm 14.5$ | $-15.9 \pm 17.3$ |
| IG | lopt | $\underline{73.1 \pm 19.2}$ | $30.6 \pm 26.3$ | $80.7 \pm 15.6$ | $47.4 \pm 19.6$ |
| LIME | k-means | $69.6 \pm 7.4$ | $16.8 \pm 4.2$ | $63.1 \pm 10.5$ | $22.6 \pm 14.1$ |
| SHAP | k-means | $55.8 \pm 23.1$ | $11.7 \pm 35.7$ | $63.6 \pm 15.8$ | $35.7 \pm 14.6$ |
| SHAP | mean | $68.3 \pm 20.5$ | $31.5 \pm 30.4$ | $57.3 \pm 18.0$ | $18.1 \pm 29.0$ |
| SHAP | zeros | $59.9 \pm 20.1$ | $-13.8 \pm 20.0$ | $63.8 \pm 14.8$ | $-18.9 \pm 12.7$ |
| SHAP | lopt | $\mathbf{82.6 \pm 13.9}$ | $\mathbf{49.4 \pm 29.1}$ | $\underline{83.7 \pm 12.8}$ | $\underline{57.7 \pm 21.7}$ |

## 5 Results

Explainers are applied to both the optimized autoencoder trained on the MAS data, as well as the applied autoencoder on the ERPSim data to obtain explanations for each explainer and explainer hyperparameter configuration. Then, we evaluate the quality of the resulting explanations using the respective ground truth, observing whether the MAS data is sufficient to identify highly performing explainers that transfer well to the ERPSim data.

At first, we examine the explanation performance on the MAS(1) data and the underlying ERPSim(1) company data in Table 3. When observing the general explanation scores across all configurations, our results show that a high performance on MAS(1) translates well to the underlying ERPSim(1) data. Especially for the SHAP and IG explainers different reference values can cause both high and low explanation quality, matching existing results that explanations are highly sensitive to reference values [21]. The optimized (lopt) references obtain the highest scores on the synthetic MAS(1) data for both SHAP and IG and would therefore be selected by our framework. These explainer choices transfer well to the ERPSim(1) data, with both choices achieving very high explanation scores. While our framework is clearly able to identify well-performing explainers, it does not find the explainer with the highest overall performance on the ERPSim(1) data, which is the Gradient×Input method in this instance, indicating that this method does not transfer from the simulated fraud cases to the cases contained in ERPSim. Nevertheless, we observe that explainers that perform highly on the synthetic MAS(1) data, and would therefore be selected for use on the underlying ERPSim(1) data in our framework, also consistently achieve high explanation scores on the ERPSim(1) data.

**Table 4.** Mean and standard variation of explainer performance on ERPSim(2) and the simulated MAS(2) data. Best and second-best results highlighted in bold and underline.

| Explainer | reference | $\text{ROC}_{\text{XAI}}^{MAS(2)}$ | $\text{COS}_{\text{XAI}}^{MAS(2)}$ | $\text{ROC}_{\text{XAI}}^{ERPSim(2)}$ | $\text{COS}_{\text{XAI}}^{ERPSim(2)}$ |
|---|---|---|---|---|---|
| Saliency | | $52.4 \pm 18.6$ | $-1.8 \pm 16.9$ | $61.4 \pm 11.9$ | $6.9 \pm 10.5$ |
| Gradient×Input | | $61.1 \pm 30.4$ | $17.9 \pm 42.1$ | $86.7 \pm 15.4$ | $\underline{65.2 \pm 18.0}$ |
| LRP | | $66.3 \pm 30.9$ | $28.7 \pm 51.1$ | $66.3 \pm 19.9$ | $28.0 \pm 22.8$ |
| IG | mean | $65.9 \pm 32.8$ | $30.5 \pm 44.9$ | $74.4 \pm 19.4$ | $34.0 \pm 28.7$ |
| IG | zeros | $67.6 \pm 21.4$ | $0.7 \pm 18.1$ | $71.1 \pm 12.9$ | $-5.2 \pm 10.6$ |
| IG | lopt | $73.1 \pm 21.1$ | $37.2 \pm 24.3$ | $\mathbf{91.1 \pm 6.1}$ | $54.7 \pm 15.2$ |
| LIME | k-means | $55.3 \pm 9.8$ | $15.7 \pm 6.2$ | $75.0 \pm 6.6$ | $31.3 \pm 9.0$ |
| SHAP | k-means | $60.8 \pm 34.6$ | $1.2 \pm 49.3$ | $76.9 \pm 13.0$ | $46.3 \pm 11.2$ |
| SHAP | mean | $75.8 \pm 25.6$ | $\underline{38.4 \pm 36.9}$ | $69.2 \pm 24.6$ | $34.7 \pm 38.8$ |
| SHAP | zeros | $\underline{77.4 \pm 25.7}$ | $-0.0 \pm 20.0$ | $75.2 \pm 12.1$ | $-8.0 \pm 12.9$ |
| SHAP | lopt | $\mathbf{86.5 \pm 16.3}$ | $\mathbf{61.7 \pm 28.1}$ | $\underline{89.3 \pm 10.0}$ | $\mathbf{66.9 \pm 19.1}$ |

Even more importantly, our analysis reveals the absence of configurations that perform well on MAS(1) yet exhibit complete failure on ERPSim(1). This mitigates concerns regarding the framework producing misleading explanations when applied to actual company data. Our analysis indicates that, while pinpointing the single most effective explainer on the company data remains challenging, the ability of our framework to identify highly effective explainers nonetheless represents a significant advancement. This outcome is particularly valuable given that, in the absence of our framework, selecting explainers and reference values for company data in real-world applications would be largely arbitrary, owing to the absence of ground truth. Observing our evaluation results, such an arbitrary choice of explainers and their hyperparameters could prove detrimental in this scenario, as our results on ERPSim(1) include explainers that perform close to entirely random noise (50.0 ROC and 0.0 COS) e.g. when using the LRP explainer or the SHAP explainer with mean reference values.

In the analysis of the second dataset, ERPSim(2), and the correspondingly generated MAS(2) data, as detailed in Table 4, we observe consistent overall patterns in the performance of explainers and the transferability of explanation scores between the two datasets. While the best explainer choice according to the MAS(2) data again does not achieve the best total scores on the ERPSim(2) data on all metrics, it still manages to perform very well compared to all other explainer choices. Other high performing explainer choices on the MAS(2) data also manage to score well on the ERPSim(2) data, avoiding particularly low explanation scores on ERPSim(2), as seen with the Saliency explainer.

Overall, our framework is able to successfully identify explainer and hyperparameter choices that provide high quality explanations on unlabeled occupational fraud detection datasets through the use of MAS-based data generation.

## 6 Conclusion

In this study, we introduced a solution to address the complexities of selecting optimal feature relevance explainers and their corresponding hyperparameters within unlabeled occupational fraud detection datasets. Our methodology extends prior research on MAS-based data generation for modeling company data by reflecting the economic state and business processes of a given company. Specifically, we advanced the simulation capabilities to generate ground truth explanations for fraud instances addressing the lack of validation of post-hoc feature relevance explainers particularly in the setting of anomaly detection with imbalanced, rare and unlabeled instances such as fraud. The efficacy of our framework was validated on two fraud detection datasets, where it consistently showed its capabilities to identify well performing explanation methods. At the same time, our findings underscore the importance of a methodological selection process such as our proposed framework for choosing suitable explainers, as arbitrary explainer choices may result in near random explanations. The application of MAS-based data generators has proven to be a useful approach in deriving ground truth explanations, thereby contributing to the improvement and validation of explainable fraud detection methods. While our proposed work relies on economics research to construct the simulation and is therefore currently limited to the task of occupational fraud detection, future research might be directed towards further application domains of XAI where foundations on data generation are available to fuel simulations. In addition, the positive results from our experiments justify further evaluation of our framework for explainable occupational fraud detection directly within company operation in future research.

## References

1. ACFE: Occupational Fraud 2022: A Report to the nations (2022). https://legacy.acfe.com/report-to-the-nations/2022/. Accessed 23 Jan 2024
2. Bach, S., Binder, A., Montavon, G., Klauschen, F., Müller, K.R., Samek, W.: On pixel-wise explanations for non-linear classifier decisions by layer-wise relevance propagation. PLoS ONE **10**(7), e0130140 (2015)
3. Bellifemine, F., Bergenti, F., Caire, G., Poggi, A.: Jade-a java agent development framework. In: Bordini, R.H., Dastani, M., Dix, J., El Fallah Seghrouchni, A. (eds.) Multi-Agent Programming. Multiagent Systems, Artificial Societies, and Simulated Organizations, vol. 15, pp. 125–147. Springer, Boston (2005). https://doi.org/10.1007/0-387-26350-0_5
4. Ciatto, G., Calegari, R., Omicini, A., Calvaresi, D.: Towards XMAS: eXplainability through Multi-Agent Systems. In: AI&IoT@AI*IA (2019)
5. Ciatto, G., Schumacher, M.I., Omicini, A., Calvaresi, D.: Agent-based explanations in AI: towards an abstract framework. In: Calvaresi, D., Najjar, A., Winikoff, M., Främling, K. (eds.) EXTRAAMAS 2020. LNCS (LNAI), vol. 12175, pp. 3–20. Springer, Cham (2020). https://doi.org/10.1007/978-3-030-51924-7_1
6. Domínguez, R., Cannella, S., Framinan, J.M.: SCOPE: a multi-agent system tool for supply chain network analysis. In: IEEE EUROCON 2015 - International Conference on Computer as a Tool (EUROCON), pp. 1–5 (2015)
7. Fuchs, A., Fuchs, K., Gwinner, F., Winkelmann, A.: A meta-model for real-time fraud detection in ERP systems. In: HICSS (2021)
8. Goodfellow, I., Bengio, Y., Courville, A.: Deep Learning. MIT Press, Cambridge (2016)
9. Hägele, M., et al.: Resolving challenges in deep learning-based analyses of histopathological images using explanation methods. Sci. Rep. **10**(1), 1–12 (2020)
10. Heuillet, A., Couthouis, F., Díaz-Rodríguez, N.: Collective eXplainable AI: explaining cooperative strategies and agent contribution in multiagent reinforcement learning with shapley values. IEEE Comput. Intell. Mag. **17**(1), 59–71 (2022)
11. Kauffmann, J., Ruff, L., Montavon, G., Müller, K.R.: The clever Hans effect in anomaly detection. arXiv preprint arXiv:2006.10609 (2020)
12. Lundberg, S.M., Lee, S.I.: A unified approach to interpreting model predictions. In: Advances in Neural Information Processing Systems, pp. 4765–4774 (2017)
13. Ohana, J.J., Ohana, S., Benhamou, E., Saltiel, D., Guez, B.: Explainable AI (XAI) models applied to the multi-agent environment of financial markets. In: Calvaresi, D., Najjar, A., Winikoff, M., Främling, K. (eds.) EXTRAAMAS 2021. LNCS (LNAI), vol. 12688, pp. 189–207. Springer, Cham (2021). https://doi.org/10.1007/978-3-030-82017-6_12
14. Ribeiro, M.T., Singh, S., Guestrin, C.: Why should i trust you?: explaining the predictions of any classifier. In: 22nd ACM SIGKDD International Conference on Knowledge Discovery and Data mining, pp. 1135–1144. ACM (2016)
15. Schreyer, M., Sattarov, T., Schulze, C., Reimer, B., Borth, D.: Detection of accounting anomalies in the latent space using adversarial autoencoder neural networks. In: 2nd KDD Workshop on Anomaly Detection in Finance. ACM (2019)
16. Shrikumar, A., Greenside, P., Shcherbina, A., Kundaje, A.: Not just a black box: learning important features through propagating activation differences. arXiv preprint arXiv:1605.01713 (2016)
17. Simonyan, K., Vedaldi, A., Zisserman, A.: Deep inside convolutional networks: visualising image classification models and saliency maps. In: ICLR (2014)

18. Sundararajan, M., Taly, A., Yan, Q.: Axiomatic attribution for deep networks. In: 34th International Conference on Machine Learning, vol. 70, pp. 3319–3328. JMLR. org (2017)
19. Takeishi, N., Kawahara, Y.: A characteristic function for shapley-value-based attribution of anomaly scores. Trans. Mach. Learn. Res. (2023)
20. Tritscher, J., Gwinner, F., Schlör, D., Krause, A., Hotho, A.: Open ERP system data for occupational fraud detection (2022)
21. Tritscher, J., Krause, A., Hotho, A.: Feature relevance XAI in anomaly detection: reviewing approaches and challenges. Front. Artif. Intell. **6** (2023)
22. Tritscher, J., Roos, A., Schlör, D., Hotho, A.: Occupational fraud detection through agent-based data generation. In: Machine Learning and Principles and Practice of Knowledge Discovery in Databases. Communications in Computer and Information Science (2023)
23. Tritscher, J., Schlör, D., Gwinner, F., Krause, A., Hotho, A.: Towards explainable occupational fraud detection. In: Koprinska, I., et al. (eds.) PKDD 2022. CCIS, vol. 1753, pp. 79–96. Springer, Cham (2023). https://doi.org/10.1007/978-3-031-23633-4_7
24. Tritscher, J., Wolf, M., Hotho, A., Schlör, D.: Evaluating feature relevance XAI in network intrusion detection. In: Longo, L. (ed.) xAI 2023. CCIS, vol. 1901, pp. 483–497. Springer, Cham (2023). https://doi.org/10.1007/978-3-031-44064-9_25
25. Yepmo, V., Smits, G., Pivert, O.: Anomaly explanation: a review. Data Knowl. Eng. **137**, 101946 (2022)

# Leveraging Weakly Supervised and Multiple Instance Learning for Multi-label Classification of Passive Acoustic Monitoring Data

Ilira Troshani[1,2(✉)], Thiago S. Gouvêa[1,2], and Daniel Sonntag[1,2]

[1] Interactive Machine Learning, German Research Centre for Artificial Intelligence (DFKI), Kaiserslautern, Germany
{ilira.troshani,thiago.gouvea,daniel.sonntag}@dfki.de
[2] Carl von Ossietzky University of Oldenburg, Oldenburg, Germany

**Abstract.** Data collection and annotation are time-consuming, resource-intensive processes that often require domain expertise. Existing data collections such as animal sound collections provide valuable data sources, but their utilization is often hindered by the lack of fine-grained labels. In this study, we examine the use of existing weakly supervised methods to extract fine-grained information from existing weakly-annotated data accumulated over time and alleviate the need for collection and annotation of fresh data. We employ TALNet, a Convolutional Recurrent Neural Network (CRNN) model and train it on 60-second sound recordings labeled for the presence of 42 different anuran species and compare it to other models such as BirdNet, a model for detection of bird vocalisation. We conduct the evaluation on 1-second segments, enabling precise sound event localization. Furthermore, we investigate the impact of varying the length of the training input and explore different pooling functions' effects on the model's performance on AnuraSet. Finally, we integrate it in an interactive user interface that facilitates training and annotation. Our findings demonstrate the effectiveness of TALNet and BirdNet in harnessing weakly annotated sound collections for wildlife monitoring. Our method not only improves the extraction of information from coarse labels but also simplifies the process of annotating new data for experts.

**Keywords:** Weakly Supervised Learning · Semi-automatic data annotation · AI Transfer · Sustainability

## 1 Introduction

Passive acoustic monitoring (PAM), has emerged as a key technology for wildlife monitoring [1] while using acoustic sensors and provides a way to promote biodiversity, assess and understand the impact of climate change, and develop intervention strategies to preserve ecosystems. However, handling the large amount


A. Hotho and S. Rudolph (Eds.): KI 2024, LNAI 14992, pp. 260–272, 2024.
https://doi.org/10.1007/978-3-031-70893-0_19

of data generated by PAM still poses a barrier for adoption by both researchers and biodiversity managers [2,3]. Although a wide range of supervised machine learning methods for analyzing PAM datasets (e.g., for sound event detection) exist [4], their application is often constrained by the availability of domain-specific annotated data. Biologists traditionally rely on museum collections for studying biodiversity [5]. In modern times, multimedia registers have become increasingly important and recognized as valuable in common practice. Among these, sound archives and collections hold significant importance [6,7]. Several such collections exist, such as FNJV[1], Macaulay library[2], and Xeno-Canto[3]. These resources serve as valuable sources of annotated data for training models to automate sound event detection in large PAM datasets. However, their potential for this task is currently limited because these sound files are weakly annotated, meaning that sound recordings are labeled only at the file level, with no information about the timestamps of specific identifying species sounds. This problem is further compounded by the presence of multiple signals in these recordings, such as other species co-occurring in the same soundscape, and the voice of the naturalist who performed the recording, often speaking into the microphone and providing metadata such as species name and a description of the recording context. Effective utilization of such knowledge sources for powering ML tools rely on isolating the meaningful, identifying portions of the sound recordings. In this paper, we propose a weakly supervised method to leverage weakly annotated data and generate training data for ML models for species level sound event detection in PAM datasets (Fig. 1).

## 2 Related Work

Deep learning methods have proven very useful for detection of sound events in PAM datasets. Among the most popular convolutional neural network (CNN) architectures applied to PAM are ResNet [8], VGG [9] and DenseNet [10]. Even though they were created for computer vision tasks, these architectures proved to be very efficient in analyzing sound data. Kahl et al. [11] developed BirdNet, an EfficientNet-based model for detection of bird vocalisations. Other popular methods include convolutional recurrent neural networks (CRNNs), that combine the advantages of both CNNs and RNNs [12–14]. [15,16] compare the performance of different models pre-trained on ImageNet [17] on different PAM datasets. They show that transfer learning can be used successfully on small PAM datasets with few samples per species.

Availability of training data is crucial for the development of supervised ML models that generalize well to new recording locations, background sounds and regional differences in the species calls. BirdNet, a popular pre-trained model for species-level classification of bird vocalizations [11], is trained on datasets that consist largely of weakly annotated focal recordings. For detecting the presence of

[1] https://www2.ib.unicamp.br/fnjv/.
[2] https://www.macaulaylibrary.org/.
[3] https://xeno-canto.org/.

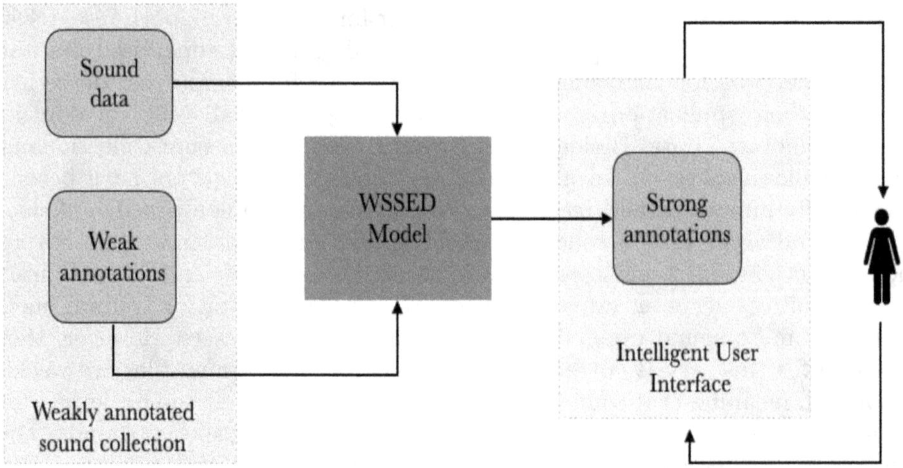


**Fig. 1.** A schematic of our proposed approach

target sounds, they used heuristic image processing methods for signal-strength estimation [20]. These recordings are often acquired with professional equipment and thus have a high sound-to-noise ratio (SNR).

In recent years, there has been a notable surge of interest within the research community in the domain of weakly supervised sound event detection (WSSED), which has been notably catalyzed by initiatives like the Detection and Classification of Acoustic Scenes and Events (DCASE) challenges and the release of extensive audio datasets such as AudioSet [21] that provide baselines for the development and evaluation of ML methods related to sound event detection (SED) and specifically WSSED. Kumar et al. [22] propose that WSSED can be treated as a problem of Multi Instance Learning; from this perspective, every audio file can be viewed as a bag $B$ of instances $x_i$ of sound events. If one or more instances have label $y_i = 1$, then the whole bag is considered to be positive and is assigned label $Y$. They explore SVM and neural network based approaches trained on weak labels for detection and achieve temporal localiza-

**Table 1.** Performance metrics on AnuraSet

| Architecture | Global | 1 s | | |
|---|---|---|---|---|
| | F1 Score | F1 Score | Precision | Recall |
| TALNet [18] | 90.00 | 64.68 | 50.82 | 88.94 |
| ResNet-50 [8] | **90.11** | 63.54 | 54.79 | 75.60 |
| ResNet-18 [8] | 88.89 | 62.85 | 56.73 | 70.46 |
| VGGish [19] | 69.64 | 41.67 | 31.36 | 59.92 |
| BirdNet [11] | 84.80 | **71.14** | 67.42 | 75.30 |

tion of sound events, harnessing richer information from the annotations within the original dataset. Xu et al. [23] introduce an attention mechanism, replacing the ReLU activation function after each convolution with GLUs. Wang et al. [18] propose TALNet, a CRNN for audio tagging and localization. They identify the best pooling function for the task. More recent approaches propose transformer-based methods for WSSED [24,25]. Current approaches combine embeddings extracted from pre-trained models such as BEATS [26] with CRNN classifiers aligning with the newest requirements of the DCASE challenges that use heterogeneous datasets that contain unlabeled, weakly labeled and synthetic datasets with strong annotations. In our work, we focus only on methods for weakly annotated datasets and how to use them to enrich annotations for PAM. Some of the deep learning methods for detection of sound events in animal sounds datasets are associated with a user interface for result inspection and easy annotation [27–29]. We contribute to this body of research by exploring how WSSED can be used for wildlife monitoring to support biodiversity conservation.

## 3 Implementation

In our implementation, we apply existing weakly supervised methods to extract detailed information from weakly-annotated data.

***Dataset***
For our experiments, we use AnuraSet, a recently released benchmark PAM dataset comprised of 1612 min of omindirectional recordings from four different sites in two Brazilian biomes: Cerrado and Atlantic Forest [30]. The dataset consists of 60 s long recording files, as well as manually created expert annotations for 42 species of anurans (frogs and toads). The annotations consist of strong labels, i.e., species identity plus on- and offset times for each call occurrence.

**Table 2.** Comparison of pooling functions for BirdNET finetuned on 60-second long inputs and evaluated on either 60-second (global) or 1-second long segments.

| Pooling Function | Global | 1 s | | |
|---|---|---|---|---|
| | F1 Score | F1 Score | Precision | Recall |
| Average (2) | 79.90 | 65.64 | 52.20 | 88.40 |
| Max pooling (1) | 81.41 | 36.70 | 74.01 | 24.40 |
| Exponential Softmax(4) | 81.57 | 64.38 | 51.10 | 86.98 |
| Linear Softmax (3) | **84.80** | **71.14** | 67.42 | 75.30 |
| Attention pooling (5) | 80.84 | 67.63 | 55.45 | 86.65 |

***Data Preprocessing***
The audio recordings are represented as Mel-frequency single channel spectrograms $S \in R^{m \times n}$, where $m = 64$ is the number of frequency bins and $n$ is

the number of frames. As "frame" we denote the minimal time segment, so $n$ depends on the length of the input files. For the 60-second long recordings $n = 2400$. To compute the spectrograms, we use a window size of 1102 and hop length 551. Raw recordings are resampled to 22kHz. For comparing performance when training is carried with inputs of different durations, we partition the 60-second audio recordings from the training set into non-overlapping 9-second and 3-second long segments. We keep the same frame length and number of frequency bins as described in TALNet [18], but adjust the number of frames according to the segment length. Considering the unbalanced nature and relatively small size of the dataset when training with 60-second long input, we perform iterative stratification to ensure balanced train and test splits, with 80% for training and 20% for test. For each segment, a vector of binary labels is generated to indicate presence of calls from each of the 42 species; each entry is set to 1 if a call of that species is present anywhere in the corresponding segment, and 0 otherwise. To create the Mel-frequency spectrograms, we use native torchaudio [31] transforms for audio processing.

***Model Architecture***

For the sound event detection and localization we use TALNet [18] a convolutional recurrent neural network developed for audio tagging and localization on AudioSet and the DCASE challenge 2017. The network consists of three convolutional layers, a pooling layers and one recurrent layer.

To perform WSSED using transfer learning on the PAM dataset, we use ResNet-50 and ResNet-18 [8] pretrained on ImageNet [17], VGGish [32] pretrained on AudioSet and BirdNet [11] pretrained on bird vocalisations, leveraging their feature extraction capabilities to capture fundamental patterns typical for spectrograms.

All presented models treat WSSED as a multiple instance learning problem; specifically, the strategy consists of training models to make predictions for each frame of a multi-frame data point, and then apply a pooling function. The pooling function combines frame level predictions into segment level ones while retaining important information. The pooling layer applies different pooling functions such as Max Pooling (Eq. 1), Average Pooling (Eq. 2), Linear Softmax (Eq. 3), Exponential Softmax (Eq. 4) and Attention pooling (Eq. 5)

$$y = max_i y_i \tag{1}$$

$$y = \frac{1}{n}\sum_i y_i \tag{2}$$

$$y = \frac{\sum_i y_i^2}{\sum_i y_i} \tag{3}$$

$$y = \frac{\sum_i y_i exp(y_i)}{\sum_i exp(y_i)} \tag{4}$$

$$y = \frac{\sum_i y_i w_i}{\sum_i w_i} \tag{5}$$

***Experimental Setup***
We train the network on samples of the AnuraSet with weak labels (60-seconds long samples) and evaluate the performance using the strong labels (1-second). In the training procedure, we use the Adam optimizer [33], and a learning rate of $3 \times 10^{-4}$. As a loss function we use the binary cross entropy loss:

$$L(y, \hat{y}) = -(y * log(\hat{y} + (1 - y) * log(1 - \hat{y})) \tag{6}$$

In Eq. 6, $y$ represents the true labels, while $\hat{y}$ the predicted probabilities. Time and frequency masking are applied as suggested in SpecAugment [34]. We create shuffled batches of size 32 samples and train for 100 epochs.

Evaluation metric is the *global F1 score* (Eq. 7, 8) assessing how well the model can identify only the presence or absence of events within an audio file, and *1-second F1 score*, an indication of how well the model can localise sound events in an audio file with a precision of 1 s.

$$F1 = 2 \times \frac{precision \times recall}{precision + recall} \tag{7}$$

where,

$$\begin{aligned} precision &= \frac{TP}{TP + FP} \\ recall &= \frac{TP}{TP + FN} \end{aligned} \tag{8}$$

To compare the model performance on different input lengths, we also conduct experiments with 9-second and 3-second long files.

## 4 Interactive Training and Annotation Tool

We create a user interface to allow the user to train a model from weakly annotated data and use it to inspect results and change weakly-supervised learning parameters at inference time, enabling them to do semi-automatic annotation of new unlabeled data.

In our interface the user can change parameters such as pooling function and prediction window size. This human intervention at test time allows for improvement of the final results. We design the tool with two user personas in mind: the expert and the novice, therefore we create a training (Fig. 2) and an annotation pane (Fig. 3a).

### 4.1 Training

The training pane has three main areas that are dynamically filled upon user action to minimize user input and collapsible for less visual clutter. Here the expert can select a weakly-annotated dataset and a pretrained model, such as BirdNet [11] or ResNet-50 [8]. As the pooling function plays an important role in the performance of weakly supervised learning models [18,35] we make it a configurable parameter.

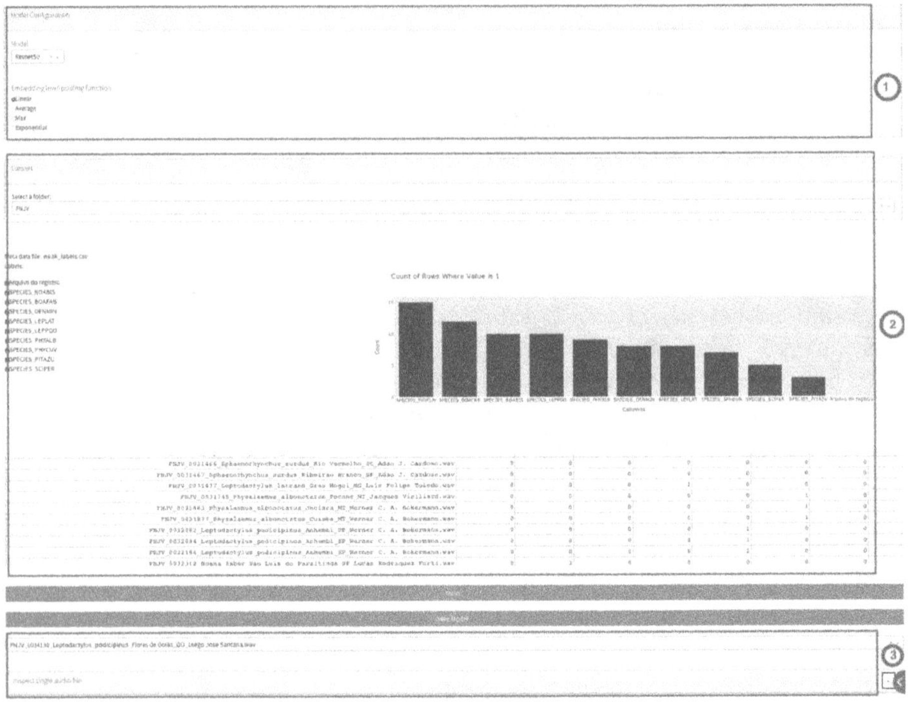

**Fig. 2.** Training pane: Domain experts can train a model with a backbone of choice of pretrained models and configure parameters such as pooling function (model configuration area (1)) using a weakly annotated dataset, such as a subset of the FNJV dataset for anurans. Upon selection of the dataset, the label distribution is computed for a better overview of the dataset (dataset area (2)). At the bottom (single file inspection area (3)), the user can inspect single spectrograms, here collapsed.

### 4.2 Annotation

In the annotation pane, the user can upload and annotate an audio file. Upon upload, inference runs in the background while the spectrogram of the audio file is computed and displayed. Upon selection, the time intervals are displayed as gray overlays on the spectrogram. The user can modify them, delete or add new ones. For the more experienced user we have created a Settings section (Fig. 3a) to modify post-hoc model parameters related to weakly supervised learning such as pooling function, prediction threshold to make the model more or less sensitive towards certain classes and prediction window. Currently, our tool is not meant for automatic dataset annotation, but to help novice and expert users, generate accurate annotations in a more efficient way. The application is implemented in Python using Dash[4] and PyTorch [36].

[4] https://dash.plotly.com.

(a) Single File annotation

(b) WSL Settings of the weakly supervised learning model.

**Fig. 3.** Annotation pane: Upon file upload, the model annotates the file with the detected species' calls and onset, offset time. The user can modify the predictions if necessary (a) using UI components such as slider, buttons and input fields. The more experienced machine learning practitioner can adjust the post-hoc parameters of the weakly supervised setting to achieve the best performance such as pooling function, prediction threshold and the precision of the prediction window for the onset-offset times (0.1 – 1 s)(b).

## 5 Results

We assess the model's performance on PAM data using the AnuraSet dataset. We start by analyzing models trained on 60-second long inputs. To compute F1 scores for both tagging and localization tasks, we use weak and strong labels. For this, we make predictions on 1-second windows by aggregating probabilities across 10 frames, followed by the application of a threshold as described in [18]. To compare the performance of TALNet with a pretrained model, we use ResNet-50, ResNet-18, VGGish and BirdNet. In Table 1 we report the global and 1-second F1 scores on AnuraSet. Since the performance of the model on 1 s segments is essential for our goal, we report the related precision and recall too. As it is evident from the table TALNet performs better than ResNet-50 in the localization task (1 s segments) and slightly worse in the tagging task

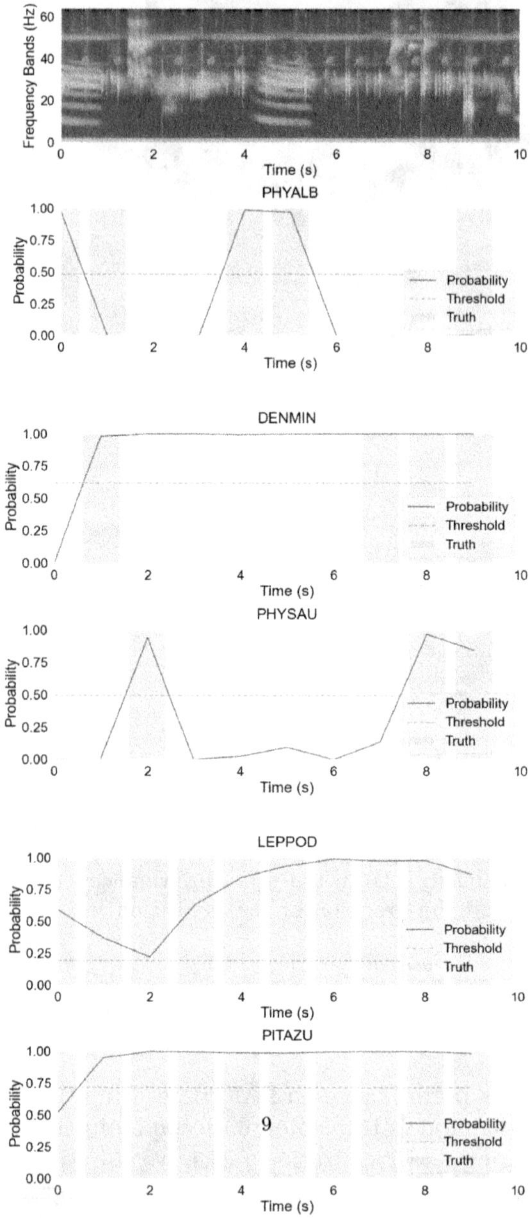


**Fig. 4.** Spectrogram of a representative 10-second audio segment, and bar plots with predicted and observed labels for five example species at 1-second resolution. Gray bars are true labels; green line is predicted probability of species occurrence; yellow line is the decision threshold. Notice that species `LEPPOD`, `PHYALB`, `PHYSAU`, and `PITAZU` are correctly localized by the model, while `DENMIN` gets mistakenly identified as occurring during the entire duration of the audio clip. (Color figure online)

**Table 3.** Micro F1 score of the BirdNet model trained on inputs of varying duration (3, 9, or 60 s), and evaluated globally and on 1-second long segments. Performance drops as duration of training samples increases.

| Length of Training Input | Micro F1 Score | |
|---|---|---|
| | Global | 1 s |
| 3 s | 90.27 | 74.50 |
| 9 s | 89.81 | 71.00 |
| 60 s | 84.80 | 71.14 |

(60 s segments) but is outperformed by BirdNet in the localization task (1 s segements). Figure 4 illustrates prediction results for a 10 s long file with five species present.

Table 2 shows the results for different pooling functions on AnuraSet. To analyze the influence of input length in performance, we finetune and evaluate BirdNet on three different input lengths (Table 3). The decrease of the input length to 9 s improves the performance by 3.36% for the 1-second F1 score and 5.47% for the global F1 score, indicating that the model's sensitivity to input length is task dependent. This finding confirms the results in [37].

## 6 Conclusion and Future Work

In this paper, we proposed the use of the existing CRNN based approach TALNet and pretrained models such as BirdNet to harness more information from weakly annotated data for wildlife monitoring and evaluated its performance on a benchmark PAM dataset. We demonstrated that domain transfer of existing models developed for different acoustic environments, such as the one in AudioSet to passive acoustic monitoring (PAM) datasets does not always require a complex model architecture and input modifications. With TALNet we achieved a 90% global F1 score in the tagging task while with BirdNet 71.14% F1 score in the localization task of animal sounds for 60-second long recordings. Further, we implemented a user interface to allow domain experts to train the model and investigate the results. Designed with the human "component" in mind, our tool makes model configuration and inference as transparent as possible and allows the user to modify the results if necessary. We plan to refine our tool by implementing more feedback-loops and include iterative model training. Future research includes applying our approach to PAM collections to generate annotated data from the weakly labelled recordings. Based on the promising results using AnuraSet, we could train BirdNet using the recordings of anuran calls and the weak annotations from museum collections such as the FNJV collection and calculate the evaluation metrics using the strong labels from AnuraSet.

**Acknowledgements.** This research is part of the Computational Sustainability & Technology project area(https://cst.dfki.de/), and has been supported by the Ministry for Science and Culture of Lower Saxony (MWK), the Endowed Chair of Applied Artificial Intelligence, Oldenburg University, and DFKI.

## References

1. Sugai, L.S.M., Silva, T.S.F., Ribeiro, J.W., Llusia, D.: Terrestrial passive acoustic monitoring: review and perspectives. Bioscience **69**(1), 15–25 (2019). https://doi.org/10.1093/biosci/biy147. Accessed 2023-03-01
2. Tuia, D., et al.: Perspectives in machine learning for wildlife conservation. Nat. Commun. **13**(1), 792 (2022)
3. Gouvêa, T.S., et al.: Interactive machine learning solutions for acoustic monitoring of animal wildlife in biosphere reserves. In: Proceedings of the Thirty-Second International Joint Conference on Artificial Intelligence, pp. 6405–6413. International Joint Conferences on Artificial Intelligence Organization, Macau, SAR, China (2023). https://doi.org/10.24963/ijcai.2023/711, https://www.ijcai.org/proceedings/2023/711. Accessed 16 Aug 2023
4. Stowell, D.: Computational bioacoustics with deep learning: a review and roadmap. PeerJ **10**, 13152 (2022). https://doi.org/10.7717/peerj.13152. Accessed 2023-08-01
5. Meineke, E.K., Davies, T.J., Daru, B.H., Davis, C.C.: Biological collections for understanding biodiversity in the Anthropocene. Philos. Trans. Royal Soc. B: Biol. Sci. **374**(1763), 20170386 (2018). https://doi.org/10.1098/rstb.2017.0386. Accessed 2023-08-01
6. Dena, S., Rebouças, R., Augusto-Alves, G., Zornosa-Torres, C., Pontes, M.R., Toledo, L.F.: How much are we losing in not depositing anuran sound recordings in scientific collections? Bioacoustics **29**(5), 590–601 (2020). https://doi.org/10.1080/09524622.2019.1633567. Accessed 2023-08-01
7. Sugai, L.S.M., Llusia, D.: Bioacoustic time capsules: using acoustic monitoring to document biodiversity. Ecol. Ind. **99**, 149–152 (2019). https://doi.org/10.1016/j.ecolind.2018.12.021. Accessed 2023-08-01
8. He, K., Zhang, X., Ren, S., Sun, J.: Deep residual learning for image recognition. In: Proceedings of the IEEE Conference on Computer Vision and Pattern Recognition, pp. 770–778 (2016)
9. Simonyan, K., Zisserman, A.: Very Deep Convolutional Networks for Large-Scale Image Recognition. arXiv (2015). https://doi.org/10.48550/arXiv.1409.1556 . http://arxiv.org/abs/1409.1556. Accessed 02 Aug 2023
10. Huang, G., Liu, Z., Van Der Maaten, L., Weinberger, K.Q.: Densely connected convolutional networks. In: Proceedings of the IEEE Conference on Computer Vision and Pattern Recognition, pp. 4700–4708 (2017)
11. Kahl, S., Wood, C.M., Eibl, M., Klinck, H.: BirdNET: a deep learning solution for avian diversity monitoring. Eco. Inform. **61**, 101236 (2021). https://doi.org/10.1016/j.ecoinf.2021.101236. Accessed 2023-05-12
12. Tzirakis, P., Shiarella, A., Ewers, R., Schuller, B.W.: Computer audition for continuous rainforest occupancy monitoring: the case of Bornean gibbons' call detection (2020)
13. Çakır, E., Parascandolo, G., Heittola, T., Huttunen, H., Virtanen, T.: Convolutional recurrent neural networks for polyphonic sound event detection. IEEE/ACM Trans. Audio Speech Lang. Process. **25**(6), 1291–1303 (2017). https://doi.org/10.1109/TASLP.2017.2690575 . Conference Name: IEEE/ACM Transactions on Audio, Speech, and Language Processing

14. Xie, J., Hu, K., Zhu, M., Guo, Y.: Bioacoustic signal classification in continuous recordings: syllable-segmentation vs sliding-window. Expert Syst. Appl. **152**, 113390 (2020)
15. Dufourq, E., Batist, C., Foquet, R., Durbach, I.: Passive acoustic monitoring of animal populations with transfer learning. Eco. Inform. **70**, 101688 (2022). https://doi.org/10.1016/j.ecoinf.2022.101688. Accessed 2023-09-19
16. Kath, H., Serafini, P.P., Campos, I.B., Gouvea, T., Sonntag, D.: Leveraging transfer learning and active learning for sound event detection in passive acoustic monitoring of wildlife. In: 3rd Annual AAAI Workshop on AI to Accelerate Science and Engineering. AAAI Workshop on AI to Accelerate Science and Engineering (AI2ASE-2024), Befindet Sich AAAI, February 26, Vancouver, BC, Canada (2024)
17. Deng, J., Dong, W., Socher, R., Li, L.-J., Li, K., Fei-Fei, L.: ImageNet: a large-scale hierarchical image database. In: 2009 IEEE Conference on Computer Vision and Pattern Recognition, pp. 248–255. IEEE (2009)
18. Wang, Y., Li, J., Metze, F.: A comparison of five multiple instance learning pooling functions for sound event detection with weak labeling. In: ICASSP 2019–2019 IEEE International Conference on Acoustics, Speech and Signal Processing (ICASSP), pp. 31–35 (2019). https://doi.org/10.1109/ICASSP.2019.8682847. ISSN: 2379-190X
19. Hershey, S., et al.: CNN architectures for large-scale audio classification. In: 2017 IEEE International Conference on Acoustics, Speech and Signal Processing (ICASSP), pp. 131–135. IEEE (2017)
20. Sprengel, E., Jaggi, M., Kilcher, Y., Hofmann, T.: Audio Based Bird Species Identification using Deep Learning Techniques. LifeCLEF 2016 (2016)
21. Gemmeke, J.F., et al.: Audio set: an ontology and human-labeled dataset for audio events. In: 2017 IEEE International Conference on Acoustics, Speech and Signal Processing (ICASSP), pp. 776–780 (2017). https://doi.org/10.1109/ICASSP.2017.7952261
22. Kumar, A., Raj, B.: Audio event detection using weakly labeled data. In: Proceedings of the 24th ACM International Conference on Multimedia, pp. 1038–1047 (2016). https://doi.org/10.1145/2964284.2964310 . arXiv:1605.02401 [cs]. http://arxiv.org/abs/1605.02401. Accessed 13 Sept 2023
23. Xu, Y., Kong, Q., Wang, W., Plumbley, M.D.: Large-scale weakly supervised audio classification using gated convolutional neural network. In: 2018 IEEE International Conference on Acoustics, Speech and Signal Processing (ICASSP), pp. 121–125 (2018). https://doi.org/10.1109/ICASSP.2018.8461975. ISSN: 2379-190X
24. Miyazaki, K., Komatsu, T., Hayashi, T., Watanabe, S., Toda, T., Takeda, K.: Weakly-supervised sound event detection with self-attention. In: ICASSP 2020 – 2020 IEEE International Conference on Acoustics, Speech and Signal Processing (ICASSP), pp. 66–70 (2020). https://doi.org/10.1109/ICASSP40776.2020.9053609 . ISSN: 2379-190X
25. Xin, Y., Yang, D., Zou, Y.: Audio pyramid transformer with domain adaption for weakly supervised sound event detection and audio classification. In: Proceedings of the Interspeech 2022, pp. 1546–1550 (2022)
26. Chen, S., et al.: BEATs: Audio Pre-Training with Acoustic Tokenizers. arXiv (2022). https://doi.org/10.48550/arXiv.2212.09058, http://arxiv.org/abs/2212.09058. Accessed 03 Aug 2023
27. Jiang, J.-J., et al.: Whistle detection and classification for whales based on convolutional neural networks. Appl. Acoust. **150**, 169–178 (2019)

28. Coffey, K.R., Marx, R.E., Neumaier, J.F.: Deepsqueak: a deep learning-based system for detection and analysis of ultrasonic vocalizations. Neuropsychopharmacology **44**(5), 859–868 (2019)
29. Cohen, Y., Nicholson, D.A., Sanchioni, A., Mallaber, E.K., Skidanova, V., Gardner, T.J.: Automated annotation of birdsong with a neural network that segments spectrograms. Elife **11**, 63853 (2022)
30. Cañas, J.S., et al.: A dataset for benchmarking neotropical anuran calls identification in passive acoustic monitoring. Sci. Data **10**(1), 771 (2023)
31. Yang, Y.-Y., et al.: TorchAudio: building blocks for audio and speech processing. arXiv preprint arXiv:2110.15018 (2021)
32. Hershey, S., et al.: CNN Architectures for Large-Scale Audio Classification. arXiv (2017). https://doi.org/10.48550/arXiv.1609.09430 . http://arxiv.org/abs/1609.09430. Accessed 11 Aug 2023
33. Kingma, D.P., Ba, J.: Adam: a method for stochastic optimization. arXiv preprint arXiv:1412.6980 (2014)
34. Park, D.S., et al.: SpecAugment: a simple data augmentation method for automatic speech recognition. arXiv preprint arXiv:1904.08779 (2019)
35. Troshani, I., Gouvea, T., Sonntag, D.: Leveraging sound collections for animal species classification with weakly supervised learning. In: 3rd Annual AAAI Workshop on AI to Accelerate Science and Engineering. AAAI Workshop on AI to Accelerate Science and Engineering (AI2ASE-2024), AAAI, Vancouver, Canada (2024)
36. Paszke, A., et al.: Pytorch: an imperative style, high-performance deep learning library. In: Advances in Neural Information Processing Systems 32, pp. 8024–8035. Curran Associates, Inc., ??? (2019). http://papers.neurips.cc/paper/9015-pytorch-an-imperative-style-high-performance-deep-learning-library.pdf
37. Shah, A., Kumar, A., Hauptmann, A.G., Raj, B.: A closer look at weak label learning for audio events. CoRR arXiv:abs/1804.09288 (2018)

# Technical Communications

# Leveraging YOLO for Real-Time Video Analysis of Animal Welfare in Pig Slaughtering Processes

Christian Beecks[1,2](✉), Anandraj Amalraj[1], Alexander Graß[1,3], Marc Jentsch[1], Felix Kitschke[4], Maximilian Norz[4], and Patric Schäffer[5]

[1] Fraunhofer FIT, Sankt Augustin, Germany
christian.beecks@fernuni-hagen.de
[2] FernUniversität in Hagen, Hagen, Germany
[3] RWTH Aachen University, Aachen, Germany
[4] elanyo GmbH, Stuttgart, Germany
[5] Ulmer Fleisch GmbH, Ulm, Germany

**Abstract.** Artificial intelligence has empowered digitalization into a new era of intelligent systems. Machine learning solutions are being tailored to various application scenarios, leading to automated functionalities along complex real-world processes. In this paper, we investigate the domain of animal welfare and present our latest findings in relation to the automated detection of animal welfare violations. To this end, we introduce three different situations of increased animal welfare risk occurring in a pig slaughtering process and elucidate YOLO-based approaches to detect these situations based on video data. Though the reported results are considered to be preliminary, our solution already detects most of the situations of increased animal welfare risk with high accuracy.

**Keywords:** YOLO · Computer Vision · Machine Learning · AI Application · Animal Welfare · Pig Slaughtering

## 1 Introduction

The rapid proliferation of artificial intelligence (AI) has revolutionized the capabilities of intelligent systems across diverse application domains. Ranging from animal welfare [6] to zoology [8], artificial intelligence has made significant progress in the last decades enabling the support and automation of complex, human-driven real-world processes. Fundamental to this success are advances [7] in the field of machine learning (ML), which have led to high-performance solutions with unprecedented efficiency and accuracy.

This paper is concerned with the application of AI-based approaches to the domain of animal welfare [2], which refers to the well-being of animals. Our focus lies on the detection of potential animal welfare violations in pig slaughtering [3, 4], where animal welfare monitoring is already carried out by qualified employees


A. Hotho and S. Rudolph (Eds.): KI 2024, LNAI 14992, pp. 275–281, 2024.
https://doi.org/10.1007/978-3-031-70893-0_20

inspecting video recordings for the detection of possible animal welfare risks and violations. Although this process is compliant with technical, organizational and legal regulations, there are periods of time when animal welfare violations may go undetected. To this end, we aim to develop an AI-based system that can assist the control process by detecting situations of increased animal welfare risk, which can then be presented to the responsible animal welfare officers. In this way, more relevant scenes can be inspected. At the same time, work efficiency of slaughterhouse workers will be improved by reducing the number of manual video inspections where animal welfare violations can be ruled out from the outset.

In this paper, we describe different YOLO-based approaches for the identification and detection of three distinct situations of increased animal welfare risk. The following situations are commonly encountered in pig slaughtering processes: (i) detection of high animal quantity prior to the stunning system, (ii) detection of conspicuous animal behavior, and (iii) detection of incorrect use of tools. We outline the developed video-based solutions, concretely, YOLO-based object detection [1] enhanced by scenario-dependent postprocessings and present preliminary results of our experimental evaluation. In summary, we make the following contributions:

- We provide different approaches for an application of state-of-the-art real-time object detection in the context of a concrete industrial scenario.
- We investigate the performance of our proposal and show its high accuracy in an empirical evaluation.

The remainder of this paper is structured as follows: Situations of increased animal welfare risk are presented in Sect. 2. Section 3 then introduces our YOLO-based approach for detecting these situations. Section 4 summarizes the results of our preliminary performance evaluation, before we conclude this paper with an outlook on future research work in Sect. 5.

## 2 Animal Welfare Risk Situations

As pointed out in a recent survey by Zhang et al. [6], artificial intelligence has the potential to enhance animal welfare across different application scenarios. In our research, we focus on the particular application domain of pig slaughtering. The technical objective comprises a detection of video scenes with increased animal welfare risk that could potentially lead to animal welfare violations. The final decision as to whether such a detected scene is an actual animal welfare violation or not will be made by the responsible animal welfare officer. Prior to outlining our methodology for identifying these video scenes, we first describe the corresponding situations.

- **High number of pigs prior to the stunning system**: These are situations where the maximum number of pigs in a particular driving area are exceeding a predefined limit, cf. Fig. 1 (left), which can cause an increase of the pigs' stress potential and the risk of injury.

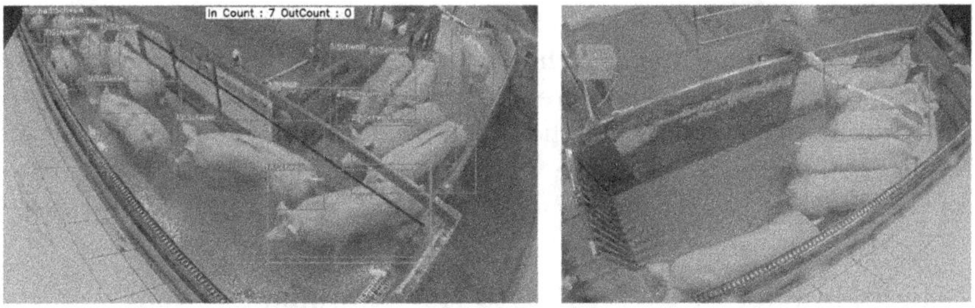


**Fig. 1.** Pigs passing through the driving corridor (left) and paddle usage (right).

- **Conspicuous pig behavior**: Pigs normally walk or stand in the driving area. Pigs that are sitting or lying down are an indication of possible health or injury problems. In these cases, employees are supposed to immediately assess the situation and take appropriate action.
- **Incorrect use of tools**: Pigs are driven in with the help of paddles, cf. Fig. 1 (right). Pigs are used to move forward when shaking paddles, producing a distinct noise. While a light tap with the paddle is allowed, strong hits constitute an animal welfare violation. Likewise, no objects other than the paddle may be used for driving. It is also not permitted to use the paddle with the narrow edge or upside down.

## 3 YOLO-Based Approach for Detecting Animal Welfare Risk Situations

In this section, we will present concrete approaches for the prediction of different animal welfare situations. All approaches have been implemented and deployed on an existing computing cluster that monitors and analyzes slaughtering processes in real-time.

We initially gathered video data from the pig slaughtering process by installing four video cameras at specific positions. Two of these video cameras are shown in Fig. 1. For legal and operational reasons, the entire video material has been pixelated via the publicly available CenterFace model [5][1].

Since in all approaches we use YOLOv8 [1] as underlying object detection model, any captured video data requires a semantic enrichment in form of data annotations provided with the help of domain experts. To this end, we define scenario-based labels for each object to be detected. Labels are determined via the open-source data labeling tool Label Studio[2], where each label consists of the object class and bounding box coordinates. As labels are only determined for some key frames, we linearly interpolate bounding boxes across all intermediate frames. Additional steps include an automatic alignment of bounding

[1] https://github.com/ORB-HD/deface.
[2] https://labelstud.io.

box information for videos with different frame rates, a validation of taggings and data augmentation to enrich the amount of available training data. Resultant labeling information are referenced by further metadata and stored in a relational database, representing the input used for model training and model validation, as described below. We will demonstrate how the generated results can be exploited for more complex approaches tailored to individual tasks.

### 3.1 High Number of Pigs Prior to the Stunning System

This particular situation can be found at two different stages in the driving corridor and is formally modelled as a multi-class prediction problem with class labels $\{pig, gate.closed, gate.open\}$. While the label *pig* is used for indicating an animal occurrence, the labels *gate.open* and *gate.closed* are used to indicate the status of the driving corridor gate, as can be seen on the left side in Fig. 1. These two additional labels are used for automating the pig counting process and achieving robustness with respect to technical deviations.

In the inference phase, the number of pigs is predicted via YOLO on each individual frame and the positions of the bounding boxes' centers are checked to be located within the regions of interest in the driving corridor. By storing the animal counts of consecutive frames in a queue of length 25, the total pig count of the region of interest is derived by the value of the third quantile (75%) of the queue, which is then compared against a threshold, which is 5 in this situation, in order to raise an alert.

When it comes to the automation of this process, pigs are tracked in each frame by using YOLO's tracking functionality (BoT-SORT) in accordance with the aforementioned gate status. The predicted bounding box centers are checked, as above, and a counter is initialized with the number of pigs in the first frame. In each subsequent frame, the pigs are tracked according to their tracking IDs and the counter is updated. This process involves detecting and tracking pigs, updating a history of positions, buffering IDs for continuity, determining object movement direction, and updating counts accordingly. Additionally, it includes the logic for handling direction changes and a reset mechanism triggered by consistent detection of the *gate.closed* status. Algorithm 1 shows pseudo-code of a simplified pig counting version. Automated pig counting enables monitoring of an additional driving corridor, with a threshold, i.e., the maximum number of pigs, of value 10.

### 3.2 Conspicuous Pig Behavior

Similar to the previous situation, this situation has been modelled as a multi-class prediction problem with class labels $\{pig.lying, pig.sitting\}$ reflecting two designated pig behaviors. Object prediction is performed on each frame using YOLO, followed by a check to determine whether the predicted objects' bounding box centers fell within predefined regions of interest for each camera. The prediction results of each frame were recorded and post-processed by applying

**Algorithm 1** YOLO-based Pig Counting

```
Require: model                                        ▷ Model for object detection and tracking
  in_cnt ← 0, out_cnt ← 0, id_buffer ← []            ▷ Init counters and tracking buffer
  for each frame do
      objs = model.track(frame)
      if not gate.closed ∈ objs then
          objs = filter_area_of_interest(objs)
          id_buffer = update_tracking_buffer(objs)
          in_cnt, out_cnt = val_line_crossing(objs, id_buffer)
      else
          emit(in_cnt − out_cnt)
          in_cnt ← 0, out_cnt ← 0, id_buffer ← []
```

a sliding window average to define start and end times of potential conspicuous behaviors, which are then reported to the responsible person. Detection robustness is achieved by compensating detection errors in individual frames via buffering.

### 3.3 Incorrect Use of Tools

This situation mainly aims to recognize the improper use of a paddle, cf. Fig. 1. From an algorithmic perspective, this situation is modelled as a binary classification problem to detect and track the paddle in combination with an estimation of its velocity. The velocity calculation is based on the maximum displacement regarding a straight movement of the tracked paddle represented by x and y coordinates between adjacent frames.

**Definition 1.** ***Paddle Velocity Estimation***
*Let $fps$ be the number of frames per second and $\Delta(x1), \Delta(y1), \Delta(x2), \Delta(y2)$ be the pixel coordinate differences of a bounding box among two subsequent frames. Paddle velocity is estimated by the maximum pixel displacement over time as:*

$$ips = max(\Delta(x1), \Delta(y1), \Delta(x2), \Delta(y2)) * fps$$

The maximum displacement is used to approximate the speed in inch per second (ips) of a paddle over multiple subsequent frames. By averaging through a sliding window, followed by only considering velocities which exceed the 90%-quantile, incorrect paddle usages are determined including their start and end times.

## 4 Preliminary Experiments

In this section, we report the results of our preliminary performance evaluation. For this purpose, we collected video data with a frame rate of 50FPS and used the labeling process as described in Sect. 3. All experiments were performed in the computing cluster of the slaughtering process based on an AMD EPYC 7313P 16-Core CPU with an NVIDIA GeForce RTX 3090 GPU.

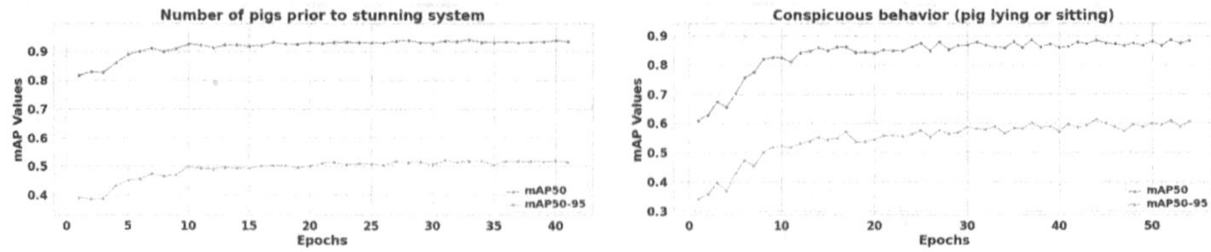


**Fig. 2.** Mean average precision for counting the number of pigs (left) and conspicuous pig behavior (right).

**Table 1.** Performance of the hyperparameter-optimized YOLOv8m model.

| Animal welfare risk situation | Precision | Recall | mAP50 | mAP50-95 |
|---|---|---|---|---|
| Number of pigs prior to stunning system | 0.9117 | 0.8897 | 0.9364 | 0.5205 |
| Conspicuous behavior (pig lying or sitting) | 0.9015 | 0.8281 | 0.8843 | 0.6113 |
| Incorrect use of tools (paddle) | 0.8361 | 0.7857 | 0.8078 | 0.4380 |

The evaluation process was carried out by first training a YOLOv8m model, suited for fast real-time object detection, and then performing genetic hyperparameter optimization with 20 iterations to optimize model quality. To this end, we split the video data into training data (75%) for initial model training, validation data (15%) for improving model quality by means of hyperparameter optimization, and test data (10%) for the evaluation of common metrics such as precision, recall, mAP50, and mAP50-95.

The results of the performance evaluation of the YOLOv8m model can be found in Table 1. The high performance is reflected in the precision, recall, and mean average precision values for an IoU of 0.5 which are in the range of 0.808 to 0.936. We thus conclude that the developed YOLO-based approach is well suited for our particular application scenario. While the mAP50-95 is lower for individual frames, the classification accuracy is further improved at a higher level via multiple frames representing a violation sequence. An illustration of mAP metrics for pig counting and conspicuous pig behavior are shown in Fig. 2.

## 5 Conclusions and Future Work

In this paper, we have proposed a YOLO-based approach in the domain of animal welfare. In addition to creating a labeled video dataset from an existing pig slaughtering process, we have trained and applied YOLO for the prediction of different animal welfare risk situations. Although the research outlined in this paper remains ongoing, the preliminary results of the conducted performance study are indicating the potential of our proposal: Predicting the number of pigs as well as conspicuous pig behavior shows and average precision of 91%, while the incorrect use of tools can be detected with an average precision of 84%.

We believe that our research contributions provide insight into an interesting application domain for artificial intelligence and that these findings are useful for both researchers and practitioners in this and adjacent research domains.

The approaches presented can be extended in a number of ways. One promising research direction is the additional utilization of audio data. In addition, we plan to expand the list of evaluated models, while simultaneously aiming for a better understanding of the model behavior by means of XAI methods.

## References


1. Cheng, T., Song, L., Ge, Y., Liu, W., Wang, X., Shan, Y.: Yolo-world: real-time open-vocabulary object detection. arXiv preprint arXiv:2401.17270 (2024)
2. Englund, M.D., Cronin, K.A.: Choice, control, and animal welfare: definitions and essential inquiries to advance animal welfare science. Front. Vet. Sci. **10**, 1250251 (2023)
3. Isbrandt, R., Wiegard, M., Meemken, D., Langkabel, N.: Impact of procedures and human-animal interactions during transport and slaughter on animal welfare of pigs: a systematic literature review. Animals **12**(23), 3391 (2022)
4. Nielsen, S.S., et al.: Welfare of pigs at slaughter. EFSA J. **18**(6), e06148 (2020)
5. Xu, Y., Yan, W., Sun, H., Yang, G., Luo, J.: Centerface: joint face detection and alignment using face as point. CoRR **abs/1911.03599** (2019). http://arxiv.org/abs/1911.03599
6. Zhang, L., et al.: Advancements in artificial intelligence technology for improving animal welfare: current applications and research progress. Anim. Res. One Health **2**(1), 93–109 (2024)
7. Zhang, Y.: New advances in machine learning. BoD–Books on Demand (2010)
8. Zhang, Y.J., Luo, Z., Sun, Y., Liu, J., Chen, Z.: From beasts to bytes: revolutionizing zoological research with artificial intelligence. Zool. Res. **44**(6), 1115 (2023)

# Early Explorations of Lightweight Models for Wound Segmentation on Mobile Devices

Vanessa Borst(✉), Timo Dittus, Konstantin Müller, and Samuel Kounev

Software Engineering Group, University of Würzburg, Würzburg, Germany
vanessa.borst@uni-wuerzburg.de

**Abstract.** The aging population poses numerous challenges to healthcare, including the increase in chronic wounds in the elderly. The current approach to wound assessment by therapists based on photographic documentation is subjective, highlighting the need for computer-aided wound recognition from smartphone photos. This offers objective and convenient therapy monitoring, while being accessible to patients from their home at any time. However, despite research in mobile image segmentation, there is a lack of focus on mobile wound segmentation. To address this gap, we conduct initial research on three lightweight architectures to investigate their suitability for smartphone-based wound segmentation. Using public datasets and UNet as a baseline, our results are promising, with both ENet and TopFormer, as well as the larger UNeXt variant, showing comparable performance to UNet. Furthermore, we deploy the models into a smartphone app for visual assessment of live segmentation, where results demonstrate the effectiveness of TopFormer in distinguishing wounds from wound-coloured objects. While our study highlights the potential of transformer models for mobile wound segmentation, future work should aim to further improve the mask contours. (An extended version of this paper is available for download from arXiv).

**Keywords:** Medical Image Segmentation · Mobile AI · Deep Learning

## 1 Introduction

Chronic wounds, such as diabetic foot ulcers, affect a significant proportion of the world's population and present a major challenge to healthcare systems worldwide. In the United States alone, it is estimated that more than 6 million people suffer from chronic wounds each year [23]. In addition to diabetes, age is a significant risk factor, with a disproportionate impact on the elderly. This is evidenced by a prevalence of 7.8% in German nursing home residents [24] compared to 1% in the German population [9]. As the world's population continues to age and the incidence of conditions such as diabetes increases, the prevalence of chronic wounds is expected to soar in the coming years. This increase poses


A. Hotho and S. Rudolph (Eds.): KI 2024, LNAI 14992, pp. 282–291, 2024.
https://doi.org/10.1007/978-3-031-70893-0_21

a significant challenge to healthcare providers in providing appropriate care to those affected. From the patient's perspective, the need for frequent visits to specialised wound care facilities to manage chronic wounds can be a significant burden. Access to such clinics may be limited, particularly in remote areas, and mobility issues exacerbate the challenge. There is an urgent need for innovative solutions to alleviate these burdens and improve access to wound care services.

Telemedicine and automation offer promising ways to address these challenges by improving efficiency and accessibility. Automated wound size monitoring has the potential to eliminate the need for patients to visit clinics solely for monitoring purposes, thereby reducing the associated logistical challenges and relieving the burden on healthcare professionals. Furthermore, automated approaches provide timely and accurate information on wound status, facilitating proactive intervention. Mobile semantic segmentation emerges as a key technology in realizing these advancements, providing real-time feedback and segmentation masks directly to users, thereby circumventing latency issues associated with server-based processing. In addition, deploying models directly on smartphones ensures enhanced privacy and security by keeping sensitive medical data local. In addition, it enables offline functionality, which is crucial for applications in remote areas with limited network coverage. Despite the advantages of mobile segmentation, there are also significant challenges associated with this endeavour. To be effective, such solutions must include small, lightweight models that can be seamlessly deployed on smartphones. At the same time, such models must be robust in the face of challenging conditions such as different backgrounds, varying lighting conditions, and different camera settings, as mobile wound monitoring is intended to take place in the patient's home environment.

However, despite the necessity for compact yet resilient models, the investigation into mobile wound segmentation remains limited to date. While several methods have been proposed for wound segmentation utilizing feature-engineering-based machine learning [14,27,35], deep learning [4,8,16,20], or even both [15,31], they often lack emphasis on lightweight architectures and mobile suitability. Some publications suggest that the analyzed architecture(s) could be used on mobile devices due to its efficiency, but such indications often rely solely on metrics such as parameter count as mobile performance indicator, lacking real-world deployment [20,33]. Of the few approaches deployed to a mobile device in practice, many apply traditional image processing rather than deep learning [7,25,30], while others evaluate only single architectures [19], or require manual annotation outlines from users [1,2]. To the best of our knowledge, a systematic comparison of different neural network-based wound segmentation techniques on mobile devices is still pending. In contrast, the number of proposed architectures for mobile segmentation beyond the wound context has increased significantly. To address this discrepancy, this work conducts initial investigations into the applicability of different architectures from non-wound domains for mobile wound segmentation, including a visual comparison in real-world usage.

# 2 Approach

## 2.1 Task Definition and Requirements

This study aims to assess the efficacy of lightweight models for mobile wound segmentation by systematically comparing approaches from beyond the wound domain. In particular, we focus on their ability to perform live segmentation from smartphone camera streams, for which we consider the following requirements paramount: (i) **accuracy**: High segmentation performance for accurate analysis of wound area (ii) **efficiency**: Small number of parameters to enable mobile processing (iii) **robustness**: Effective performance under varying conditions for use in non-standardized home environments (iv) **practicality**: Effective segmentation in real-world deployment to enhance medical practice.

## 2.2 Model Selection

The number of proposed models for segmentation on mobile devices has increased significantly beyond wound segmentation. Notable advances include CNN-based methods, such as encoder-decoder architectures with lightweight backbones such as MobileNets [10,11] combined with different decoders, ENet [21], UNeXt [29], or Fast-SCNN [22], which have proven successful in mobile environments. In addition, vision transformer-based models, such as Mobile-Former [3], MobileViT [17], SeaFormer [32], or TopFormer [38], have been explored increasingly. From this spectrum, our model selection aims to cover different network types (CNN, vision transformer) and application domains (general-purpose, (bio)medicine), while setting reasonable limits on training effort. Therefore, we limit the inclusion of architectures to four, prioritizing the assessment of different-sized variants of selected methods where feasible. Specifically, we selected TopFormer [38], a representative (hybrid) vision transformer among recent general-purpose state-of-the-art (SOTA) methods, and UNeXt [29], a compact CNN-based SOTA technique for medical segmentation. The selection of these two SOTA models was based on the following criteria: (i) competitive performance on respective benchmark datasets (ii) publication in prestigious journals or conferences (iii) efficient resource utilization, as indicated by reported parameter counts and GFLOPs (iv) code availability on GitHub. Additionally, ENet [21], a widely referenced general-purpose CNN for real-time semantic segmentation from 2016, was included due to its remarkable efficiency in terms of trainable parameters and promising outcomes in medical segmentation studies [28,37], given its size. Lastly, to establish a baseline for wound segmentation, we integrated UNet [26], a prominent model in biomedical segmentation. Table 1 outlines the selection, with parameters reported in million (M) for binary segmentation (one out-channel).

**Table 1.** Overview of the included segmentation techniques.

| | UNet [26] | ENet [21] | UNeXt [29] | | TopFormer [38] | | |
|---|---|---|---|---|---|---|---|
| | | | Base | Small | Base | Small | Tiny |
| **Params** | 31.03M | 0.35M | 1.47M | 0.25M | 5.03M | 3.01M | 1.39M |
| **Type** | CNN | CNN | CNN | | Hybrid CNN-ViT | | |
| **Domain** | Biomedicine | General purpose | Medicine | | General purpose | | |

# 3 Experimental Setup

## 3.1 Dataset

We perform experiments on a combination of two prominent datasets: the *Foot Ulcer Segmentation Challenge 2021 (FuSeg)* [34] and the *Diabetic Foot Ulcer Challenge 2022 (DFUC 2022)* [13,36] datasets. FuSeg encompasses 1,210 foot ulcer images, each with a resolution of 512 × 512 pixels, sourced from 889 patients' medical visits spanning from October 2019 to April 2021. Meanwhile, the DFUC 2022 dataset comprises 4,000 foot ulcer images with a resolution of 640 × 480 pixels, captured approximately 30 to 40 cm away from the ulcer. Our custom dataset, *Combined Foot Ulcers (CFU)*, integrates FuSeg and DFUC 2022, comprising images with mixed resolutions of either 512 × 512 or 640 × 480 pixels. Due to the privacy of test set annotations (200 out of 1,210 for FuSeg and 2,000 out of 4,000 for DFUC 2022), only publicly accessible training and validation images are utilized for CFU. To prevent data leakage, duplicate images within FuSeg and DFUC 2022 were identified and removed. A similarity analysis, employing the perceptual hashing algorithm [18], was conducted. Image sets with identical raw bytes or perceptual hash values exhibiting a Hamming distance of 11 or less were deemed duplicates. For illustrative purposes, a number of examples are provided in the appendix of the extended version of this paper. Eliminating duplicate image-annotation pairs, we obtained a final dataset with 2,887 samples after discarding 123 pairs. Next, these samples were randomly partitioned into training (60%), validation (20%), and test (20%) set.

## 3.2 Evaluation Protocol

To assess model performance, we employed established segmentation metrics, such as Dice score (DSC), Intersection over Union (IoU), precision (Prc), and recall (Rec), utilizing the micro-averaging method while excluding the background class. Our analysis covered two training modalities: training models from scratch and utilizing pre-training on either the ImageNet [6] or Cityscapes [5] dataset. To facilitate this, we downloaded pre-trained weights where available: ENet (Cityscapes), UNet (ImageNet), and TopFormer (ImageNet). Since pre-trained weights were unavailable for UNeXt, we trained both variants ourselves for 100 epochs on Cityscapes, using the best checkpoints as initial weights.

### 3.3 Implementation Details

We implemented this approach using Python 3.10.9 and PyTorch 2.0.1 on a server with two partitioned Nvidia A100 GPUs, reserving a large-sized instance with 40 GB VRAM for our experiments. All models were trained end-to-end using the AdamW optimizer with an initial learning rate of 0.0001, a batch size of four, and the binary cross entropy (BCE) loss. We used a *ReduceLROn-Plateau* scheduler conditioned on the Dice score, with a minimum learning rate of 0.000001. All models were trained for 200 epochs from scratch and for at least 130 epochs in the pre-training setting. For both, the best checkpoints, according to the mean IoU on the validation set, were used for the ensuing test set evaluation. All images were resized to $512 \times 512$ pixels and normalized during training. We applied thresholding to annotated masks to eliminate out-of-class pixels and used the following data augmentations for each training set sample: Gaussian blur with a kernel size of 25 and a randomly sampled standard deviation from a uniform distribution of [0.001, 2.0], random affine transforms (incl. translations by up to 12.5%, rotations by up to 180°, scaling between 50% and 150% of the original size, and shear of up to 22.5°), color jitter, as well as random horizontal and vertical flips, each with a 50% probability. Further details regarding the augmentations are given in the extended version of this paper, using a sample configuration file.

## 4 Evaluation

### 4.1 Results

We compare the selected models in Table 2, where we report the micro-averaged metric scores of each variant for both training modalities. Our findings reveal that, without pretraining, UNeXt-B demonstrates the most favorable segmentation performance in terms of DSC and IoU scores, effectively balancing parameter count and segmentation quality. Notably, all lightweight models, except UNeXt-S, achieve results within a similar size range as the larger-sized UNet, with the base variants of UNeXt and TopFormer, and ENet even slightly surpassing UNet. Pretraining considerably enhances performance across all models regarding IoU and DSC metrics. Specifically, all lightweight networks show comparable performance, with IoU scores ranging from 69% to 72% and DSC values from 82% to 84%, except for UNeXt-S, which performs worse.

**Table 2.** Comparison of all models on the CFU dataset. The micro-averaged metrics (in %) were calculated on the test set. Parameters are reported in millions.

| | Without pretraining | | | | With pretraining | | | | |
|---|---|---|---|---|---|---|---|---|---|
| Model | IoU | DSC | Prc | Rec | IoU | DSC | Prc | Rec | Params↓ |
| U-Net | 65.69 | 79.30 | 81.57 | **77.14** | **74.69** | **85.56** | 87.74 | 83.39 | 31.03M |
| TopFormer-B | 66.10 | 79.59 | 86.63 | 73.61 | 72.41 | 84.00 | 84.56 | **83.44** | 5.03M |
| TopFormer-S | 63.36 | 77.57 | 83.31 | 72.57 | 71.56 | 83.42 | 85.95 | 81.04 | 3.01M |
| UNeXt-B | **68.96** | **81.63** | 88.49 | 75.76 | 70.12 | 82.43 | 87.40 | 78.01 | 1.47M |
| TopFormer-T | 62.51 | 76.93 | 84.27 | 70.77 | 69.62 | 82.09 | 83.90 | 80.36 | 1.39M |
| ENet | 67.27 | 80.43 | **89.21** | 72.23 | 71.71 | 83.52 | **89.98** | 77.93 | 0.35M |
| UNeXt-S | 59.16 | 74.34 | 77.52 | 71.41 | 64.13 | 78.14 | 87.52 | 70.58 | **0.25M** |

### 4.2 Real-World Deployment

To assess the models' real-world capabilities, we developed a Flutter prototype. For seamless integration within the app, all PyTorch models except UNet were converted to TorchScript by tracing. The deployed networks performed live segmentation of camera streams, with center-cropped images to $224 \times 224$ pixels for efficient processing. Following preliminary trials, a prediction threshold of 0.75 was established for the derivation of binary masks, with the objective of achieving more precise segmentations and reduced jitter during inference. We rigorously tested the architectures on an *OnePlus 7 Pro* Android phone from 2019. All models ran smoothly without stalling, demonstrating the feasibility of mobile segmentation using Flutter. To enable visual assessments, we devised six distinct scenes and captured live segmentations via screenshots, utilizing the variants with pretraining. The scenes encompass: (i) a neutral environment devoid of objects (ii) two diverse wound images not from the CFU dataset (iii) two scenes featuring everyday objects with colors resembling wounds (iv) a scene with two everyday objects, one of which has a wound-like color. The resulting segmentation masks generated by the models are depicted in Fig. 1. Visually, ENet exhibits the worst results among the tested models, with pixelated and holey masks as well as inaccuracies such as falsely identifying everyday objects, including the blue sponge. In contrast, the TopFormer and UNeXt variants show more promising results, identifying wounds with fewer imperfections, albeit with somewhat indistinct edges and contours. Notwithstanding the recognition difficulties encountered by the UNeXt models when presented with objects of wound-like colours, there is an overall improvement for both architectures compared to ENet. Overall, the TopFormer-based networks appear to be superior, as they proved quite effective at distinguishing between wounds and wound-coloured objects. Notably, the smallest variant (TF-T) performs exceptionally well in our example scenes, while the CNN-based UNeXt-B, despite exhibiting promising performance in Table 2, displays significantly poorer performance.

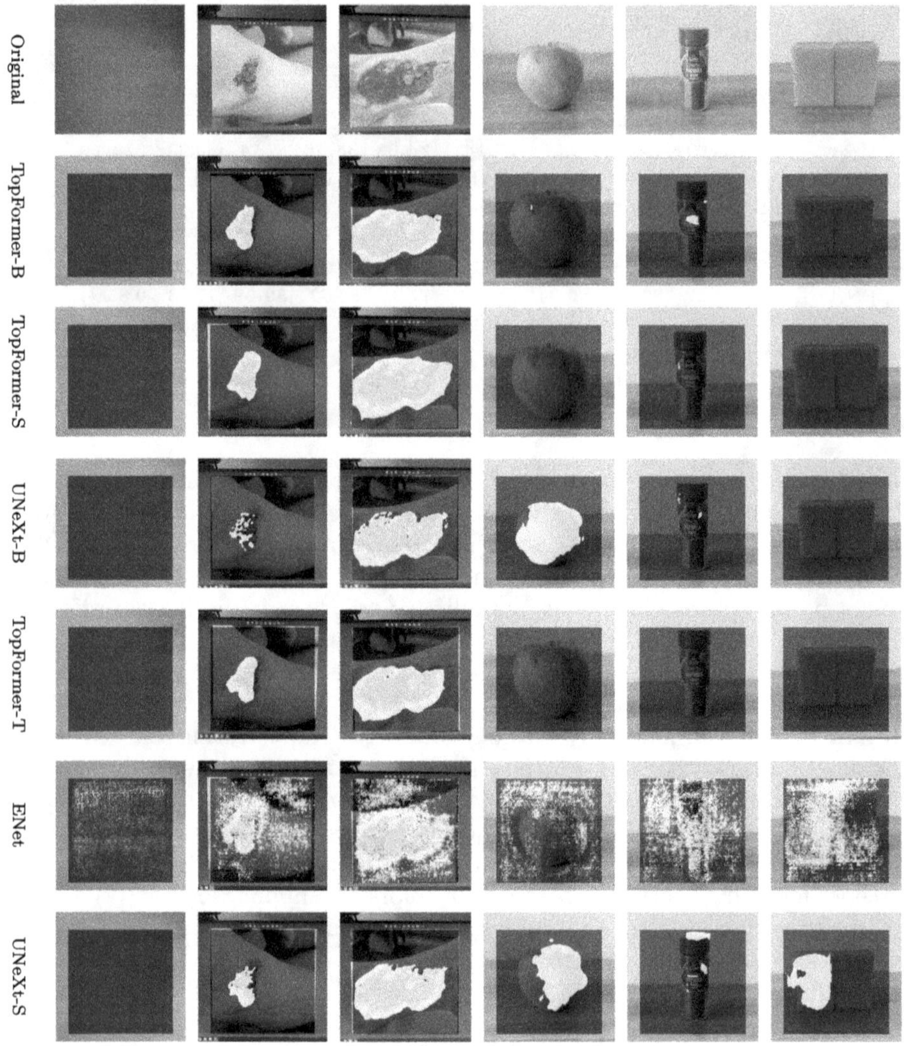


**Fig. 1.** Live segmentation of different model variants in our mobile app, arranged in descending order of parameter count. (Color figure online)

## 5 Conclusion and Future Work

In conclusion, our preliminary investigations into lightweight neural networks demonstrate their potential for mobile wound segmentation. The analyzed architectures, especially TopFormer, yield promising results in real-world application scenarios, although some refinements are still necessary, such as clearer edge detection. Future work will include extending the comparison to further promising architectures such as MBSNet [12], Mobile-Former [3], or SeaFormer [32]. Furthermore, retraining the networks on an expanded dataset with more images, including those featuring objects with wound-similar colors, is expected to enhance the robustness and generalization ability of the models, thereby improving the accessibility and effectiveness of future wound care.

**Acknowledgements.** We thank Adrian Strutt for conducting preliminary experiments with ENet prior to the start of this work.

**Disclosure of Interests.** The authors have no competing interests to declare.

## References

1. Cazzolato, M.T., et al.: Semi-automatic ulcer segmentation and wound area measurement supporting telemedicine. In: 2020 IEEE 33rd International Symposium on Computer-Based Medical Systems (CBMS), pp. 356–361. IEEE (2020)
2. Cazzolato, M.T., et al.: The UTrack framework for segmenting and measuring dermatological ulcers through telemedicine. Comput. Biol. Med. **134**, 104489 (2021)
3. Chen, Y., et al.: Mobile-former: bridging MobileNet and transformer. In: Proceedings of the IEEE/CVF Conference on Computer Vision and Pattern Recognition, pp. 5270–5279 (2022)
4. Chino, D.Y., Scabora, L.C., Cazzolato, M.T., Jorge, A.E., Traina-Jr, C., Traina, A.J.: Segmenting skin ulcers and measuring the wound area using deep convolutional networks. Comput. Meth. Programs Biomed. **191**, 105376 (2020)
5. Cordts, M., et al.: The cityscapes dataset for semantic urban scene understanding. In: Proceedings of the IEEE Conference on Computer Vision and Pattern Recognition, pp. 3213–3223 (2016)
6. Deng, J., Dong, W., Socher, R., Li, L.J., Li, K., Fei-Fei, L.: ImageNet: a large-scale hierarchical image database. In: 2009 IEEE Conference on Computer Vision and Pattern Recognition, pp. 248–255. IEEE (2009)
7. Ferreira, F., et al.: Experimental study on wound area measurement with mobile devices. Sensors **21**(17), 5762 (2021)
8. Goyal, M., Yap, M.H., Reeves, N.D., Rajbhandari, S., Spragg, J.: Fully convolutional networks for diabetic foot ulcer segmentation. In: 2017 IEEE International Conference on Systems, Man, and Cybernetics (SMC), pp. 618–623. IEEE (2017)
9. Heyer, K., Herberger, K., Protz, K., Glaeske, G., Augustin, M.: Epidemiology of chronic wounds in Germany: analysis of statutory health insurance data. Wound Repair Regeneration **24**(2), 434–442 (2016)
10. Howard, A., et al.: Searching for mobileNetV3. In: Proceedings of the IEEE/CVF International Conference on Computer Vision, pp. 1314–1324 (2019)
11. Howard, A., Zhmoginov, A., Chen, L.C., Sandler, M., Zhu, M.: Inverted residuals and linear bottlenecks: mobile networks for classification, detection and segmentation. In: Proceedings of CVPR, pp. 4510–4520 (2018)
12. Jin, S., Yu, S., Peng, J., Wang, H., Zhao, Y.: A novel medical image segmentation approach by using multi-branch segmentation network based on local and global information synchronous learning. Sci. Rep. **13**(1), 6762 (2023)
13. Kendrick, C., et al.: Translating clinical delineation of diabetic foot ulcers into machine interpretable segmentation. arXiv preprint arXiv:2204.11618 (2022)

14. Kolesnik, M., Fexa, A.: Multi-dimensional color histograms for segmentation of wounds in images. In: Kamel, M., Campilho, A. (eds.) ICIAR 2005. LNCS, vol. 3656, pp. 1014–1022. Springer, Heidelberg (2005). https://doi.org/10.1007/11559573_123
15. Li, F., Wang, C., Liu, X., Peng, Y., Jin, S.: A composite model of wound segmentation based on traditional methods and deep neural networks. Comput. Intell. Neurosci. **2018** (2018)
16. Liu, X., Wang, C., Li, F., Zhao, X., Zhu, E., Peng, Y.: A framework of wound segmentation based on deep convolutional networks. In: 2017 10th International Congress on Image and Signal Processing, Biomedical Engineering and Informatics (CISP-BMEI), pp. 1–7. IEEE (2017)
17. Mehta, S., Rastegari, M.: MobileViT: light-weight, general-purpose, and mobile-friendly vision transformer. In: International Conference on Learning Representations (2021)
18. Micikevicius, P., et al.: Mixed precision training. In: International Conference on Learning Representations (2018)
19. Ong, E.P., Yin, C.T.K., Lee, B.H.: Efficient deep learning-based wound-bed segmentation for mobile applications. In: 2020 42nd Annual International Conference of the IEEE Engineering in Medicine & Biology Society (EMBC), pp. 1654–1657. IEEE (2020)
20. Oota, S.R., Rowtula, V., Mohammed, S., Liu, M., Gupta, M.: WSNet: towards an effective method for wound image segmentation. In: Proceedings of the IEEE/CVF Winter Conference on Applications of Computer Vision, pp. 3234–3243 (2023)
21. Paszke, A., Chaurasia, A., Kim, S., Culurciello, E.: ENet: a deep neural network architecture for real-time semantic segmentation. arXiv preprint arXiv:1606.02147 (2016)
22. Poudel, R.P.K., Liwicki, S., Cipolla, R.: Fast-SCNN: fast semantic segmentation network. In: 30th British Machine Vision Conference 2019, BMVC 2019, Cardiff, UK, 9-12 September 2019. BMVA Press (2019)
23. Powers, J.G., Higham, C., Broussard, K., Phillips, T.J.: Wound healing and treating wounds: chronic wound care and management. J. Am. Acad. Dermatol. **74**(4), 607–625 (2016)
24. Raeder, K., Jachan, D.E., Müller-Werdan, U., Lahmann, N.A.: Prevalence and risk factors of chronic wounds in nursing homes in Germany: a cross-sectional study. Int. Wound J. **17**(5), 1128–1134 (2020)
25. Rocha, C.D.F.D., Carvalho, B.S., Marques, V.G., Silva, B.M.: WoundArch: a hybrid architecture system for segmentation and classification of chronic wounds. In: Proceedings of the 14th International Joint Conference on Biomedical Engineering Systems and Technologies, pp. 651–658. INSTICC, SciTePress (2021)
26. Ronneberger, O., Fischer, P., Brox, T.: U-Net: convolutional networks for biomedical image segmentation. In: Navab, N., Hornegger, J., Wells, W.M., Frangi, A.F. (eds.) MICCAI 2015. LNCS, vol. 9351, pp. 234–241. Springer, Cham (2015). https://doi.org/10.1007/978-3-319-24574-4_28
27. Song, B., Sacan, A.: Automated wound identification system based on image segmentation and artificial neural networks. In: 2012 IEEE International Conference on bioinformatics and biomedicine, pp. 1–4. IEEE (2012)
28. Song, E., Zhan, B., Liu, H.: Combining external-latent attention for medical image segmentation. Neural Netw. **170**, 468–477 (2024)
29. Valanarasu, J.M.J., Patel, V.M.: UNext: MLP-based rapid medical image segmentation network. In: Wang, L., Dou, Q., Fletcher, P.T., Speidel, S., Li, S. (eds.)

MICCAI 2022. LNCS, vol. 13435, pp. 23–33. Springer, Cham (2022). https://doi.org/10.1007/978-3-031-16443-9_3
30. Varma, A., Varma, A., Jakkampudi, H.M., Raj, A.N.J.: Vision based pus segmentation and area estimation of wound using android application. Indian J. Sci. Technol. (2016)
31. Wagh, A., et al.: Semantic segmentation of smartphone wound images: comparative analysis of AHRF and CNN-based approaches. IEEE Access **8**, 181590–181604 (2020)
32. Wan, Q., Huang, Z., Lu, J., Yu, G., Zhang, L.: SeaFormer: squeeze-enhanced axial transformer for mobile semantic segmentation. In: International Conference on Learning Representations (ICLR) (2023)
33. Wang, C., et al.: Fully automatic wound segmentation with deep convolutional neural networks. Sci. Rep. **10**(1), 21897 (2020)
34. Wang, C., et al.: FUSeg: the foot ulcer segmentation challenge. Information **15**(3), 140 (2024)
35. Wang, L., Pedersen, P.C., Agu, E., Strong, D.M., Tulu, B.: Area determination of diabetic foot ulcer images using a cascaded two-stage SVM-based classification. IEEE Trans. Biomed. Eng. **64**(9), 2098–2109 (2016)
36. Yap, M.H., et al.: Diabetic foot ulcers segmentation challenge report: benchmark and analysis. Med. Image Anal. **94**, 103153 (2024)
37. Yuan, F., Zhang, Z., Fang, Z.: An effective CNN and transformer complementary network for medical image segmentation. Pattern Recogn. **136**, 109228 (2023)
38. Zhang, W., et al.: TopFormer: token pyramid transformer for mobile semantic segmentation. In: Proceedings of the IEEE/CVF Conference on Computer Vision and Pattern Recognition, pp. 12083–12093 (2022)

# Automated Design in Hybrid Action Spaces by Reinforcement Learning and Differential Evolution

Quirin Göttl(✉), Haris Asif, Alexander Mattick, Robert Marzilger, and Axel Plinge

Fraunhofer Institute for Integrated Circuits (IIS), Fraunhofer IIS, Nuremberg, Germany
{Quirin.Goettl,Haris.Asif,Alexander.Mattick, Robert.Marzilger,Axel.Plinge}@iis.fraunhofer.de

**Abstract.** Many real world applications of artificial intelligence and machine learning require to solve a given task inside a hybrid action space. While it is possible to approach these situations with frameworks based solely on reinforcement learning (RL), it is also possible to only use RL for the discrete part of the action space and let an optimization method at hand take care of the continuous actions. In the present work, a framework suitable for hybrid action spaces is derived that uses RL for the discrete decisions alongside a genetic algorithm for the continuous decisions. The RL methodology is based on hierarchical decomposition of the action space and Gumbel AlphaZero. The genetic algorithm is based on differential evolution and can be applied inside continuous action spaces of arbitrary dimension. The flexibility of the approach is shown in two real world use-cases, namely, synthesis of chemical process flowsheets and design of optical multi-layer films. The framework shows promising results compared to solely RL-based methodologies. The versatile framework introduced here could be applied in many other domains, such as, control of energy systems or navigation tasks.

**Keywords:** Reinforcement Learning · Genetic Algorithm · Automated Design

## 1 Introduction

In recent years it has been shown that reinforcement learning (RL) is capable of training agents to achieve superhuman performance in various tasks, e.g., board games [22], robotics [18], and discovery of algorithms [5]. In particular, RL algorithms excel at tasks that are well defined inside a homogeneous action space, i.e., an action space that is either discrete or continuous.

However, in many real world applications, hybrid action spaces (i.e., action spaces that consist of discrete and continuous actions) have to be considered, which require specifically adjusted RL frameworks [4,14,20]. For example, RL


A. Hotho and S. Rudolph (Eds.): KI 2024, LNAI 14992, pp. 292–299, 2024.
https://doi.org/10.1007/978-3-031-70893-0_22

agents are trained inside hybrid actions spaces for control of energy systems [7], vehicle routing [30], and stock trading [21]. To approach these kinds of problems, it is possible to discretize the continuous actions to obtain a solely discrete action space or to relax the discrete actions into a continuous action space [12]. Another possibility is to divide the task on several agents in a cooperative setting [6].

However, hybrid action spaces can also be tackled by RL in combination with a classical optimization approach, e.g., a genetic algorithm (GA) or a gradient-based method [24]. In this setup, the RL agent operates in a solely discrete action space, while all continuous decisions are determined by an optimization method at hand.

In this work, we present a framework that combines a RL agent and a GA to operate in hybrid action spaces. An overview is provided in Fig. 1. The RL agent consists of an artificial neural network (ANN) and a Monte-Carlo tree search (MCTS) based on the framework presented in [9] and Gumbel AlphaZero [2]. In [9], the RL agent operates in a hybrid action space and was trained to synthesize chemical process flowsheets. In this work, we adjust the methodology in a way that the RL agent only takes care of the discrete decisions, while a GA is used for all continuous decisions.

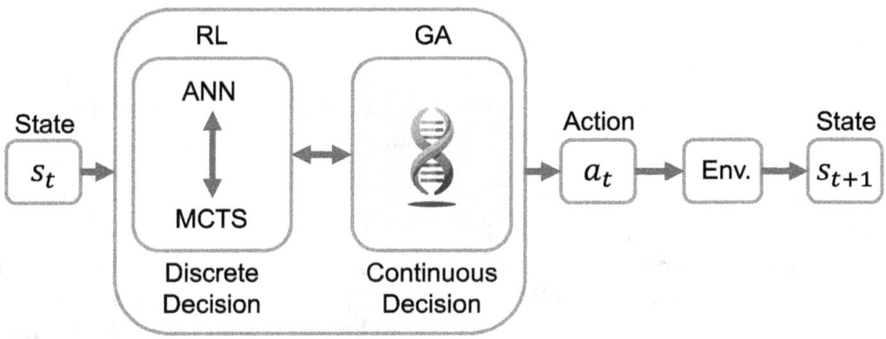


**Fig. 1.** Overview of the framework. Based on the current state $s_t$, the RL agent and the GA generate an action $a_t$ that consists of discrete and continuous decisions. The environment (Env.) simulates $a_t$ and provides the subsequent state $s_{t+1}$.

We show the flexibility of the framework by applying it to two real world use-cases: automated synthesis of chemical process flowsheets and automated optical multi-layer design.

Flowsheet synthesis in chemical engineering, is defined as the act where one invents the structure and operating levels for a new chemical manufacturing process [29]. For this purpose, a flowsheet topology has to be generated alongside the process parameters, i.e., the specifications for the individual devices, so called unit operations, contained in the process scheme. For a detailed overview regarding automated process synthesis in general, we refer to [11, 19]. In RL-based flowsheet synthesis, the agent is trained by interaction with a process simulation to combine unit operations (e.g., distillation columns, reactors) and determine their continuous specifications (e.g., pressure, temperature) in a sequential way [8]. There exist several approaches that structure the hybrid action space hierarchically to separate process synthesis into several sub-tasks [9,10,16,23].

The design of optical multi-layer films is a combinatorial optimization problem with applications in several engineering disciplines [27]. The overall goal is to determine a combination of materials and layer thicknesses that has optimal properties w.r.t. a given objective function, e.g., absorption rate. There exists a multitude of approaches for this problem statement as for example branch-and-bound [1] and the utilization of GAs [17]. It is also possible to train agents by RL to design optical multi-layer films [15,27,28].

We apply our framework to both of those use-cases to show the flexibility of the methodology. In both cases, we can show comparable or improved performances compared to the existing solely RL-based approaches.

## 2 Methodology

### 2.1 RL Framework and Decision Process

As mentioned beforehand, the RL methodology is based on [9], but, compared to that approach, the operating space of the RL agent is reduced to discrete decisions. The structure of the RL agent will be explained alongside Fig. 2 and the flowsheet synthesis use-case. For the details we refer to [9].

The goal of the RL agent is to maximize a predefined reward function by setting up a discrete flowsheet topology. The current state $s_t$, i.e., the current flowsheet, is encoded by an MLP-Mixer [26], an ANN architecture that was originally developed for computer vision. The resulting state vector is processed by several actor-critic [25] networks that generate policy and value for the subsequent MCTS. In Fig. 2, two hierarchy levels are shown to visualize the concept: at level 1, a policy for choosing a location inside the current flowsheet is generated; at level 2, a policy for choosing a unit type (e.g., a reactor or a distillation column) that is to be added at that location is generated. The MCTS is based on Gumbel AlphaZero [2] and explores promising actions (promising is quantified by the outputs of the actor-critic networks). During the tree search, the GA optimizes the continuous parameters that have to be set alongside the discrete actions (e.g., when a distillation column is placed at a certain location, the GA optimizes all continuous parameters inside the whole process with respect to the reward function). The result of the MCTS is an improved policy that is the basis for the final decision of the agent. Furthermore, the resulting improved policy is used as training target for the actor-critic networks.

The framework is flexible regarding the number of discrete hierarchy levels. For example, for the optical multi-layer design use-case, only one hierarchy level (for determination of the next material type) is required.

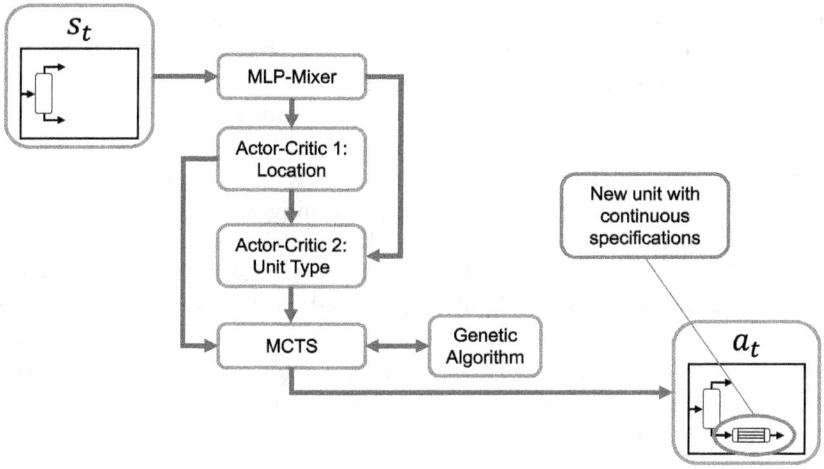


**Fig. 2.** Hierarchical decision process of the RL agent combined with the GA for the use-case flowsheet synthesis.

The core difference between our approach and [9] is that we use a GA instead of RL to determine all continuous decisions. The main advantage of this is that this can be done in any continuous action space without further adjustments to the ANN architecture.

### 2.2 Genetic Algorithm

As explained above, the RL agent operates in a hierarchical and purely discrete action space, e.g., it decides on the discrete topology of a chemical process flowsheet. The continuous parameters (e.g., pressure inside a distillation column and thickness of a material layer for use-case 1 and use-case 2, respectively) are found by a GA, which traverses the continuous (often multidimensional) action space in an effective way and yields a parameter set that is optimized with respect to the reward function. In this work, we chose the differential evolution algorithm described in [3,13] for the determination of the continuous parameters.

This part of the framework could also be replaced by other GAs [13] or gradient-based, classical optimization approaches at hand [24].

## 3 Results

### 3.1 Use Case 1: Automated Synthesis of Chemical Process Flowsheets

For this use-case, the agent interacts with the chemical process simulation that is provided in [9] as environment. The task of the agent is to synthesize process flowsheets that separate the feed stream into pure components using unit operations as for example distillation or decantation. The feed stream compositions are sampled randomly from several chemical systems. In [9], two versions for a reward function, namely 'generic' and 'literature', are presented. Reward 'generic' measures the achieved separation of the resulting processes while reward 'literature' is based on economical analysis of chemical processes. Both reward functions are scaled to the interval $[0, 1]$ and a reward close to 1 is desirable.

For both reward versions, we do a full training cycle with our framework and also reproduce the results from [9]. The resulting average rewards on the evaluation set (the same as provided in [9]) are shown in Table 1. As can be seen, our framework shows a slightly improved performance for both reward versions. It is notable that our framework can be trained in shorter time, but also important to mention that the training procedure in [9] was not optimized for such comparisons. For this reason, we omit listing the exact training times.

**Table 1.** Performance comparison for automated synthesis of chemical process flowsheets. We report the average reward (rounded) on the evaluation set provided in [9].

| Version reward function | Average reward (framework from [9]) | Average reward (our framework) |
|---|---|---|
| Reward 'generic' | 0.92 | **0.95** |
| Reward 'literature' | 0.88 | **0.90** |

### 3.2 Use Case 2: Automated Optical Multi-layer Design

For this use-case, the agent interacts with a material simulation for multi-layer films provided in [27]. The goal is to synthesize an ultra-wideband absorber by sequentially choosing a material layer (discrete decision) and specifying its thickness (continuous decision). The maximum number of material layers is specified at initialization of the environment (the agent can choose to terminate the synthesis before reaching this many layers). The performance of the agent is measured by the achieved absorption rate (in the interval $[0, 1]$, thus an absorption rate close to 1 is desirable) of the final structure.

As in [27], we repeat the training cycle for two different choices for the maximum number of allowed layers and report the results in Table 2. We reproduced the results achieved by the RL agent from [27] for multiple seeds. Additionally, we provide the absorption rates provided in [27] that were achieved after finetuning the best found design out of 10 training cycles with a gradient-based optimization method.

As can be seen in Table 2, our framework provides a slightly better result than the RL methodology from [27] in both cases and can even compare to the performance after the finetuning step from [27].

**Table 2.** Performance comparison for automated optical multi-layer design. We report the absorption rate (rounded) for the best found design (average over 5 seed runs). For [27], we report the absorption rate achieved by the RL framework and the absorption rate after finetuning (in brackets).

| Maximum number of allowed material layers | Best found absorption rate (framework from [27], avg. over multiple seeds) | Best found absorption rate (our framework, avg. over 5 seeds) |
|---|---|---|
| 6 | 0.94 (0.98) | **0.97** |
| 15 | 0.94 (0.99) | **0.97** |

## 4 Discussion and Conclusion

The present work combines a RL agent and a GA to obtain a general framework that is applicable to planning problems in hybrid action spaces. The methodology is successfully applied to two real world use-cases, namely flowsheet synthesis and optical multi-layer design.

The RL methodology adapts several elements from [9], a framework that was originally developed for automated flowsheet synthesis. The core advantage of using a GA for the continuous action space is the resulting flexibility of the framework since it can be applied independently of the number of required continuous actions. Additionally, less training time is needed, since the GA does not require any training for the specification of the continuous actions.

The results give a proof-of-concept that the framework can be applied flexibly to different problems. While in flowsheet synthesis the goal is to obtain a trained agent that can handle several initial situations (i.e., feed streams from several chemical systems) sufficiently well, the goal in optical multi-layer design is to obtain an optimal sequence of materials and thicknesses for a fixed initial situation. Herein, the problem is much more static and RL serves merely as a replacement for other optimization tools. It is important to mention that methods solely based on RL already provide good results in both use-cases. Therefore, in further research, there should be a focus on application fields that provide more room for improvements because of their complexity. For example, one could consider stochastic environments as in energy system control or navigation.

In summary, the present framework shows promising results for application in domains with hybrid action spaces. It enables to combine the creative capabilities of a RL agent in discrete action spaces with the efficiency of a GA for determining continuous parameters and thus can be applied to problems with varying structure in the underlying action space.

## References

1. Azunre, P., Jean, J., Rotschild, C., Bulovic, V., Johnson, S.G., Baldo, M.A.: Guaranteed global optimization of thin-film optical systems. New J. Phys. **21**, 073050 (2019). https://doi.org/10.1088/1367-2630/ab2e19

2. Danihelka, I., Guez, A., Schrittwieser, J., Silver, D.: Policy improvement by planning with Gumbel. In: International Conference on Learning Representations (2022)
3. Das, S., Suganthan, P.N.: Differential evolution: a survey of the state-of-the-art. IEEE Trans. Evol. Comput. **15**, 4–31 (2011). https://doi.org/10.1109/TEVC.2010.2059031
4. Dulac-Arnold, G., Mankowitz, D., Hester, T.: Challenges of real-world reinforcement learning. In: International Conference on Machine Learning (2019)
5. Fawzi, A., Balog, M., Huang, A., et al.: Discovering faster matrix multiplication algorithms with reinforcement learning. Nature **610**, 47–53 (2022). https://doi.org/10.1038/s41586-022-05172-4
6. Fu, H., Tang, H., Hao, J., Lei, Z., Chen, Y., Fan, C.: Deep multi-agent reinforcement learning with discrete-continuous hybrid action spaces. In: International Joint Conference on Artificial Intelligence (2019)
7. Gao, Y., Matsunami, Y., Miyata, S., Akashi, Y.: Multi-agent reinforcement learning dealing with hybrid action spaces: a case study for off-grid oriented renewable building energy system. Appl. Energy **326**, 120021 (2022). https://doi.org/10.1016/j.apenergy.2022.120021
8. Gao, Q., Schweidtmann, A.M.: Deep reinforcement learning for process design: review and perspective. Curr. Opin. Chem. Eng. **44**, 101012 (2024). https://doi.org/10.1016/j.coche.2024.101012
9. Göttl, Q., Pirnay, J., Burger, J., Grimm, D.G.: Deep reinforcement learning uncovers processes for separating azeotropic mixtures without prior knowledge. arXiv:2310.06415 (2023)
10. Göttl, Q., Tönges, Y., Grimm, D.G., Burger, J.: Automated flowsheet synthesis using hierarchical reinforcement learning: proof of concept. Chem. Ing. Tec. **93**(12), 2010–2018 (2021). https://doi.org/10.1002/cite.202100086
11. Grossmann, I.E., Harjunkoski, I.: Process systems engineering: academic and industrial perspectives. Comput. Chem. Eng. **126**(12), 474–484 (2019). https://doi.org/10.1016/j.compchemeng.2019.04.028
12. Hausknecht, M.J., Stone, P.: Deep reinforcement learning in parameterized action space. In: International Conference on Learning Representations (2016)
13. Hu, Z., Li, P., Liu, Y.: Enhancing the performance of evolutionary algorithm by differential evolution for optimizing distillation sequence. Molecules **27**(12), 3802 (2022). https://doi.org/10.3390/molecules27123802
14. Hubert, T., Schrittwieser, J., Antonoglou, I., Barekatain, M., Schmitt, S., Silver, D.: Learning and planning in complex action spaces. In: International Conference on Machine Learning (2021)
15. Jiang, A., Osamu, Y., Chen, L.: Multilayer optical thin film design with deep Q learning. Sci. Rep. **10**, 12780 (2020). https://doi.org/10.1038/s41598-020-69754-w
16. Khan, A.A., Lapkin, A.A.: Designing the process designer: hierarchical reinforcement learning for optimisation-based process design. Chem. Eng. Process. Process Intensification **180**, 108885 (2022). https://doi.org/10.1016/j.cep.2022.108885
17. Li, D.G., Watson, A.C.: Optical thin film optimization design using genetic algorithms. In: IEEE International Conference on Intelligent Processing Systems (1997). https://doi.org/10.1109/ICIPS.1997.672752
18. Mahmood, A.R., Korenkevych, D., Vasan, G., Ma, W., Bergstra, J.: Benchmarking reinforcement learning algorithms on real-world robots. In: Conference on Robot Learning (2018)

19. Martin, M., Adams II, T.A.: Challenges and future directions for process and product synthesis and design. Comput. Chem. Eng. **128**, 421-436 (2019). https://doi.org/10.1016/j.compchemeng.2019.06.022
20. Neunert, M., et al.: Continuous-discrete reinforcement learning for hybrid control in robotics. In: Conference on Robot Learning (2019)
21. Pan, F., Zhang, T., Luo, L., He, J., Liu, S.: Learn continuously, act discretely: hybrid action-space reinforcement learning for optimal execution. In: Proceedings of the Thirty-First International Joint Conference on Artificial Intelligence (2022). https://doi.org/10.48550/ARXIV.2207.11152
22. Silver, D., Schrittwieser, J., Simonyan, K., et al.: Mastering the game of go without human knowledge. Nature **550**, 354–359 (2017). https://doi.org/10.1038/nature24270
23. Stops, L., Leenhouts, R., Gao, Q., Schweidtmann, A.M.: Flowsheet generation through hierarchical reinforcement learning and graph neural networks. AIChE J. **69**(1), e17938 (2023). https://doi.org/10.1002/aic.17938
24. Sun, S., Cao, Z., Zhu, H., Zhao, J.: A survey of optimization methods from a machine learning perspective. IEEE Trans. Cybern. **50**, 3668–3681 (2019). https://doi.org/10.1109/TCYB.2019.2950779
25. Sutton, R.S., Barto, A.G.: Reinforcement Learning: An Introduction, 2nd edn. The MIT Press, Cambridge (2018)
26. Tolstikhin, I.O., et al.: MLP-mixer: an all-MLP architecture for vision. Neural Inf. Process. Syst. (2021)
27. Wang, H., Zheng, Z., Ji, C., Guo, L.J.: Automated multi-layer optical design via deep reinforcement learning. Mach. Learn. Sci. Technol. **2**, 025013 (2021). https://doi.org/10.1088/2632-2153/abc327
28. Wankerl, H., Stern, M.L., Mahdavi, A., Eichler, C., Lang, E.W.: Parameterized reinforcement learning for optical system optimization. J. Phys. D Appl. Phys. **54**, 305104 (2021). https://doi.org/10.1088/1361-6463/abfddb
29. Westerberg, A.: A retrospective on design and process synthesis. Comput. Chem. Eng. **28**(4), 447–458 (2004). https://doi.org/10.1016/j.compchemeng.2003.09.029
30. Xu, R., Xu, J., Liu, X., Lin, Y.: Safe hybrid-action reinforcement learning-based decision and control for discretionary lane change. Machines **12**, 252 (2024). https://doi.org/10.3390/machines12040252

# Instance Segmentation with a Novel Tree Log Detection Dataset

Julian Haasis(✉), Christopher Bonenberger, and Markus Schneider

Institute of Artificial Intelligence, University of Applied Sciences Ravensburg-Weingarten, Weingarten, Germany
Julian.Haasis@rwu.de
https://www.rwu.de

**Abstract.** Reliable tree log detection is a key requirement for automation of forestry operations. Despite the substantial progress regarding object detection in general, tree log detection lags behind due to the lack of well-annotated datasets. In order to address this gap, we introduce the *Tree Log Detection Dataset* (TLDD). This real-world dataset is collected using a combination of 360° multiline LIDAR and stereo cameras. It offers a wide range of annotated segmentation masks for over 1000 images of about 22000 tree logs. We assess the quality of the presented data set by comparing it to existing data sets using state-of-the-art architectures such as MaskDINO and Mask2Former. Our experiments demonstrate the quality of TLDD and confirm the efficiency of attention-based, transformer-like networks.

**Keywords:** Wood Log Detection · Forestry · Semantic Segmentation

## 1 Introduction

In recent years many tasks were automated with the help of deep learning, robotics, and sensor availability. Forestry, however, has not made huge steps forward due to the advances in digitization. This is mainly because this industry is propelled by a large mechanization effort and therefore there is a high potential concerning automating this domain. Here, the detection of tree logs represents a crucial aspect, offering potential benefits for automated log picking.

The development of effective tree log detection algorithms depends on access to well-annotated datasets. However, there is a gap in terms of data availability compared to other well-established domains such as autonomous driving (KITTI [10]), and object detection in different surroundings (Pascal VOC [8], COCO [17], etc. [1–5,7,11,18,19,21–23]). To address this gap and advance the automation of forestry operations, the *Tree Log Detection Dataset* (TLDD) is introduced. Inspired by the success of benchmark data sets like Pascal VOC, the TLDD hopefully provides a valuable resource to empower researchers in developing tree log detection algorithms and helps to drive progress in automating forestry operations, similar to well-established datasets in other computer vision domains.

The data set under consideration is available under https://www.kaggle.com/datasets/julejan/treelogdetectiondataset.


A. Hotho and S. Rudolph (Eds.): KI 2024, LNAI 14992, pp. 300–307, 2024.
https://doi.org/10.1007/978-3-031-70893-0_23

Similar to our work, the authors of the TimberSeg1.0 data set [9] suggest evaluating the data set quality using state-of-the-art neural networks. This work aims to improve the detection of logs, by using available log datasets and using the recent state-of-the-art network architectures. In the following, we evaluate the difference between the examined datasets and between current state-of-the-art networks. Mask2Former [6], which showed to be effective for long elongated objects [9], is used as a baseline. Furthermore, MaskDINO [16] is used for comparison, in order to evaluate the impact of different transformer networks for the given task. Finally, we evaluate how the combination of different datasets affects the overall performance.

## 2 Related Work

Although there is just a small number of log detection data sets available, there are plenty of other publicly available datasets in the context of forestry environment. Many forestry-related data sets are concerned with automating forestry applications, tree detection and segmentation, while others also target navigation tasks in the forest environment. As an example, [13] published a data set containing 100 real images with over 900 annotated trees in Canadian forests. [15] provide a data set of 4226 manually annotated objects, out of which 2562 objects correspond to tree trunks classified into three different instance categories. The data consists of RGB stereo images, point clouds, sparse depth maps, and ground truth manual annotations for semantic, instance, and panoptic segmentation.

**Table 1.** Related work of datasets in the forest environment

| Dataset Name | BBox | Inst. Seg | Sem. Seg | Panoptic Seg | Point Cloud | Optical Flow | Depth |
|---|---|---|---|---|---|---|---|
| CanaTree100 | ✓ | ✓ | ✗ | ✗ | ✗ | ✗ | ✓ |
| Finn Woodlands | ✓ | ✓ | ✓ | ✓ | ✗ | ✗ | ✗ |
| SynthTree43k | ✓ | ✓ | ✗ | ✗ | ✗ | ✗ | ✓ |
| TimberSeg1.0 | ✓ | ✓ | ✗ | ✗ | ✗ | ✗ | ✗ |
| **TLDD(ours)** | ✓ | ✓ | ✗ | ✗ | ✓ | ✗ | ✓ |

SYNTHTREE43K is a synthetic dataset collected from a simulated forest environment for tree detection with 17 different types of trees [13]. It consists of 43000 synthetic images and over 190000 annotated trees (produced using the Unity game engine [14]). Finally, the TimberSeg 1.0 [9] dataset consists of 220 images containing 2500 instances of wood logs in different environments and dispositions from the viewpoint of forwarders and wood loaders.

Table 1 compares the discussed datasets and their properties. Only the TimberSeg 1.0 and our TLDD data set are used for training and evaluation since, these are the only ones that have logs as a category. In the other data sets, only different types of trees are labeled.

**Fig. 1.** Samples images from the introduced *Tree Log Detection Dataset* (TLDD)

# 3 Dataset Features

Our data set is split into 10 different sequences provided as *rosbags*. This data format belongs to the Robot Operation System (ROS) and is a common way to record data in the autonomous robots domain. The duration of these rosbags varies from $600s$ to $1000s$. The annotated dataset contains about 1050 images and more than 22000 annotated segmentation masks of the *log* class. To create the annotated dataset out of the collected raw data, a selection was chosen to reduce too similar scenes. The image stream has a framerate of 15 FPS, where every tenth frame has associated ground truth annotations. Further refinement was done by eliminating very similar pictures. Blurry and unusable images were removed manually. Some sample images are shown in Fig. 1. The annotation was carried out using pixelwise mask segmentation with the CVAT [20] annotation tool. A trained Mask2Former network (see [9]) was used to pre-annotate data set. Subsequently data was processed according to the above mentioned principles.

## 3.1 Data Recording and Labeling

Data recordings were conducted under a variety of different environment conditions using a 360° , 16 line Velodyne VLP16 LIDAR sensor (Field of view: $\pm 15°$, up to $100m$), and a ZED2i stereo camera (15 FPS, 720p). The ZED2i stereo camera complements the LIDAR, providing both RGB and 3D data. The ZED2 onboard Inertial Measurement Unit (IMU) provides additional information about orientation, acceleration, and angular velocity (e.g., for crane movement detection). This adds further depth to the dataset. The two different depth cameras might make the data set also useful for Sensor Fusion applications.

In order to synchronize the different sensors and data types we use Robot Operating System (ROS). The data set consists of 10 rosbags. The two-shift working cycle of the sawmill allowed it to capture scenes from the early morning hours to the late evening, presenting the full spectrum of a typical workday. This real-world setting captures different scenarios and weather conditions, offering a wide range of perspectives for research. Moreover, the sensor rig used for data collection was positioned at various angles, capturing scenes from different viewpoints. This diversity of viewing angles ensures that the dataset encompasses a wide range of perspectives, allowing to investigate the impact of sensor placement on object detection and scene comprehension.

Additionally, 34 other test images are included for the evaluation of 2D image classification algorithms. This test dataset contains no additional 3D data. These recordings were made with several smartphones and RGB cameras (see 3). In order to show generalization abilities of the training data, the test data set is collected by retrieving images from different locations and viewing angles, as well as different camera sensors. The test data does not include pictures of scenes or locations from the training data.

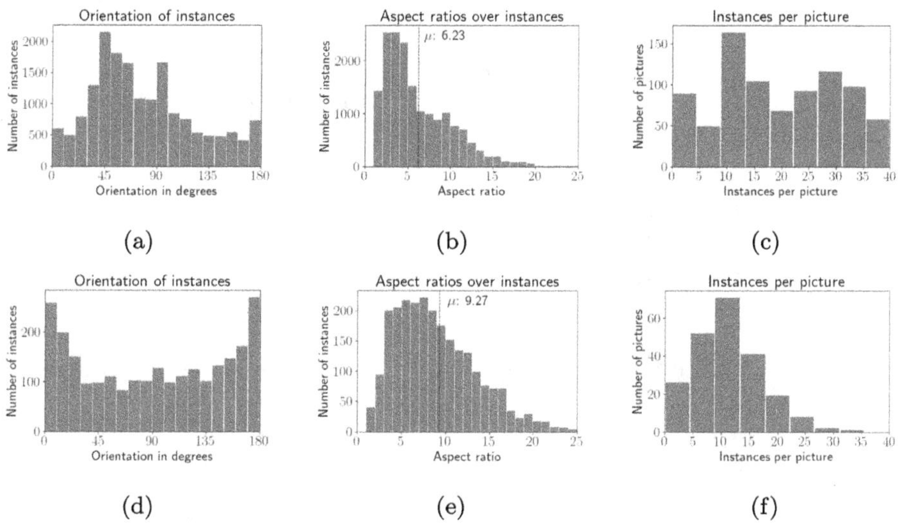


**Fig. 2.** Dataset statistics: The TimberSeg 1.0 data set is illustrated at the bottom, our TLDD data set at the top. Figures 2a and 2d shows the distribution of log annotations throughout the data sets. This graph is calculated by iterating over all annotations, calculating a rotated bounding box for each of these annotations (rotation angles are aggregated to bins). Figures 2b and 2e shows aspect ratios of all logs and Figs. 2c and 2f shows the instances per picture

### 3.2 Dataset Comparison

An overview about the two data sets under consideration is shown in 2. Regarding orientation, in TLDD the slight peak around 45° to 90° means that a lot of logs are parallel to the horizon. The logs in TimberSeg have a peak around

0/180°, which means that the orientation of these logs is mostly vertical. The average aspect ratio of all logs $\mu = 6.23$ in TLDD can be explained by the fact that the sorting crane often moves close to the logs next to it. In TimberSeg logs are further away, leading to a higher sample mean $\mu = 9.27$. Together these two datasets add up at this point and increase the variety.

**Fig. 3.** Test data (left), Best performing MaskDINO 44.55AP (middle), Best performing Mask2Former 38.81AP (right)

TimberSeg 1.0 also contains already four different environment types such as, lumberyard, roadside, trailer and forest. TLDD contains a not yet existing environment type sawmill.

# 4 Experiments and Results

**Table 2.** Results of the networks on the test data

| Model | Dataset | Backbone | Epochs | *MaskAP* | *BoxAP* |
|---|---|---|---|---|---|
| Mask2Former | TimberSeg | SwinB | 10 | $28.37 \pm 1.65$ | |
| Mask2Former | TLDD | SwinB | 10 | $31.41 \pm 2.75$ | |
| MaskDINO[1] | TLDD&TimberSeg | SwinL | 10 | $32.23 \pm 5.33$ | $32.42 \pm 5.54$ |
| MaskDINO | TLDD | R50 | 10 | $31.62 \pm 2.64$ | $32.11 \pm 3.29$ |
| MaskDINO | TimberSeg | R50 | 10 | $34.51 \pm 4.49$ | $36.01 \pm 2.73$ |
| Mask2Former | TLDD | SwinB(IN21K) | 10 | $36.14 \pm 1.71$ | |
| MaskDINO | TimberSeg | SwinB | 10 | $36.73 \pm 1.51$ | $37.46 \pm 2.24$ |
| Mask2Former | TLDD&TimberSeg | SwinB | 10 | $36.2 \pm 3.06$ | |
| MaskDINO | TimberSeg | R50 | 10 | $37.44 \pm 0.61$ | $37.91 \pm 0.79$ |
| Mask2Former | TLDD&TimberSeg | SwinB(IN21K) | 10 | $38.81 \pm 0.43$ | |
| MaskDINO | TLDD&TimberSeg | R50 | 10 | $39.85 \pm 2.7$ | $40.36 \pm 2.79$ |
| MaskDINO | TLDD&TimberSeg | SwinB | 10 | $44.09 \pm 2.7$ | $42.66 \pm 2.47$ |
| MaskDINO | TLDD | SwinL | 10 | $\mathbf{44.55 \pm 1.26}$ | $43.62 \pm 0.69$ |

[1]no maskenhance

Multiple experiments were conducted to compare and analyze the performances of state-of-the-art networks on TLDD and TimberSeg 1.0. Previous research [9] have shown that neural networks designed like Mask R-CNN perform less effectively in detecting elongated and rotated objects compared to attention-based, transformer-like networks. As the dataset size is too small, pretrained models have to be taken into account. For comparability, the pretrained network models should ideally be trained on similar datasets. As stated in the methodology section, Mask2Former and MaskDINO are further used to evaluate the performance of the dataset on state-of-the-art networks. It is assumed that MaskDINO is achieving better results than Mask2Former due to the unified query selection. For the evaluation of the predicted masks, the standard COCO metrics are being used. All networks were evaluated on the test dataset of TLDD, which includes images of color sensors not included in the TLDD or TimberSeg 1.0 training dataset, as well as different resolutions and locations. To have a fair comparison between the datasets, they are trained on each network with the identical training settings. For each network, three training runs are conducted, each initialized randomly.

Here, 10 epochs are chosen to train all networks for a better comparability. The final model performed better than the one trained with 10 Epochs. Another job for future work could be if overfitting occurs on such a high epoch number.

For each network, the same data augmentation is used (large-scale jittering (LSJ) method, following [12]). Mask2Former was trained according to [6] (several backbones with a window size of 12, pretrained on ImageNet-1k and COCO). The number of object queries for the decoder is set to 100. Both networks are trained for 10 Epochs and with a batch size of 3 using a NVIDIA A40 GPU with a VRAM of 48GB. For MaskDINO we followed the recommendations of [16].

### 4.1 Main Results

As can be seen from Table 2 MaskDINO clearly outperforms Mask2Former under the same configurations. The instance segmentation results surpass the baseline of Mask2Former by a large margin (+5.7 AP) on the 10 Epoch settings. Another important finding is that networks trained with TimberSeg 1.0 achieve similar or better performance on the test data set than TLDD, although it contains just a quarter of the annotated images compared to TLDD. This could originate from the fact that TimberSeg 1.0 was collected in a much greater time span, at different seasons using different sensors.

## 5 Conclusion

We presented a new dataset, which consists over 1000 images of over 22000 annotated tree log instance segmentations as rosbags. This allows to replicate a real-world scenario in a sawmill. We provide images collected from different, scenes, days, viewing angles and image sensors. Using MaskDINO and Mask2Former we have demonstrated the utility of the data set and its generalization ability.

## References


1. Behley, J., et al.: SemanticKITTI: a dataset for semantic scene understanding of LiDAR sequences (2019)
2. Brostow, G.J., Fauqueur, J., Cipolla, R.: Semantic object classes in video: a high-definition ground truth database. Pattern Recogn. Lett. **30**(2), 88–97 (2009). https://doi.org/10.1016/j.patrec.2008.04.005, https://www.sciencedirect.com/science/article/pii/S0167865508001220, video-based Object and Event Analysis
3. Cabon, Y., Murray, N., Humenberger, M.: Virtual KITTI 2 (2020)
4. Caesar, H., et al.: nuScenes: a multimodal dataset for autonomous driving. In: 2020 IEEE/CVF Conference on Computer Vision and Pattern Recognition (CVPR), pp. 11618–11628. IEEE Computer Society, Los Alamitos, CA, USA (2020). https://doi.org/10.1109/CVPR42600.2020.01164
5. Che, Z., et al.: D2-city: a large-scale dashcam video dataset of diverse traffic scenarios. ArXiv abs/1904.01975 (2019). https://api.semanticscholar.org/CorpusID:102486513
6. Cheng, B., Misra, I., Schwing, A.G., Kirillov, A., Girdhar, R.: Masked-attention mask transformer for universal image segmentation (2022)
7. Cordts, M., et al.: The cityscapes dataset for semantic urban scene understanding. In: 2016 IEEE Conference on Computer Vision and Pattern Recognition (CVPR), pp. 3213–3223. IEEE Computer Society, Los Alamitos, CA, USA (2016). https://doi.org/10.1109/CVPR.2016.350
8. Everingham, M., Van Gool, L., Williams, C.K.I., Winn, J., Zisserman, A.: The pascal visual object classes (VOC) challenge. Int. J. Comput. Vision **88**(2), 303–338 (2010)
9. Fortin, J.M., Gamache, O., Grondin, V., Pomerleau, F., Giguère, P.: Instance segmentation for autonomous log grasping in forestry operations. In: 2022 IEEE/RSJ International Conference on Intelligent Robots and Systems (IROS), pp. 6064–6071 (2022). https://doi.org/10.1109/IROS47612.2022.9982286
10. Geiger, A., Lenz, P., Stiller, C., Urtasun, R.: Vision meets robotics: the KITTI dataset. Int. J. Robot. Res. (IJRR) (2013)
11. Geyer, J., et al.: A2D2: Audi autonomous driving dataset (2020)
12. Ghiasi, G., et al.: Simple copy-paste is a strong data augmentation method for instance segmentation. In: 2021 IEEE/CVF Conference on Computer Vision and Pattern Recognition (CVPR), pp. 2917–2927 (2021). https://doi.org/10.1109/CVPR46437.2021.00294
13. Grondin, V., Fortin, J.M., Pomerleau, F., Giguère, P.: Tree detection and diameter estimation based on deep learning. For. Int. J. For. Res. (2022)
14. Juliani, A., et al.: Unity: a general platform for intelligent agents. arXiv preprint arXiv:1809.02627 (2018)
15. Lagos, J., Lempiö, U., Rahtu, E.: Finnwoodlands dataset. In: Gade, R., Felsberg, M., Kämäräinen, J.K. (eds.) SCIA 2023. LNCS, vol. 13885, pp. 95–110. Springer, Cham (2023). https://doi.org/10.1007/978-3-031-31435-3_7
16. Li, F., et al.: Mask Dino: Towards a unified transformer-based framework for object detection and segmentation (2022)
17. Lin, T.-Y., et al.: Microsoft COCO: common objects in context. In: Fleet, D., Pajdla, T., Schiele, B., Tuytelaars, T. (eds.) ECCV 2014. LNCS, vol. 8693, pp. 740–755. Springer, Cham (2014). https://doi.org/10.1007/978-3-319-10602-1_48

18. Neuhold, G., Ollmann, T., Bulò, S.R., Kontschieder, P.: The mapillary vistas dataset for semantic understanding of street scenes. In: 2017 IEEE International Conference on Computer Vision (ICCV), pp. 5000–5009 (2017). https://doi.org/10.1109/ICCV.2017.534
19. Russakovsky, O., et al.: ImageNet large scale visual recognition challenge (2015)
20. Sekachev, B., et al.: opencv/cvat: v1.1.0 (2020). https://doi.org/10.5281/zenodo.4009388
21. Shao, S., et al.: Objects365: a large-scale, high-quality dataset for object detection. In: 2019 IEEE/CVF International Conference on Computer Vision (ICCV), pp. 8429–8438 (2019). https://doi.org/10.1109/ICCV.2019.00852
22. Silberman, N., Hoiem, D., Kohli, P., Fergus, R.: Indoor segmentation and support inference from RGBD images. In: Fitzgibbon, A., Lazebnik, S., Perona, P., Sato, Y., Schmid, C. (eds.) ECCV 2012. LNCS, vol. 7576, pp. 746–760. Springer, Heidelberg (2012). https://doi.org/10.1007/978-3-642-33715-4_54
23. Yu, F., et al.: BDD100K: a diverse driving dataset for heterogeneous multitask learning (2020)

# LaFAM: Unsupervised Feature Attribution with Label-Free Activation Maps

Aray Karjauv[1,2(✉)] and Sahin Albayrak[1]

[1] Technische Universität Berlin, 10623 Berlin, Germany
[2] GT-ARC gGmbH, 10587 Berlin, Germany
Aray.Karjauv@tu-berlin.de

**Abstract.** Convolutional Neural Networks (CNNs) are known for their ability to learn hierarchical structures, naturally developing detectors for objects, and semantic concepts within their deeper layers. Activation maps (AMs) reveal these saliency regions, which are crucial for many Explainable AI (XAI) methods. However, the direct exploitation of raw AMs in CNNs for feature attribution remains underexplored in literature. This work revises Class Activation Map (CAM) methods by introducing the Label-free Activation Map (LaFAM), a streamlined approach utilizing raw AMs for feature attribution without reliance on labels. LaFAM presents an efficient alternative to conventional CAM methods, demonstrating particular effectiveness in saliency map generation for self-supervised learning while maintaining applicability in supervised learning scenarios.

**Keywords:** XAI · Self-supervised learning · CNN

## 1 Introduction

Self-supervised learning (SSL) has emerged as a flexible approach in computer vision [5,12], demonstrating the ability to learn the underlying structure from data without manual annotations. This property enables SSL models to be applied to a variety of downstream tasks, making them often referred to as general purpose or foundation models [3]. However, the absence of labels presents a challenge in evaluating whether these models have learned relevant representations or if they captured biases and shortcuts [16]. Therefore, it is crucial to identify and address potential shortcomings, particularly in light of the EU AI Act [15], which requires transparency in high-risk areas such as healthcare.

While Vision Transformers [8] show impressive results, convolutional neural networks (CNNs) continue to be extensively used due to their well-established effectiveness in extracting spatially coherent features and their interpretability [22,25,26]. Evidence suggests that CNNs can inherently develop detectors for semantic objects without explicit supervision [2,11,27], making Activation


A. Hotho and S. Rudolph (Eds.): KI 2024, LNAI 14992, pp. 308–315, 2024.
https://doi.org/10.1007/978-3-031-70893-0_24

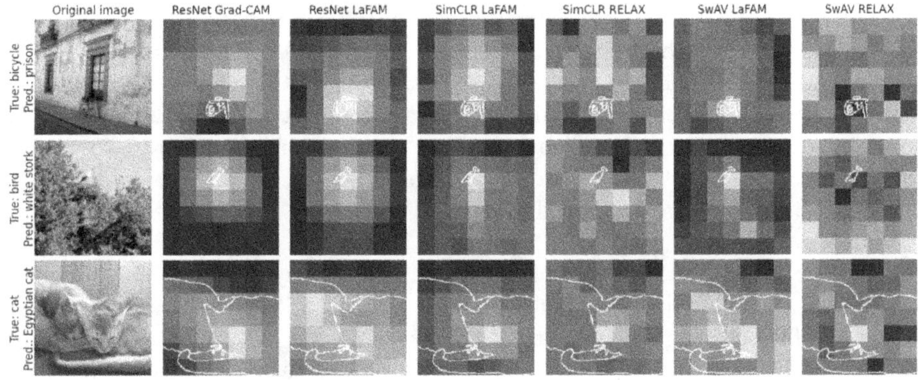


**Fig. 1.** Saliency maps comparison. The contours are overlaid to provide a reference for the ground truth. The title on the left displays the true labels from PASCAL VOC 2012 [9] alongside the ImageNet labels predicted by ResNet50. Grad-CAM utilizes the predicted labels to generate saliency maps. Noticeably, the LaFAM saliency maps are very similar to those produced by Grad-CAM, while RELAX produces noisy saliency maps. The first row demonstrates a misclassification example, showcasing a situation where Grad-CAM fails to highlight the correct region. A detailed comparison is presented in Sect. 4.

Maps (AMs) in deeper layers a key factor in the success of current XAI methods [4,6,17,19,21,23,26,28].

Surprisingly, while AMs are acknowledged for their ability to capture abstract concepts the utilization of raw AMs has been largely overlooked in the literature. Addressing this gap, we propose the Label-free Activation Map (LaFAM) as an alternative to traditional occlusion and gradient-based techniques. LaFAM does not require labels, making it ideally suited for SSL models. It may further benefit supervised learning due to its ability to highlight salient regions that traditional methods might ignore since they primarily focus on generating class-specific saliency maps. For reproducibility and further examination, both the source code[1] and a live demonstration[2] are made available.

## 2 Related Work

Early foundational work by Zeiler and Fergus (2014) demonstrates the use of occlusion sensitivity to show how CNN predictions decline when key image features are masked, thereby localizing critical features for classification [26]. Building on this concept, the Randomized Input Sampling for Explanation (RISE) [18] method employs random masks to create occlusions, facilitating the generation of saliency maps that identify salient regions in image classification tasks.

[1] https://github.com/karray/LaFAM.

[2] https://karay.me/examples/lafam.

Zhou et al. (2016) enhanced interpretability by introducing Class Activation Maps (CAMs) which are a linear combination of AMs in the last convolutional layer, multiplied by the weights of the output layer [28]. Nevertheless, this method is limited by its architecture-specific requirements, which include a global average pooling followed by a fully connected output layer. To mitigate this architectural constraint, various gradient-based [6,17,21] and gradient-free methods [4,19,23] were proposed that aim to obtain a class-specific score to weigh the AMs.

However, these methods do not apply to SSL, as they require labels to calculate the class scores. Given the lack of annotations in SSL, there is a requirement for a label-free scoring function that can measure the output. Representation learning explainability (RELAX) [24] method addresses this issue by adapting RISE to SSL models. It utilizes cosine similarity as a replacement for the class score function. It measures the difference between the embedding of the original image and the embeddings of the occluded images, which is then used to weigh the occlusion masks.

Nevertheless, it suffers from several limitations including computational inefficiency due to multiple model inference and noisy results. Additionally, using solid color patches for occlusion may produce out-of-distribution samples, which can produce misleading explanations that do not reflect the behavior of a model on typical in-distribution data.

We propose to utilize the raw AMs directly for generating saliency maps without further manipulations, as they already encapsulate essential semantic and spatial information [11,27]. Given that this approach does not require labeled data, it is particularly well-suited for SSL. Additionally, this method may significantly benefit self-supervised models by revealing all learned concepts, including those not explicitly labeled as shown in Fig. 2.

## 3 Method

LaFAM is a post hoc analytical label and gradient-free method that generates saliency maps by averaging the activation maps at a selected convolutional layer. Unlike fully connected layers, convolutional layers preserve spatial information, which is essential for constructing spatially coherent saliency maps. LaFAM requires a single forward pass during inference to obtain the activation of the target convolutional layer, which makes it computationally efficient.

Let $A_{i,j}^{l,k}$ denote the activation at spatial position $(i, j)$ in the $k$-th channel of the $l$-th convolutional layer. The average activation $\bar{A}_{i,j}^{l}$ over channels at position $(i, j)$ for the $l$-th convolutional layer is computed as follows:

$$\bar{A}_{i,j}^{l} = \frac{1}{K}\sum_{k=1}^{K} A_{i,j}^{l,k}, \tag{1}$$

where $K$ is the number of channels in the $l$-th convolutional layer.

The ResNet used in this work incorporates Rectified Linear Units (ReLUs), which ensures that the AMs contain only non-negative values, representing a positive contribution to the output. However, these values are unbounded in the positive direction, requiring normalization. Therefore, the saliency map $M_{\text{LaFAM}}$ is constructed by first applying min-max normalization to the averaged map $\bar{A}^l$ scaling the values to the range $[0, 1]$. The resulting map is then upsampled to match the input image size:

$$M_{\text{LaFAM}} = \text{Up}\left(N\left(\bar{A}^l\right)\right), \tag{2}$$

where $N$ denotes the min-max normalization function, and Up represents the operation of upsampling the normalized map to the input image size.

## 4 Experiment

We systematically evaluate and compare LaFAM with RELAX in SSL settings. For additional details, please refer to the extended version of this paper available on arXiv[3].

This evaluation is conducted using segmentation masks derived from the validation sets of ImageNet-S [10]. To ensure a fair and consistent comparison, all methods employ nearest neighbor interpolation for upscaling the saliency maps. Furthermore, we isolate pertinent samples by selecting distinct categories and discarding masks with multiple classes. The refined dataset consists of 11,294 samples.

Quantitative analysis is performed using the Quantus framework [14], with a particular focus on saliency map quantification through segmentation masks. Visual examples of the saliency maps are presented in Fig. 1.

We employ SimCLR [7] and SwAV [5] models with a ResNet50 [13] backbone, pretrained[4] on ImageNet-1k [20]. Given that the final convolutional layer produces AMs with a $7 \times 7$ spatial dimension and 2048 channels, RELAX is set to generate an equal number of masks with the same spatial resolutions. These masks are produced based on a Bernoulli distribution with a p-parameter of 0.5. Moreover, we compare the performance of LaFAM and Grad-CAM in supervised settings as a baseline for reference using a pretrained ResNet50 classifier[5].

Table 1 presents the results for the ImageNet-1k validation set, comparing LaFAM with RELAX for SimCLR and SwAV in SSL settings, and with Grad-CAM in the supervised setting as a point of comparison.

In the SSL scenario, LaFAM outperforms RELAX across all metrics. Specifically, the higher Pointing-Game and Sparseness scores indicate more precise and focused saliency maps. RELAX shows a low Sparseness score, indicating that it tends to produce noticeably noisy saliency maps, especially for small objects. This observation is supported by the qualitative findings, as depicted in Fig. 1.

[3] Extended version available at https://arxiv.org/abs/2407.06059.

[4] Pretrained models from PyTorch Lightning Bolts.

[5] Pretrained model from Torchvision.

**Table 1.** Saliency maps performance comparison on ImageNet-S (higher values are better).

| Metric | Supervised (ResNet50) | | SSL (SimCLR) | | SSL (SwAV) | |
|---|---|---|---|---|---|---|
| | Grad-CAM | LaFAM | RELAX | LaFAM | RELAX | LaFAM |
| Pointing-Game | **94.00** | 90.67 | 88.29 | **92.14** | 85.47 | **89.90** |
| Sparseness | **42.74** | 34.82 | 35.26 | **49.70** | 31.49 | **39.92** |
| Relevance Mass Accuracy | **50.28** | 45.89 | 46.19 | **53.32** | 42.96 | **50.13** |
| Relevance Rank Accuracy | **62.22** | 59.50 | 58.13 | **61.44** | 53.64 | **64.69** |
| Top-K Intersection | **75.07** | 69.09 | 71.21 | **76.59** | 63.83 | **71.68** |
| AUC | **83.12** | 80.45 | 76.49 | **81.28** | 70.13 | **83.03** |

In supervised settings, LaFAM is competitive with Grad-CAM across several metrics. As expected, Grad-CAM demonstrates superior results in the Sparseness metric. It implies that Grad-CAM generates less scattered explanations, as its design inherently highlights discriminative regions associated with specific class labels. However, CAM methods require labels and it can be a limiting factor since class scores are computed with respect to the most probable prediction. This can lead to wrong or misleading explanations when the prediction is incorrect.

In contrast, a lower Sparseness score for LaFAM suggests that it captures a broader range of learned features, which can be beneficial compared to CAM methods, especially when an image contains multiple objects in a scene. It, therefore, may not be necessarily interpreted as a shortcoming but rather as a more universal approach as shown in Fig. 2.

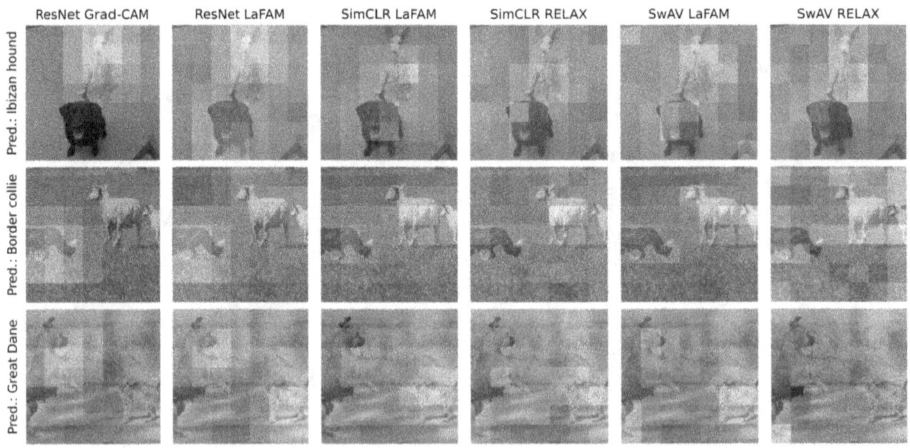


**Fig. 2.** Saliency maps comparison for scenes with two distinct objects. Left-hand labels indicate ImageNet labels predicted by ResNet50 classifier.

Figure 2 depicts scenes with two distinct objects, demonstrating the ability of LaFAM to highlight multiple concepts simultaneously. This capability is crucial not only for SSL but may also benefit supervised models by providing insights into their behavior when processing complex scenes. Unlike traditional CAM methods, which focus on a single object due to their class-specific approach, LaFAM allows for a more comprehensive understanding of learned concepts. This can be particularly beneficial when predictions are incorrect or when assessing model robustness in real-world situations, where scenes often contain multiple classes. By aggregating AMs without relying on specific predictions, LaFAM can help identify and address model misclassifications.

## 5 Discussion

This study evaluates methods XAI methods in the context of CNNs and compares the performance of LaFAM and RELAX within the SSL scenario and LaFAM and Grad-CAM under supervised conditions as a reference point.

Our findings reveal that LaFAM consistently outperforms RELAX across all metrics for SSL models, particularly for scenes with small-sized objects. While the performance of RELAX is suboptimal, it can be enhanced by decreasing cell sizes and increasing the number of masks to generate more detailed saliency maps, although this increment raises computational costs. In contrast, LaFAM is more computationally efficient and requires no additional hyperparameters, making it a more practical choice for saliency map generation in SSL scenarios.

A common limitation of traditional CAM methods, including LaFAM, is low resolution. However, it might be possible to address this limitation by redistributing the AM values to earlier layers through Layer-wise Relevance Propagation (LRP) [1]. This approach could potentially mitigate the issue, but its effectiveness requires further investigation.

In conclusion, the results highlight that LaFAM emerges as a robust and flexible method, contributing to the diversification of the XAI toolbox. Its computational efficiency and label-free saliency map generation make it valuable for understanding model decisions across both supervised and self-supervised learning paradigms.

**Acknowledgements.** This work is based on the research conducted in the gemeinwohlorientierter KI-Anwendungen (Go-KI) project (Offenes Innovationslabor KI zur Förderung gemeinwohlorientierter KI-Anwendungen), funded by the German Federal Ministry of Labour and Social Affair (BMAS) under the funding reference number DKI.00.00032.21.

## References

1. Bach, S., Binder, A., Montavon, G., Klauschen, F., Müller, K.R., Samek, W.: On pixel-wise explanations for non-linear classifier decisions by layer-wise relevance propagation. PLoS ONE **10**(7), e0130140 (2015)

2. Bau, D., Zhou, B., Khosla, A., Oliva, A., Torralba, A.: Network dissection: Quantifying interpretability of deep visual representations. In: Proceedings of the IEEE Conference on Computer Vision and Pattern Recognition, pp. 6541–6549 (2017)
3. Bommasani, R., et al.: On the opportunities and risks of foundation models. arXiv preprint arXiv:2108.07258 (2021)
4. Byun, S.Y., Lee, W.: Recipro-CAM: gradient-free reciprocal class activation map. arXiv preprint arXiv:2209.14074 (2022)
5. Caron, M., Misra, I., Mairal, J., Goyal, P., Bojanowski, P., Joulin, A.: Unsupervised learning of visual features by contrasting cluster assignments. Adv. Neural. Inf. Process. Syst. **33**, 9912–9924 (2020)
6. Chattopadhay, A., Sarkar, A., Howlader, P., Balasubramanian, V.N.: Grad-CAM++: generalized gradient-based visual explanations for deep convolutional networks. In: 2018 IEEE Winter Conference on Applications of Computer Vision (WACV), pp. 839–847. IEEE (2018)
7. Chen, T., Kornblith, S., Norouzi, M., Hinton, G.: A simple framework for contrastive learning of visual representations. In: International Conference on Machine Learning, pp. 1597–1607. PMLR (2020)
8. Dosovitskiy, A., et al.: An image is worth 16x16 words: transformers for image recognition at scale. arXiv preprint arXiv:2010.11929 (2020)
9. Everingham, M., Van Gool, L., Williams, C., Winn, J., Zisserman, A.: The PASCAL visual object classes challenge 2012 (VOC2012) results (2012). http://www.pascal-network.org/challenges. In: VOC/voc2012/workshop/index. html
10. Gao, S., Li, Z.Y., Yang, M.H., Cheng, M.M., Han, J., Torr, P.: Large-scale unsupervised semantic segmentation. IEEE Trans. Pattern Anal. Mach. Intell. (2022)
11. Gonzalez-Garcia, A., Modolo, D., Ferrari, V.: Do semantic parts emerge in convolutional neural networks? Int. J. Comput. Vision **126**, 476–494 (2018)
12. Grill, J.B., et al.: Bootstrap your own latent-a new approach to self-supervised learning. Adv. Neural. Inf. Process. Syst. **33**, 21271–21284 (2020)
13. He, K., Zhang, X., Ren, S., Sun, J.: Deep residual learning for image recognition. corr abs/1512.03385 (2015) (2015)
14. Hedström, A., et al.: Quantus: an Explainable AI toolkit for responsible evaluation of neural network explanations and beyond. J. Mach. Learn. Res. **24**(34), 1–11 (2023), http://jmlr.org/papers/v24/22-0142.html
15. Madiega, T.: Artificial intelligence act. European Parliamentary Research Service, European Parliament (2021)
16. Meehan, C., Bordes, F., Vincent, P., Chaudhuri, K., Guo, C.: Do SSL models have Déjà Vu? a case of unintended memorization in self-supervised learning. In: Thirty-seventh Conference on Neural Information Processing Systems (2023). https://openreview.net/forum?id=lkBygTc0SI
17. Omeiza, D., Speakman, S., Cintas, C., Weldermariam, K.: Smooth grad-CAM++: an enhanced inference level visualization technique for deep convolutional neural network models. arXiv preprint arXiv:1908.01224 (2019)
18. Petsiuk, V., Das, A., Saenko, K.: Rise: randomized input sampling for explanation of black-box models. arXiv preprint arXiv:1806.07421 (2018)
19. Ramaswamy, H.G., et al.: Ablation-CAM: visual explanations for deep convolutional network via gradient-free localization. In: Proceedings of the IEEE/CVF Winter Conference on Applications of Computer Vision, pp. 983–991 (2020)
20. Russakovsky, O., et al.: Imagenet large scale visual recognition challenge. Int. J. Comput. Vis. **115**, 211–252 (2015)

21. Selvaraju, R.R., Cogswell, M., Das, A., Vedantam, R., Parikh, D., Batra, D.: Grad-CAM: visual explanations from deep networks via gradient-based localization. In: Proceedings of the IEEE International Conference on Computer Vision, pp. 618–626 (2017)
22. Smith, S.L., Brock, A., Berrada, L., De, S.: ConvNets match vision transformers at scale. arXiv preprint arXiv:2310.16764 (2023)
23. Wang, H., et al.: Score-CAM: score-weighted visual explanations for convolutional neural networks. In: Proceedings of the IEEE/CVF Conference on Computer Vision and Pattern Recognition Workshops, pp. 24–25 (2020)
24. Wickstrøm, K.K., et al.: RELAX: representation learning explainability. Int. J. Comput. Vis. **131**(6), 1584–1610 (2023)
25. Wightman, R., Touvron, H., Jégou, H.: Resnet strikes back: an improved training procedure in timm. arXiv preprint arXiv:2110.00476 (2021)
26. Zeiler, M.D., Fergus, R.: Visualizing and understanding convolutional networks. In: Fleet, D., Pajdla, T., Schiele, B., Tuytelaars, T. (eds.) ECCV 2014. LNCS, vol. 8689, pp. 818–833. Springer, Cham (2014). https://doi.org/10.1007/978-3-319-10590-1_53
27. Zhou, B., Khosla, A., Lapedriza, A., Oliva, A., Torralba, A.: Object detectors emerge in deep scene CNNs. arXiv preprint arXiv:1412.6856 (2014)
28. Zhou, B., Khosla, A., Lapedriza, A., Oliva, A., Torralba, A.: Learning deep features for discriminative localization. In: Proceedings of the IEEE Conference on Computer Vision and Pattern Recognition, pp. 2921–2929 (2016)

# Uli-RL: A Real-World Deep Reinforcement Learning Pedagogical Agent for Children

Anna Riedmann[1(✉)], Julia Götz[1], Carlo D'Eramo[2,3], and Birgit Lugrin[1]

[1] Socially Interactive Agents, University of Würzburg, Würzburg, Germany
anna.riedmann@uni-wuerzburg.de

[2] Center for Artificial Intelligence and Data Science, University of Würzburg, Würzburg, Germany

[3] Technical University of Darmstadt, Darmstadt, Germany

**Abstract.** Deep Reinforcement Learning (DRL) has proven its usefulness in various fields, such as robotic control systems, recommendation algorithms, and natural language dialogue interfaces. Recently, we have been witnessing a growing interest in applying DRL in education, with early results suggesting beneficial effects. However, the majority of research in educational applications apply methods in simulation without evaluation with real learners, thus providing scarce evidence of its effectiveness on real-world problems. Arguably, real-world applications are crucial to properly assess the validity of DRL methods. To this end, we present an approach for integrating DRL into an empirically validated digital reading application for second graders in the form of an adaptive pedagogical agent. We use DRL to tailor the agent's feedback behavior to each child's individual learning needs. We evaluate our approach with second graders, investigating their performance and overall motivation, and compare it to a control version of the app. Through this work, we contribute an innovative approach to the use of DRL within the context of primary education, showcasing promising results in a real-world evaluation.

**Keywords:** deep reinforcement learning · deep q network · pedagogical agent

## 1 Introduction

Reinforcement Learning (RL) is a method for learning an optimal state-action-mapping while interacting with an unknown environment by leveraging a reward signal [25]. Deep Reinforcement Learning (DRL) further combines neural networks with classic RL algorithms and serves as a powerful tool that has demonstrated its usefulness in different fields, such as robotic control systems, recommendation algorithms, and natural language dialogue interfaces [26]. There


A. Hotho and S. Rudolph (Eds.): KI 2024, LNAI 14992, pp. 316–323, 2024.
https://doi.org/10.1007/978-3-031-70893-0_25

seems to be growing interest in applying (D)RL in education and recent research has shown promising learning related outcomes when applying it in educational contexts [8,10]. However, the amount of real-world applications of RL is generally rather small, which can be attributed to the complexity of real-world environments, posing challenges for implementing these systems in practical settings [9]. Comparably to the pitfalls of recommender systems described by Dulac-Arnold et al. [9], RL-based educational applications suffer from a huge amount of diverse optimization goals (increase motivation and learning gain, reduce drop out, improve algorithm performance) while real-time interaction requires the system to react and decide immediately. Especially in an educational context, offline data is often not available in large quantities, however, RL-based systems rely heavily on offline data logs, since online experimentation is costly.

Despite the challenges, a RL policy must be evaluated with real students to determine its effectiveness [8]. A review by Doroudi et al. [8] showed that half of the included studies applying RL for instructional sequencing in education demonstrated that RL policies significantly outperformed all baselines, with most of them reporting a large Cohen's $d$ effect size of 0.8. However, significantly superior RL approaches often used weak baselines (e.g. random or poorly designed baselines), while expert-crafted baselines are applied less frequently. Further, RL in education seems to be most efficient when guided by concepts and theories from cognitive psychology and the learning sciences [8].

In summary, the field of RL in education, while being increasingly studied especially with adult learners (e.g. [1,3,14]), seems to be lacking approaches framed by knowledge from cognitive sciences and evaluated in a real-world setting in comparison to sophisticated baselines. To address this gap, we present our approach Uli-RL and report the results of a real-world experiment with second graders to demonstrate its feasibility, comparing it to an expert-designed baseline. We thus contribute a novel concept of applying DRL in the context of primary education, guided by knowledge from educational psychology.

## 2 Concept

In general, RL has been applied in a few areas within primary education, such as tailoring interaction behavior to a learner's knowledge mastery level [6], recommending exercises [2] and selecting sentence structure and complexity in educational stories [20]. These applications have generally led to significant improvements in learning outcomes and emotional engagement [6,20]. Nonetheless, as far as we know, this is the first time RL techniques have been employed to personalize the feedback behavior of a pedagogical agent (PA) in primary education. PAs, "defined as a virtual character [...] that seeks to promote learning, enhance motivation, and provide support to engage in an educational activity" [16, p. 307], have demonstrated their potential to support student learning [7] and are able to provide motivational benefits in certain contexts [24]. Building on our previous work [21], we present the final version of our DRL-powered PA that delivers personalized feedback in the context of a reading intervention. With this approach, we aim to address individual differences in children's learning behaviors and abilities while enhancing motivation and knowledge retention.

The environment for implementing and testing our concept is based on an empirically validated digital reading application [12,22], which is itself derived from a validated analogue reading intervention [19]. The app includes 21 games that target different reading aspects, aiming to improve phonological and orthographic comparison processes, and reading comprehension. It features a PA named "Uli, the owl" (see Fig. 1) that offers feedback and support after a number of incorrect attempts and follows a predetermined behavioral sequence. The PA either comments on the accuracy of the child's response, provides the correct solution, or offers help by referring to a tutorial video.

In previous research, we showed the advantages of incorporating adaptive mechanisms rooted in educational psychology into this learning environment and the PA's behavior [23]. Our previous approach used a fixed feedback implementation that we now aim to improve by automating the agent's behavior through DRL. This allows the PA to adjust its feedback to each child's learning process in real-time. For this purpose we selected three games that cover all central aspects of the digital reading training: *Syllable Salad*, *Bee Flight* and *Uschi's Mail*. *Syllable Salad* requires children to place syllables in correct order to form a word (see Fig. 1 right), *Bee Flight* involves separating words into syllables by drawing arcs beneath the words, and *Uschi's Mail* addresses reading comprehension by challenging children to read and understand texts to answer questions about the content.

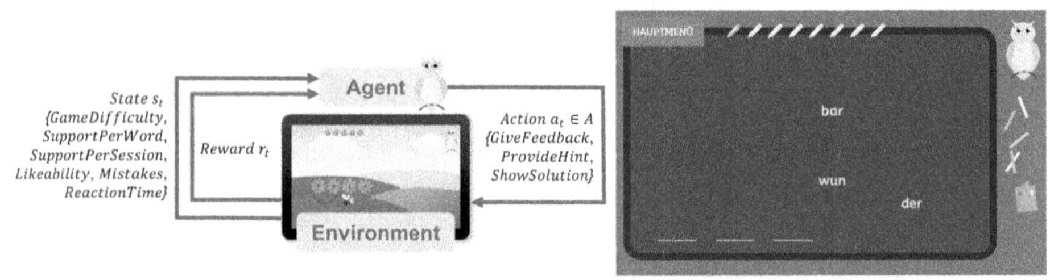


**Fig. 1.** Left: Agent-environment interaction at time-step t; Right: Game *Syllable Salad* with the pedagogical agent "Uli, the owl" in the top right corner.

Children receive feedback by the PA for each attempt to solve the current task. When a child answers a task incorrectly, the DRL-based PA chooses the suitable feedback behavior, either providing encouraging audio feedback (e.g. "You'll get it next time."), providing a hint by referring to the tutorial video, or displaying the correct solution. By personalizing the DRL-based agent's feedback behavior to cater to each child's individual learning process, we aim for a more effective and engaging educational experience. This could result in improved reading skills and enhanced motivation among children, potentially yielding long-term advantages for their academic achievement and overall well-being.

## 3 Uli-RL

We consider our user-centered environment as a Markov Decision Process (MDP), comprising a set of states, actions, rewards, and probabilities that describe the dynamics of the decision-making process [25]. To represent the state, state feature variables are utilized to encapsulate the history of the learning session, retaining all relevant details about past and current session interactions. The state features are selected based on relevant factors impacting the perception and effectiveness of feedback in educational science (see [4,11,27]). They involve the current game difficulty (1–5) and the support provided by the PA (i.e. referring to the tutorial video), both for the word item currently worked on and for the overall game session. Further, the likeability (i.e. the extent to which children are currently motivated to interact with the app) is queried regularly (every five tasks) within the app using a pop-up where children can adjust a slider between 0–1 (with 1 as highest likeability), aiming to include motivational aspects into the state space, as suggested by Doroudi et al. [8]. Additionally, the learner's amount of mistakes is tracked for the current game session as well as the reaction time for each interaction. Together, all specified features represent the current state $s = \{GameDifficulty, SupportPerWord, SupportPerSession, Likeability, Mistakes, ReactionTime\}$.

For each incorrect answer given by the child while interacting with the app, the DRL-powered PA selects one out of three feedback behaviors using a Deep Q-Network (DQN) algorithm [18] to account for the continuous state space. Thus, we define the set of actions as $A = \{GiveFeedback, ProvideHint, ShowSolution\}$. Rewards are assigned depending on the correctness of the given answer (either 1 for correct or 0 for incorrect answers), with a discount $\xi \in \{0, 1\}$ for previously given incorrect answers, where the value of $\xi$ is computed as the accuracy divided by the total number of answers given in the current game. This results in $r = \xi * accuracy$. Training is performed on a previously collected dataset of the app's target group, containing interaction data of 144 children working through the reading training in the app in 20 sessions, playing the three games multiple times, involving 40–60 interactions per game. The data collected encompasses accuracy, reaction time, difficulty level, and specific word characteristics. All other necessary information for training, such as actions and states, was inferred from the dataset.

For our experiment, we integrated the pre-trained DRL model into the existing reading app using the Chaquopy plugin [5], allowing the PA to select the respective feedback behavior whenever a child makes a mistake while interacting with the app. We employ two neural networks with identical architectures to train the DQN algorithm. For our Q-Network, we used a two-layer sequential neural network, with the $Tanh$ function as activation function and the final layer aligning with the action space dimension. Input feature values are min-max normalized and our algorithm is fine-tuned using the Adam stochastic optimization technique [15]. To derive the optimal action-value function, we follow the Bellman Optimality Equation in that:

$$Q^*(s,a) = r + \gamma \max_{a'} Q^*(s',a') \tag{1}$$

In this context, $r$ represents the expected immediate reward for choosing action $a$ in state $s$ (as defined earlier), $\gamma$ is the discount factor, and $Q^*(s',a')$ is the optimal action-value function for taking an action $a'$ in the subsequent state $s'$ and then adhering to policy $\pi$ afterwards.

The Q-Network is updated during live interactions with the target group (see Fig. 1 left), while the target network is updated in accordance to the next interaction a child takes after it has made a mistake and received feedback by the DRL-powered PA. We chose the Huber loss function as a robust alternative to MSELoss. To balance exploration and exploitation, the DRL-based PA makes use of the epsilon-greedy policy, selecting the action with the highest estimated reward, with $\varepsilon = 6e - 1$ and linear epsilon decrement for each interaction with $\varepsilon_{decay} = 1e - 3$. We set the learning rate as $lr = 9e - 1$. We chose this rather high value due to the real-world evaluation setting where children interact with the app only for a short time, performing a limited number of interactions, which does not allow for a long training process.

## 4 Real-World Experiment

We evaluated the performance of the DRL-powered PA in a real-world setting by investigating whether the DRL-induced policy significantly outperforms the expert baseline in an applied school context in terms of students' motivation and performance. We expected higher motivation (H1) and higher performance (H2) in the DRLApp condition.

### 4.1 Method

We investigated the motivation of the children using the Reduced Instructional Materials Motivation Survey (RIMMS) [17], slightly adapted to the target group as described in Riedmann et al. [23]. The questionnaire comprises four subscales, measured on a five-point Likert scale ($\alpha = .85$). Further, the questionnaire included an additional question regarding the children's perception of the given feedback ("Did you like the feedback from Uli, the owl?"), answered on a five-point Likert scale. The children's performance was assessed as the accuracy ratio for all games, as a value between 0 and 1 that is calculated as the number of correct answers given in relation to the overall number of interactions. Additionally, prior reading skills were assessed by the teacher beforehand.

Thirty-eight primary school students (aged 8–10 years; $M = 8.03$, $SD = 0.75$), whose parents provided informed consent, voluntarily agreed to take part in the experiment and were randomly assigned to either the group with the adaptive PA (DRLApp, $n = 19$) or the control group using the expert-designed version with the non-adaptive PA (ExpertApp, $n = 19$). All children played the games *Syllable Salad*, *Bee Flight* and *Uschi's Mail* on a 10.5-inch Android tablet. Both the sequence, the type of games, and the content of the agent's feedback in the learning app were identical for both conditions, while the feedback selection varied. During the interaction, data on the children's performance in the app and reaction time was tracked. Afterwards, children filled out the questionnaire. Working on the app and the subsequent questionnaire lasted about 35 min. Our experiment was approved by the responsible institutional ethics committee (application number 100124).

### 4.2 Results

All analyses were conducted with JASP [13] and the Bonferroni corrected alpha was set at .025. We used Mann-Whitney U tests for group comparisons. Regarding the children's prior reading skills, there was no significant difference between the two groups ($U = 152.00$, $p = .389$). Table 1 displays all results.

**Table 1.** Means for both conditions with SD in parentheses and results of U tests.

| | Scale | DRLApp | ExpertApp | $U$ | $p$ |
|---|---|---|---|---|---|
| Motivation | 1–5 | 4.70 (0.45) | 4.65 (0.39) | 149.00 | .176 |
| Performance | 0–1 | 0.80 (0.10) | 0.80 (0.12) | 184.50 | .552 |
| Feedback | 1–5 | 4.37 (0.96) | 4.42 (0.69) | | |

## 5 Discussion and Conclusion

In our work, we aimed to address the gap in real-world evaluations of DRL methods within the context of education. We investigated the effect of our Uli-RL approach in a reading app for second graders and evaluated its usefulness in a user study in a real-world setting. We did not find significantly higher motivation and performance in the DRLApp condition and thus reject H1 and H2. Notably, motivation and performance values were high in both conditions, suggesting that the DRL-powered PA was equally effective as the expert-designed version employing concepts from cognitive psychology. This seems to be the case even though the experiment lasted only 35 min, allowing just a relatively small number of interactions children had with the app to be used as training data. Further, the expert-designed app already has inherent high motivational potential and provided varying feedback in terms of a fixed feedback schedule. Due to this aspect and the relatively short time working on the app, potential benefits of individualized feedback might not have become apparent, requiring a longer time working on the app. Reviewing the interaction data, the DRL model initially followed the pre-trained feedback schedule. Over time, the PA's feedback was adapted based on the children's learning behavior: children who struggled received solutions or explanations more often, while those who succeeded in previous tasks were given more opportunities to try again. Thus, our approach might have the potential to retain the children's motivation in the long-term while reducing drop-out, requiring future work.

In conclusion, our study contributes an innovative approach to applying DRL in a real-world setting in the context of education. We demonstrated the feasibility of a DRL model running on an average Android tablet, while being equally effective compared to an expert-designed version, with the feedback provided by DRL-powered agent being perceived very well. Future work should investigate the promising potential of long-term applications of a DRL-powered PA.

**Acknowledgments.** Partly funded by the German Federal Ministry of Education and Research (BMBF) under the grants MobiLe (03VP07080) and TriFORCE (01IS22078).

## References

1. Abdelshiheed, M., Hostetter, J.W., Barnes, T., Chi, M.: Leveraging deep reinforcement learning for metacognitive interventions across intelligent tutoring systems. In: Wang, N., Rebolledo-Mendez, G., Matsuda, N., Santos, O.C., Dimitrova, V. (eds.) AIED 2023. LNAI, vol. 13916, pp. 291–303. Springer, Cham (2023). https://doi.org/10.1007/978-3-031-36272-9_24
2. Ai, F., Chen, Y., Guo, Y., Zhao, Y., Wang, Z., Fu, G., Wang, G.: Concept-aware deep knowledge tracing and exercise recommendation in an online learning system. Int. Educ. Data Min. Soc. (2019)
3. Ausin, M.S., Maniktala, M., Barnes, T., Chi, M.: Tackling the credit assignment problem in reinforcement learning-induced pedagogical policies with neural networks. In: Roll, I., McNamara, D., Sosnovsky, S., Luckin, R., Dimitrova, V. (eds.) AIED 2021. LNCS (LNAI), vol. 12748, pp. 356–368. Springer, Cham (2021). https://doi.org/10.1007/978-3-030-78292-4_29
4. Bimba, A.T., Idris, N., Al-Hunaiyyan, A., Mahmud, R.B., Shuib, N.L.B.M.: Adaptive feedback in computer-based learning environments: a review. Adapt. Behav. **25**(5), 217–234 (2017). https://doi.org/10.1177/1059712317727590
5. Chaquo Ltd: Chaquopy (2022). https://chaquo.com/chaquopy/
6. Chen, H., Park, H.W., Breazeal, C.: Teaching and learning with children: impact of reciprocal peer learning with a social robot on children's learning and emotive engagement. Comput. Educ. **150**, 103836 (2020). https://doi.org/10.1016/j.compedu.2020.103836
7. Dai, L., Jung, M.M., Postma, M., Louwerse, M.M.: A systematic review of pedagogical agent research: similarities, differences and unexplored aspects. Comput. Educ. **190**, 104607 (2022). https://doi.org/10.1016/j.compedu.2022.104607, https://www.sciencedirect.com/science/article/pii/S0360131522001786
8. Doroudi, S., Aleven, V., Brunskill, E.: Where's the reward? Int. J. Artif. Intell. Educ. **29**(4), 568–620 (2019). https://doi.org/10.1007/s40593-019-00187-x
9. Dulac-Arnold, G., et al.: Challenges of real-world reinforcement learning: definitions, benchmarks and analysis. Mach. Learn. **110**(9), 2419–2468 (2021). https://doi.org/10.1007/s10994-021-05961-4
10. Fahad Mon, B., Wasfi, A., Hayajneh, M., Slim, A., Abu Ali, N.: Reinforcement learning in education: a literature review. Informatics **10**(3), 74 (2023). https://doi.org/10.3390/informatics10030074
11. Hattie, J., Timperley, H.: The power of feedback. Rev. Educ. Res. **77**(1), 81–112 (2007). https://doi.org/10.3102/003465430298487

12. Heß, J., et al.: Mobile: Konzeption einer digitalen silbenbasierten leseförderung für die grundschule. Psychol. Erzieh. Unterr. **71**(1), 41–51 (2024). https://doi.org/10.2378/peu2024.art05d
13. JASP Team: Jasp (2021). https://jasp-stats.org/
14. Ju, S., Zhou, G., Abdelshiheed, M., Barnes, T., Chi, M.: Evaluating critical reinforcement learning framework in the field. In: Roll, I., McNamara, D., Sosnovsky, S., Luckin, R., Dimitrova, V. (eds.) AIED 2021. LNCS (LNAI), vol. 12748, pp. 215–227. Springer, Cham (2021). https://doi.org/10.1007/978-3-030-78292-4_18
15. Kingma, D.P., Ba, J.: Adam: a method for stochastic optimization. http://arxiv.org/pdf/1412.6980
16. Lane, H.C., Schroeder, N.L.: Pedagogical agents. In: Lugrin, B., Pelachaud, C., Traum, D.R. (eds.) The Handbook on Socially Interactive Agents, pp. 307–330. ACM books, Association for Computing Machinery, New York, NY (2022). https://doi.org/10.1145/3563659.3563669
17. Loorbach, N., Peters, O., Karreman, J., Steehouder, M.: Validation of the instructional materials motivation survey (IMMS) in a self-directed instructional setting aimed at working with technology. Br. J. Educ. Technol. **46**(1), 204–218 (2015). https://doi.org/10.1111/bjet.12138
18. Mnih, V., et al.: Human-level control through deep reinforcement learning. Nature **518**(7540), 529–533 (2015). https://doi.org/10.1038/nature14236
19. Müller, B., Richter, T., Karageorgos, P.: Syllable-based reading improvement: effects on word reading and reading comprehension in grade 2. Learn. Instr. **66** (2020). https://doi.org/10.1016/j.learninstruc.2020.101304
20. Park, H.W., Grover, I., Spaulding, S., Gomez, L., Breazeal, C.: A model-free affective reinforcement learning approach to personalization of an autonomous social robot companion for early literacy education. In: Proceedings of the AAAI Conference on Artificial Intelligence, vol. 33, pp. 687–694 (2019). https://doi.org/10.1609/aaai.v33i01.3301687
21. Riedmann, A., Lugrin, B.: Towards an adaptive pedagogical agent in a reading intervention using reinforcement learning. In: Lugrin, B., Latoschik, M., von Mammen, S., Kopp, S., Pécune, F., Pelachaud, C. (eds.) Proceedings of the 23rd ACM International Conference on Intelligent Virtual Agents, pp. 1–3. ACM Digital Library, Association for Computing Machinery, Erscheinungsort nicht ermittelbar (2023). https://doi.org/10.1145/3570945.3607320
22. Riedmann, A., et al.: Iteratively digitizing an analogue syllable-based reading intervention. Interact. Comput. **33**(4), 411–425 (2022)
23. Riedmann, A., Schaper, P., Jakob, B., Lugrin, B.: A theory based adaptive pedagogical agent in a reading app for primary students - a user study. In: Crossley, S., Popescu, E. (eds.) ITS 2022. LNCS, vol. 13284, pp. 276–292. Springer, Cham (2022). https://doi.org/10.1007/978-3-031-09680-8_26
24. Schroeder, N.L., Adesope, O.O.: A systematic review of pedagogical agents' persona, motivation, and cognitive load implications for learners. J. Res. Technol. Educ. **46**(3), 229–251 (2014). https://doi.org/10.1080/15391523.2014.888265
25. Sutton, R.S., Barto, A.: Reinforcement Learning: An Introduction Adaptive Computation and Machine Learning, 2nd edn. The MIT Press, Cambridge, London, (2018)
26. Wang, H.N., et al.: Deep reinforcement learning: a survey. Front. Inf. Technol. Electron. Eng. **21**(12), 1726–1744 (2020). https://doi.org/10.1631/FITEE.1900533
27. Wisniewski, B., Zierer, K., Hattie, J.: The power of feedback revisited: a meta-analysis of educational feedback research. Front. Psychol. **10**, 3087 (2019). https://doi.org/10.3389/fpsyg.2019.03087

# Explanatory Interactive Machine Learning with Counterexamples from Constrained Large Language Models

Emanuel Slany(✉), Stephan Scheele, and Ute Schmid

Fraunhofer Institute for Integrated Circuits IIS, Comprehensible AI, Erlangen, Germany
{emanuel.slany,stephan.scheele,ute.schmid}@iis.fraunhofer.de

**Abstract.** In Explanatory Interactive Machine Learning (XIML), counterexamples refine machine learning models by augmenting human feedback. Traditionally created through random sampling or data augmentation, the emergence of Large Language Models (LLMs) now allows an infinite amount of new training instances to be queried through simple natural language prompts. However, validation of LLM results becomes crucial as they may produce potentially inaccurate or "hallucinated" content, which has led to an increased incorporation of logical reasoning with LLMs in recent literature. We present LlmXiml, a framework that integrates logically constrained LLMs into XIML. Our results indicate that LLM-generated counterexamples improve the model performance and logical reasoning increases the counterexamples' correctness.

**Keywords:** Explanatory Interactive Machine Learning · Counterexamples · Large Language Models · Probabilistic Logic Programming

## 1 Introduction

**Motivation.** Counterexamples in Explanatory Interactive Machine Learning (XIML) [11] are novel training instances that have the objective to unlearn indecisive and enforce decisive correlations from a human annotator's perspective. Large Language Models (LLMs) such as `gpt-3.5-turbo`[1] enable users to query counterexamples by natural language.

**Problem.** LLM-generated counterexamples might be invalid [2], which causes an incorrect projection of the human domain knowledge into the ML model. The model's decision boundary might not reach a human-defined optimum.

**Solution.** We enrich the generation procedure by logical reasoning that acts as a gatekeeper to filter out invalid counterexamples and refines the LLM instruction.

[1] https://platform.openai.com/docs/models/gpt-3-5-turbo.


A. Hotho and S. Rudolph (Eds.): KI 2024, LNAI 14992, pp. 324–331, 2024.
https://doi.org/10.1007/978-3-031-70893-0_26

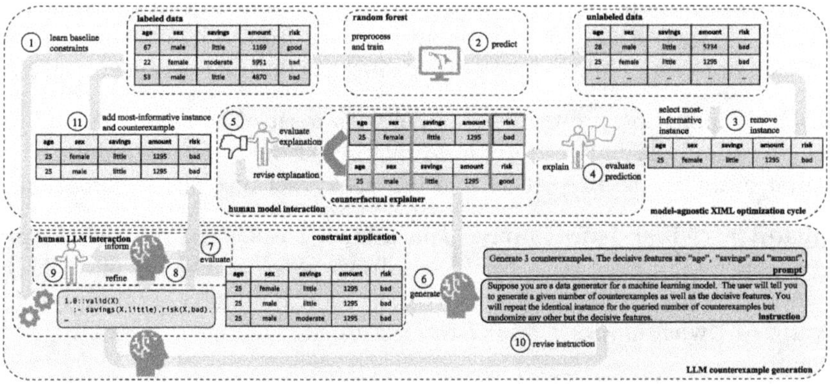


**Fig. 1.** Counterexample generation with constrained LLMs. We learn a set of constraints (1) and train a random forest on the labeled data set and predict an unlabeled data set (2). Next, we select the most-informative instance (3). A user evaluates the prediction (4) and the explanation (5). A LLM generates counterexamples (6), which are evaluated by a probabilistic logical program (7). The LLM translates violated validity constraints to humans (8) and humans generate new validity constraints (9). The violated constraints revise the instruction (10). The validated counterexamples serve as new training data (11).

**Related Work.** Several model-agnostic XIML algorithms propose counterexample generation methods. The objective of the XIML framework CAIPI [11] is to unlearn spurious correlations on image data, where counterexamples randomize spurious correlations induced by decoy pixels. Others apply CAIPI to medicine [9], where physicians can train a ML model for tissue classification. Counterexamples are augmented decisive features of the tissue. Recently, FAIR-CAIPI optimizes fairness metrics of ML models [3]. Counterexamples randomize the protected attribute. Closely related, is the construction of counterexamples by a Variational Autoencoder [10]. There exist approaches that generate data from LLMs [1,7], methods that apply logical reasoning to LLMs [5], and methods that use LLMs to translate between logic and natural language [12].

**Contributions.** We contribute LLMXIML (Fig. 1) – an integration of logically constrained LLMs in the XIML optimization cycle.

**Research Questions.** We demonstrate our approach by unlearning a spurious correlation on various tabular data sets and ask the following research questions: (**R1**) Does reasoning enhance the correctness of LLM-generated counterexamples? (**R2**) How do LLM-generated counterexamples with and without reasoning affect the predictive and explanatory quality of XIML optimization?

## 2 Constrained LLMs for XIML

This section presents LLMXIML (Fig. 1) in the spirit of [10].

**Notation.** Let $f : \mathcal{X}^n \rightarrow \mathcal{Y} \in \{0,1\}$ be a binary classification model with $(x_1, ..., x_i, ..., x_n)^T \in \mathcal{X}$ and the identifier set $\mathcal{F} = \{1, ..., i, ..., n\}$. Let $l : \mathcal{X} \rightarrow \mathcal{Y}$

be a labeling function. Further, let $x^{(n)}$ ($y^{(n)}$) be the $n$-th instance in $\mathcal{X}$ ($\mathcal{Y}$). Finally, let $(x_{\mathcal{L}}, y_{\mathcal{L}}) \in \mathcal{L} \subseteq \mathcal{X} \times \mathcal{Y}$ be an instance from a labeled set and $x_{\mathcal{U}}$ be an instance from an unlabeled set $\mathcal{U} \subseteq \mathcal{X}$. A procedure FIT trains or updates $f$.

XIML optimization pre-trains a model with a set of labeled data and selects in each iteration the most-informative instance from a set of unlabeled data. Users first evaluate the prediction and if it is correct also the explanation.

**Definition 1 (Most-Informative Instance).** Here, $f(x)$ returns the prediction score. Let $m = \arg\min_n \{|f(x^{(n)}) - \alpha| | x^{(n)} \in \mathcal{U}\}$ be the index of the *most-informative instance* that is closest to a decision boundary $\alpha$. Suppose that a procedure MII with inputs $f$ and $\mathcal{U}$ returns $m$.

**Definition 2 (Correct Decision Making).** Let *correct decision making* imply robustness against indecisive feature permutations. EXP randomly alters each $x_i$ if $i \in \bar{v}$ and returns `True` if $f(x) = y$ and `False` otherwise, where $v \subseteq \mathcal{F}$ and $\bar{v} = \mathcal{F} \setminus v$ denote the sets of decisive and indecisive features respectively. We say that a *prediction* is *correct* if $f(x) = l(x)$.

Each optimization iteration yields one of three XIML outcomes, summarized in Table 1: **Right** for the **R**ight **R**easons (**RRR**) if both prediction and explanation are correct, **W**rong (**W**) if the prediction is incorrect, or **R**ight for the **W**rong **R**easons (**RWR**) if the prediction is correct but the explanation reveals erroneous decision making. In the last case, counterexamples are added. The final action $\mathcal{U} = \mathcal{U} \setminus x^{(m)}$ is independent from the outcome case.

**Table 1.** XIML Outcome Cases [10,11].

| Case | Prediction | Explanation | Action |
|---|---|---|---|
| RRR | correct | correct | $\mathcal{L} = \mathcal{L} \cup \{(x^{(m)}, \hat{y}^{(m)})\}$ |
| RWR | correct | incorrect | $\mathcal{L} = \mathcal{L} \cup \{\mathcal{X}', \mathcal{Y}'\} \cup \{(x^{(m)}, \hat{y}^{(m)})\}$ |
| W | incorrect | incorrect | $\mathcal{L} = \mathcal{L} \cup \{(x^{(m)}, l(x^{(m)}))\}$ |

**Definition 3 (Counterexamples).** Let a counterexample $(x', y') \in \mathcal{X}' \times \mathcal{Y}'$ replicate $y$ and $x$ for each $i \in v$ and randomize any other feature. Suppose a procedure GEN with inputs $x$, $y$, and $c$ that utilizes `gpt-3.5-turbo`[2] with the **generate** prompts (Fig. 2) and returns $c$ counterexamples $(\mathcal{X}', \mathcal{Y}')$.

**Definition 4 (Probabilistic Rule Induction [6]).** The INDUCE procedure with input $\mathcal{X} \times \mathcal{Y}$ returns a set of *probabilistic rules* $R$ *induced* by PROBFOIL$^+$.

**Definition 5 (Reasoning [4]).** Let *reasoning* select valid instances $\mathcal{X}^* \times \mathcal{Y}^* \subseteq \mathcal{X} \times \mathcal{Y}$ s.t. each instance satisfies the rules of a probabilistic rule set $R$: Let REASON denote a procedure with inputs $(\mathcal{X}, \mathcal{Y})$ and $R$ that returns $(\mathcal{X}^*, \mathcal{Y}^*)$.

[2] https://platform.openai.com/docs/models/gpt-3-5-turbo.

LLMXIML (Algorithm 1) generates counterexamples with an LLM. A logical reasoning step using ProbLog [4] validates their correctness. The LLM semantically translates between ProbLog rules and natural language (Fig. 2) to **inform** the user why counterexamples are (in)valid. The combination of ProbLog with the LLM **revises** the LLM instruction by appending natural language constraints. Users **refine** the logical program by formulating validity constraints. The set of baseline constraints is induced from the labeled data set using PROBFOIL$^+$ [6].

**Algorithm 1:** LLMXIML($\mathcal{X}_\mathcal{L}, \mathcal{Y}_\mathcal{L}, \mathcal{U}, c, n$)

**Input:** Data sets $\mathcal{X}_\mathcal{L}$, $\mathcal{Y}_\mathcal{L}$ and $\mathcal{U}$, number of counterexamples $c$, number of iterations $n$
**Output:** Model $f$

$R \leftarrow$ INDUCE($\mathcal{X}_\mathcal{L}, \mathcal{Y}_\mathcal{L}$)
**for** $i \leftarrow$ **to** $n$ **do**
  $f \leftarrow$ FIT($\mathcal{X}_\mathcal{L}, \mathcal{Y}_\mathcal{L}$)
  $m \leftarrow$ MII($f, \mathcal{U}$)
  $\hat{y}^{(m)} \leftarrow f(x_\mathcal{U}^{(m)})$
  **if** $\hat{y}^{(m)} \neq l(x_\mathcal{U}^{(m)})$ **then** $\mathcal{L} \leftarrow \mathcal{L} \cup \{(x_\mathcal{U}^{(m)}, l(x_\mathcal{U}^{(m)}))\}$ ▷ *Case:* ***W***
  **else**
    **if** EXP($f, x_\mathcal{U}^{(m)}$) **then** $\mathcal{L} \leftarrow \mathcal{L} \cup \{(x_\mathcal{U}^{(m)}, \hat{y}^{(m)})\}$ ▷ *Case:* ***RRR***
    **else**
      $\mathcal{X}^*, \mathcal{Y}^* \leftarrow$ REASON(GEN($x_\mathcal{U}^{(n)}, \hat{y}^{(m)}, c$), $R$)
      $\mathcal{L} \leftarrow \mathcal{L} \cup \{(x_\mathcal{U}^{(m)}, \hat{y}^{(m)})\} \cup \mathcal{X}^* \times \mathcal{Y}^*$ ▷ *Case:* ***RWR***
  $\mathcal{U} \leftarrow \mathcal{U} \setminus x_\mathcal{U}^{(m)}$
**return** $f$

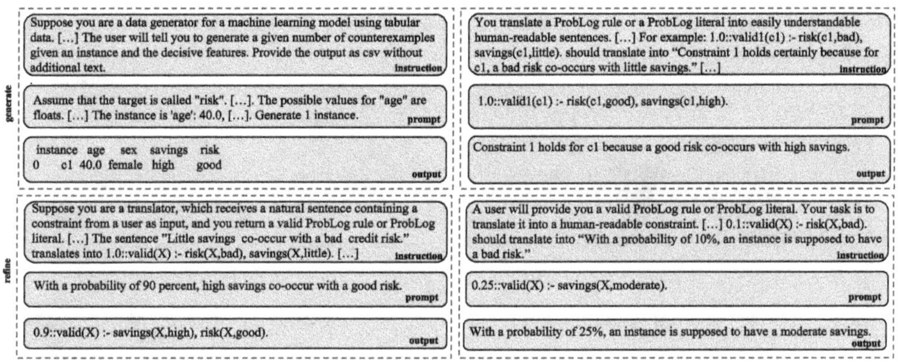


**Fig. 2.** Instructions, prompts, and (expected) LLM outputs. The LLM serves as a semantic translator: to **generate** counterexamples, to **inform** the user, to **refine** the set of ProbLog rules, and to **revise** the generation instruction.

## 3 Experimental Evaluation

**Setup.** We prefer an automated evaluation with induced spurious correlations over human annotators. In five experimental iterations, we investigate the impact of adding a single LLM-generated counterexample compared to co-active learning [8], which is equal to zero counterexamples. We assess whether the reasoning component yields additional benefits. We train a random forest[3] for 250 XIML iterations, where the initial labeled data set contains 100 random instances. A spurious correlation inserted into the labeled data set replicates the targets of the Adult, Credit, Diabetes, Diagnostic, and Heart data sets[4] with a probability of 90%. The preprocessing pipeline removes rows with missing values, standardizes numerical features, and replaces categorical features by natural numbers.

Counterexamples randomize the spurious correlation. We instruct the LLM to select a value from the set $\{0, 1, 2, ..., 10\}$ for the spurious correlation feature, where only the values 0 and 1 are valid choices for binary classification tasks. A counterexample is correct if (i) its value is valid and (ii) its value is not equal to the target. For invalid counterexamples, the reasoning and revision components remove the generated spurious correlation value from the instruction set.

**Results.** Table 2 indicates that reasoning is beneficial for the correctness of counterexamples. Contrary, Table 3 shows a stable predictive and explanatory quality. Except for the Adult data set, we observe that LLM-generated counterexamples are beneficial compared to co-active learning.

Note that `gpt-3.5-turbo` is a probabilistic model. Even if its instruction receives multiple revisions, its results might still be wrong. Similarly, learned spurious correlations in statistical ML models might persist for some instances, despite bias-mitigating counterexamples have been added multiple times.

**Table 2.** Ratio of correct counterexamples. We compare the ratio of correct counterexamples across **Data** sets and the **Modes** with and without (w/o) reasoning. Standard deviations are given in brackets. Superior results are written as bold numbers. We highlight Welch's test results as $p < 0.01$: $x^{**}$, $p < 0.05$: $x^{*}$.

| Data | Mode: $c = 1$ w/o reasoning | $c = 1$ with reasoning |
|---|---|---|
| Adult | 0.0967 (0.0819) | **0.8858**$^{*}$ (0.1413) |
| Credit | 0.1300 (0.0751) | **0.9063**$^{**}$ (0.0557) |
| Diabetes | 0.1304 (0.1068) | **0.4197** (0.0702) |
| Diagnostic | 0.0000 (0.0000) | **0.0291** (0.0482) |
| Heart | 0.0659 (0.0990) | **0.6057**$^{*}$ (0.1564) |

[3] https://scikit-learn.org/stable/modules/generated/sklearn.ensemble.RandomForestClassifier.html.

[4] All data sets from https://www.kaggle.com/datasets.

**Table 3.** Results. We compare the weighted average (Ø) **Precision**, **Recall**, and ratio of correct explanations (**Corr. Expl.**) across **Data** sets and **Modes**.

| Data | Mode | Ø Precision | Ø Recall | Ø Corr. Expl. |
|---|---|---|---|---|
| Adult | $c=0$ | 0.7431 (0.0179) | **0.7660** (0.0156) | **0.4166** (0.0122) |
| | $c=1$ w/o r. | 0.7434 (0.0157) | 0.7564 (0.0364) | 0.4064 (0.4064) |
| | $c=1$ r. | **0.7455** (0.0222) | 0.7660 (0.0273) | 0.4125 (0.0444) |
| Credit | $c=0$ | 0.6106 (0.0200) | 0.6090 (0.0208) | 0.3451 (0.0429) |
| | $c=1$ w/o r. | 0.6005 (0.0197) | 0.6013 (0.0215) | 0.3687 (0.0387) |
| | $c=1$ r. | **0.6217** (0.0328) | **0.6192** (0.0303) | **0.4560** (0.0278) |
| Diabetes | $c=0$ | 0.7290 (0.0251) | 0.7191 (0.0317) | 0.3929 (0.0521) |
| | $c=1$ w/o r. | **0.7398** (0.0222) | **0.7470** (0.0216) | 0.4628 (0.0186) |
| | $c=1$ r. | 0.7393 (0.0207) | 0.7452 (0.0205) | **0.4670** (0.0172) |
| Diagnostic | $c=0$ | 0.9074 (0.0085) | 0.9059 (0.0083) | 0.4987 (0.0018) |
| | $c=1$ w/o r. | **0.9083** (0.0078) | **0.9071** (0.0077) | **0.5000** (0.0000) |
| | $c=1$ r. | 0.9074 (0.0085) | 0.9059 (0.0083) | 0.4987 (0.0018) |
| Heart | $c=0$ | 0.8949 (0.0217) | 0.8932 (0.0232) | 0.4974 (0.0010) |
| | $c=1$ w/o r. | 0.8997 (0.0197) | 0.8977 (0.0209) | 0.4981 (0.0014) |
| | $c=1$ r. | **0.9090** (0.0214) | **0.9068** (0.0238) | **0.4989** (0.0016) |

The latter finding is confirmed by the graphical assessment (Fig. 3) with two observations: First, the explanatory quality on the Credit data set, having the worst predictive performance, benefits the most from including reasoning. And second, the Diagnostic data set with the best predictive performance barely has observable benefits from adding counterexamples. Let us answer our research

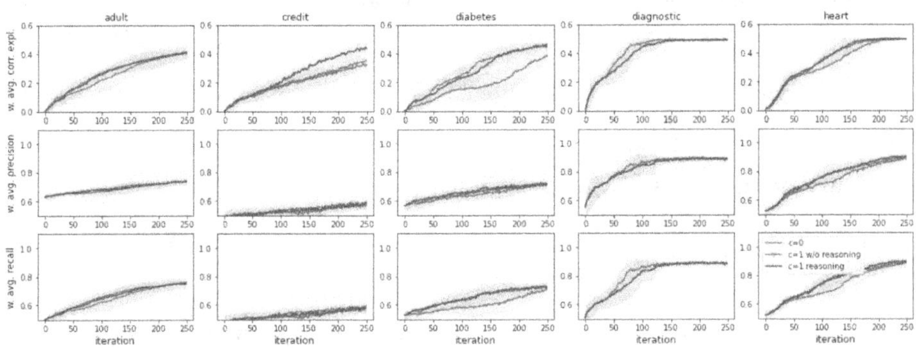


**Fig. 3.** Optimization. We compare the optimization wrt. the weighted average (w. avg.) precision, recall, and correct explanations (corr. expl.). We depict the absence of counterexamples c = 0 (red), the presence of a counterexample c = 1 without (w/o) reasoning (green), and a counterexample with reasoning (blue). (Color figure online)

questions: (**R1**) Reasoning improves the correctness of LLM-generated counterexamples. (**R2**) Such counterexamples have a beneficial impact on the predictive and explanatory quality of XIML optimization. The reasoning component does not result in additional performance benefits.

## 4 Discussion

**Conclusion.** LLMXIML generates counterexamples with ProbLog-constrained LLMs. We infer that LLM-generated counterexamples are beneficial for the predictive and explanatory performance of ML models. Although the logical post-processing increases the correctness of counterexamples, it does not outperform XIML with counterexamples that have not undergone reasoning.

**Related Results.** This article exploits the data generation ability of LLMs [1] and proposes another CAIPI [11] counterexample generation variant. It is closely related to [10], who use a Variational Autoencoder for image data. It also touches the research area that connects LLMs with logic programming [12]. The experimental setup and findings are in line with other CAIPI variants [3,9–11], which all infer an increased explanatory performance for including counterexamples.

**Limitations.** Some limitations arise from the experiments: The reasoning component does not yield measurable benefits. This is due to the circumstance that our counterexamples randomize a correlation rather than formulating one. Other publications [9–11] additionally evaluate the amount of counterexamples.

**Future Work.** Most importantly, we aim to increase the experimental support for our method. Besides a generalized experimental evaluation on tabular data sets, we will transfer LLMXIML to image data. Algorithmically, we will extract more sophisticated theories from structured causal models that act as a baseline for the reasoning step [13]. This will put us into the position to overcome the limitations, as the logical component now contains knowledge about causalities of the original data, rather than just about the absence of a spurious correlation.

**Acknowledgments.** Funded by BMBF Germany, Project hKI-Chemie (# 01IS210 23A).

**Disclosure of Interests.** The authors have no competing interests to declare.

## References

1. Chung, J.J.Y., Kamar, E., Amershi, S.: Increasing diversity while maintaining accuracy: text data generation with large language models and human interventions. In: Proceedings of the 61st Annual Meeting of the Association for Computational Linguistics, vol. 1, pp. 575–593 (2023). https://aclanthology.org/2023.acl-long.34.pdf

2. Hammond, K.J., Leake, D.B.: Large language models need symbolic AI. In: d'Avila Garcez, A.S., Besold, T.R., Gori, M., Jiménez-Ruiz, E. (eds.) Proceedings of the 17th International Workshop on Neural-Symbolic Learning and Reasoning, La Certosa di Pontignano, Siena, Italy, 3–5 July 2023. CEUR Workshop Proceedings, vol. 3432, pp. 204–209. CEUR-WS.org (2023). https://ceur-ws.org/Vol-3432/paper17.pdf
3. Heidrich, L., Slany, E., Scheele, S., Schmid, U.: FairCaipi: a combination of explanatory interactive and fair machine learning for human and machine bias reduction. Mach. Learn. Knowl. Extr. **5**, 1519–1538 (2023). https://doi.org/10.3390/make5040076
4. Kimmig, A., Demoen, B., Raedt, L.D., Costa, V.S., Rocha, R.: On the implementation of the probabilistic logic programming language ProbLog. Theory Pract. Log. Program. **11**(2–3), 235–262 (2011). https://doi.org/10.1017/S1471068410000566
5. Nye, M.I., Tessler, M.H., Tenenbaum, J.B., Lake, B.M.: Improving coherence and consistency in neural sequence models with dual-system, neuro-symbolic reasoning. In: Ranzato, M., Beygelzimer, A., Dauphin, Y.N., Liang, P., Vaughan, J.W. (eds.) Advances in Neural Information Processing Systems 34: Annual Conference on Neural Information Processing Systems 2021, NeurIPS 2021, 6–14 December 2021, Virtual, pp. 25192–25204 (2021). https://proceedings.neurips.cc/paper/2021/hash/d3e2e8f631bd9336ed25b8162aef8782-Abstract.html
6. Raedt, L.D., Dries, A., Thon, I., den Broeck, G.V., Verbeke, M.: Inducing probabilistic relational rules from probabilistic examples. In: Proceedings of the Twenty-Fourth International Joint Conference on Artificial Intelligence, IJCAI 2015, Buenos Aires, Argentina, 25–31 July 2015, pp. 1835–1843. AAAI Press (2015). http://ijcai.org/Abstract/15/261
7. Ribeiro, M.T., Lundberg, S.M.: Adaptive testing and debugging of NLP models. In: Muresan, S., Nakov, P., Villavicencio, A. (eds.) Proceedings of the 60th Annual Meeting of the Association for Computational Linguistics (Volume 1: Long Papers), ACL 2022, Dublin, Ireland, 22–27 May 2022, pp. 3253–3267. Association for Computational Linguistics (2022). https://doi.org/10.18653/V1/2022.ACL-LONG.230
8. Shivaswamy, P., Joachims, T.: Coactive learning. J. Artif. Intell. Res. **53**, 1–40 (2015). https://doi.org/10.1613/jair.4539
9. Slany, E., Ott, Y., Scheele, S., Paulus, J., Schmid, U.: CAIPI in practice: towards explainable interactive medical image classification. In: Maglogiannis, I., Iliadis, L., Macintyre, J., Cortez, P. (eds.) AIAI 2022, vol. 652, pp. 389–400. Springer, Cham (2022). https://doi.org/10.1007/978-3-031-08341-9_31
10. Slany, E., Scheele, S., Schmid, U.: Bayesian CAIPI: a probabilistic approach to explanatory and interactive machine learning. In: Nowaczyk, S., et al. (eds.) ECAI 2023, vol. 1947, pp. 285–301. Springer, Cham (2024). https://doi.org/10.1007/978-3-031-50396-2_16
11. Teso, S., Kersting, K.: Explanatory interactive machine learning. In: Proceedings of the 2019 AAAI/ACM Conference on AI, Ethics, and Society, AIES 2019, Honolulu, HI, USA, 27–28 January 2019, pp. 239–245. ACM (2019). https://doi.org/10.1145/3306618.3314293
12. Yang, Z., Ishay, A., Lee, J.: Coupling large language models with logic programming for robust and general reasoning from text. In: Findings of the Association for Computational Linguistics: ACL 2023, pp. 5186–5219 (2023). https://aclanthology.org/2023.findings-acl.321.pdf
13. Zecevic, M., Willig, M., Dhami, D.S., Kersting, K.: Causal Parrots: Large Language Models May Talk Causality But are Not Causal (2023). https://doi.org/10.48550/arXiv.2308.13067

# Extended Abstracts of Papers from other AI Conferences

# MIND Your Language: A Multilingual Dataset for Cross-Lingual News Recommendation (Extended Abstract)

Andreea Iana[1(✉)], Goran Glavaš[2], and Heiko Paulheim[1]

[1] University of Mannheim, Mannheim, Germany
andreea.iana@uni-mannheim.de
[2] University of Würzburg, Würzburg, Germany

**Abstract.** We present xMIND, an *open, multilingual* news recommendation dataset derived from the English MIND dataset using machine translation, covering 14 linguistically and geographically diverse languages, with digital footprints of varying sizes. Using xMIND, we benchmark several content-based neural news recommenders (NNRs) in zero-shot (`ZS-XLT`) and few-shot (`FS-XLT`) cross-lingual transfer scenarios, considering both monolingual and bilingual news consumption patterns. Our findings reveal that (i) current NNRs, even when based on a multilingual language model, suffer from substantial performance losses under `ZS-XLT` and that (ii) inclusion of target-language data in `FS-XLT` training has limited benefits, particularly when combined with bilingual news consumption. We release xMIND at https://github.com/andreeaiana/xMIND.

**Original Venue: SIGIR 2024** [15].

**Keywords:** multilingual news dataset · cross-lingual recommendation

## 1 Introduction

The digitalization of news consumption has established news platforms as the prevalent medium of information for Internet users. This, in turn, propelled personalized recommender systems into the main vehicle used by news websites to cater the individual information needs of readers. The global expansion of the Internet's outreach has vastly increased the language diversity of its users [20, 36], a non-negligible fraction of whom speak and consume news in two or more languages. However, the majority of online content remains available mostly in few resource-rich languages, while the majority of languages spoken worldwide appear in less than 0.1% of the websites or are not represented online [16].

News media play a central role in democratic societies, by ensuring informed citizens and providing a public forum for disseminating ideas [2,11]. In this context, recommender systems shape people's worldviews through the way in which they filter and propagate news [22]. On the one hand, research on personalized neural news recommenders (NNRs) has focused on improving their accuracy and diversity [32]. On the other hand, recent progress in areas such as


A. Hotho and S. Rudolph (Eds.): KI 2024, LNAI 14992, pp. 335–340, 2024.
https://doi.org/10.1007/978-3-031-70893-0_27

**Table 1.** The 14 languages of xMIND. Languages classified as low-resource according to [5] are highlighted in gray

| Code (ISO 693-3) | Language | Script | Macro-area | Family | Genus |
|---|---|---|---|---|---|
| SWH | Swahili | Latin | Africa | Niger-Congo | Bantu |
| SOM | Somali | Latin | Africa | Afro-Asiatic | Lowland East Cushitic |
| CMN | Mandarin Chinese | Han | Eurasia | Sino-Tibetan | Sinitic |
| JPN | Japanese | Japanese | Eurasia | Japonic | Japanesic |
| TUR | Turkish | Latin | Eurasia | Altaic | Turkic |
| TAM | Tamil | Tamil | Eurasia | Dravidian | Dravidian |
| VIE | Vietnamese | Latin | Eurasia | Austro-Asiatic | Vietic |
| THA | Thai | Thai | Eurasia | Tai-Kadai | Kam-Tai |
| RON | Romanian | Latin | Eurasia | Indo-European | Romance |
| FIN | Finnish | Latin | Eurasia | Uralic | Finnic |
| KAT | Georgian | Georgian | Eurasia | Kartvelian | Georgian-Zan |
| HAT | Haitian Creole | Latin | North America | Indo-European | Creoles and Pidgins |
| IND | Indonesian | Latin | Papunesia | Austronesian | Malayo-Sumbawan |
| GRN | Guarani | Latin | South America | Tupian | Maweti-Guarani |

neural machine translation [5–7, 10, 18] and multilingual large language models (mPLMs) [3, 4, 25, 27–29, 35] for low(er)-resource languages has begun to democratize access to information for underrepresented communities [16].

Nonetheless, the ever-growing language-diversity of online news readers has not been reflected in the news recommendation research, which focuses nearly entirely on resource-rich languages [12, 33], particularly English, and monolingual news consumption [8, 9, 17, 21, 24, 34]. In practice, this translates into less relevant, less balanced and less diverse recommendations for multilingual news consumers [20]. Overall, these limitations pose significant challenges for developing efficient multilingual news recommendation and effective cross-lingual transfer techniques to resource-lean and/or underrepresented languages.

We introduce xMIND, a new public and large-scale multilingual news dataset for multi- and cross-lingual recommendation. xMIND is derived by translating the 130,379 unique articles of the English-only dataset MIND [34] into a diverse selection of 14 high- and low-resource languages, spanning five geographical macro-areas and 13 distinct language families (cf. Table 1) using the open-source machine translation system NLLB 3.3B [5]. Compared to existing multilingual news recommendation datasets, xMIND is: **(1)** much more *diverse* – we include both resource-rich and resource-poor languages, covering a wide variety of geographical areas, family languages, and scripts, with some of the languages being out-of-sample for existing mPLMs (i.e., not present in the mPLM's pretraining); **(2)** *parallel* – the same set of news has been translated into all target languages, enabling the direct comparison of the performance of multilingual news recommenders and cross-lingual transfer approaches across target languages; **(3)** *open source* – we release xMIND in TSV format and provide scripts for integration with the corresponding click logs from MIND in the NewsRecLib library [13].

## 2 Cross-Lingual News Recommendation

We combine the xMIND news with the corresponding click logs and additional news annotations (i.e., categories and named entities) from MIND [34] based

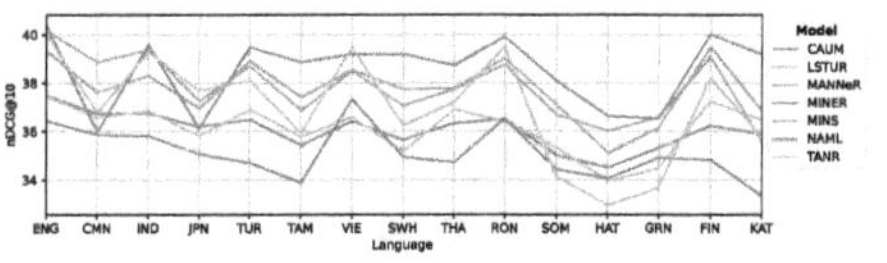


**Fig. 1.** ZS-XLT ranking performance in English and the xMIND languages

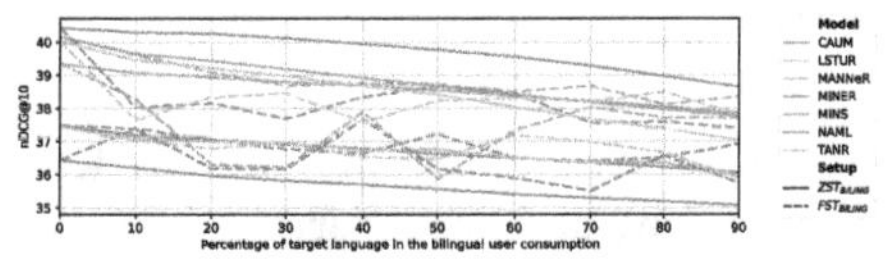


**Fig. 2.** FS-XLT ranking performance averaged over the xMIND languages

on the news IDs. We use the *small variant* of the resulting dataset to benchmark a range of content-based NNRs: (1) *NAML* [30], (2) *LSTUR* [1], (3) *MINS* [26], (4) *CAUM* [23], (5) *TANR* [31], (5) *MINER* [19], and (6) *MANNeR* [14] (only the `CR-Module`). We replace the original news encoder in all models with XLM-RoBERTa [3]. We evaluate the NNRs in two cross-lingual recommendation evaluation setups: (i) **zero-shot (ZS-XLT)** (i.e., the user history and candidates during training are *monolingual*, in English only) and (ii) **few-shot (FS-XLT)** (i.e., we increasingly replace a portion of the English training set with target-language news). We couple the two training scenarios, with two simulated types of *news consumption patterns* during inference: (i) **monolingual** (MONO) – the user reads news and receives suggestions only in the target language, and (ii) **bilingual** (BILING) – the user consumes news in English and in another language, and recommendations are also provided in the same two languages.

We find that NNRs trained monolingually on English news suffer significant performance drops when evaluated on the target languages under ZS-XLT, even when paired with a massively multilingual language model (cf. Fig. 1). These effects are stronger for languages being out-of-sample for the mPLM (i.e., HAT, GRN). Moreover, our results indicate that target-language injection during training has a limited effect in mitigating these gaps in cross-lingual transfer, especially in the context of bilingual news consumption, as shown in Fig. 2. Overall, these variations in the performance of models and the limited benefits of few-shot target-language injection – particularly when considering bilingual news consumption patterns – emphasize the need for a deeper understanding of the specific factors that drive cross-lingual NNR performance in order to inform the design of architectures that are robust to multilingual user histories.

## 3 Conclusion

We introduced xMIND, an open multilingual news recommendation dataset comprising 14 linguistically and geographically diverse languages, derived from the English MIND dataset using machine translation. Our findings from systematically benchmarking several content-based NNRs on xMIND show that current NNRs suffer considerable performance losses under ZS-XLT, while the inclusion of target-language data in FS-XLT training brings limited gains to recommenders, especially in the context of bilingual news consumption. We hope that xMIND will foster much more research on multilingual and cross-lingual news recommendation, for speakers of both high- and low-resource languages.

## References


1. An, M., Wu, F., Wu, C., Zhang, K., Liu, Z., Xie, X.: Neural news recommendation with long-and short-term user representations. In: Proceedings of the 57th Annual Meeting of the Association for Computational Linguistics, pp. 336–345 (2019). https://doi.org/10.18653/v1/P19-1033
2. Balkin, J.M.: Free speech in the algorithmic society: big data, private governance, and new school speech regulation. UCDL Rev. **51**, 1149 (2017)
3. Conneau, A., et al.: Unsupervised cross-lingual representation learning at scale. In: Proceedings of the 58th Annual Meeting of the Association for Computational Linguistics, pp. 8440–8451 (2020). https://doi.org/10.18653/v1/2020.acl-main.747
4. Conneau, A., Lample, G.: Cross-lingual language model pretraining. In: Proceedings of the 33rd International Conference on Neural Information Processing Systems, pp. 7059–7069 (2019). https://arxiv.org/abs/1901.07291
5. Costa-jussà, M.R., et al.: No language left behind: Scaling human-centered machine translation. arXiv preprint arXiv:2207.04672 (2022)
6. Dabre, R., Chu, C., Kunchukuttan, A.: A survey of multilingual neural machine translation. ACM Comput. Surv. (CSUR) **53**(5), 1–38 (2020). https://doi.org/10.1145/3406095
7. Fan, A., et al.: Beyond English-centric multilingual machine translation. J. Mach. Learn. Res. **22**(107), 1–48 (2021)
8. Gabriel De Souza, P.M., Jannach, D., Da Cunha, A.M.: Contextual hybrid session-based news recommendation with recurrent neural networks. IEEE Access **7**, 169185–169203 (2019). https://doi.org/10.1109/ACCESS.2019.2954957
9. Gulla, J.A., Zhang, L., Liu, P., Özgöbek, Ö., Su, X.: The Adressa dataset for news recommendation. In: Proceedings of the International Conference on Web Intelligence, pp. 1042–1048 (2017). https://doi.org/10.1145/3106426.3109436
10. Haddow, B., Bawden, R., Barone, A.V.M., Helcl, J., Birch, A.: Survey of low-resource machine translation. Comput. Linguist. **48**(3), 673–732 (2022). https://doi.org/10.1162/coli_a_00446
11. Helberger, N.: On the democratic role of news recommenders. In: Algorithms, Automation, and News, pp. 14–33. Routledge (2021)
12. Iana, A., et al.: Nemig-a bilingual news collection and knowledge graph about migration. In: Proceedings of the Workshop on News Recommendation and Analytics co-located with RecSys 2023 (2023)
13. Iana, A., Glavaš, G., Paulheim, H.: Newsreclib: a pytorch-lightning library for neural news recommendation. In: Proceedings of the 2023 Conference on Empirical Methods in Natural Language Processing: System Demonstrations, pp. 296–310 (2023). https://doi.org/10.18653/v1/2023.emnlp-demo.26
14. Iana, A., Glavaš, G., Paulheim, H.: Train once, use flexibly: a modular framework for multi-aspect neural news recommendation. arXiv preprint arXiv:2307.16089 (2023)
15. Iana, A., Glavas, G., Paulheim, H.: Mind your language: a multilingual dataset for cross-lingual news recommendation. In: Proceedings of the 47th International ACM SIGIR Conference on Research and Development in Information Retrieval (2024). https://doi.org/10.1145/3626772.3657867
16. Joshi, P., Santy, S., Budhiraja, A., Bali, K., Choudhury, M.: The state and fate of linguistic diversity and inclusion in the NLP world. In: Proceedings of the 58th Annual Meeting of the Association for Computational Linguistics, pp. 6282–6293 (2020). https://doi.org/10.18653/v1/2020.acl-main.560

17. Kille, B., Hopfgartner, F., Brodt, T., Heintz, T.: The plista dataset. In: Proceedings of the 2013 International News Recommender Systems Workshop and Challenge, pp. 16–23 (2013). https://doi.org/10.1145/2516641.2516643
18. Kudugunta, S., et al.: Madlad-400: a multilingual and document-level large audited dataset. In: Thirty-seventh Conference on Neural Information Processing Systems Datasets and Benchmarks Track (2023)
19. Li, J., et al.: Miner: multi-interest matching network for news recommendation. In: Findings of the Association for Computational Linguistics: ACL 2022, pp. 343–352 (2022). https://doi.org/10.18653/v1/2022.findings-acl.29
20. Ling, C., Steichen, B., Figueira, S.: Multilingual news-an investigation of consumption, querying, and search result selection behaviors. Int. J. Hum.-Comput. Interact. **36**(6), 516–535 (2020). https://doi.org/10.1080/10447318.2019.1662636
21. Lucas, J.P., da Silva, J.F.G., Figueiredo, L.F.: NPR: a news portal recommendations dataset. In: Proceedings of the The First Workshop on the Normative Design and Evaluation of Recommender Systems (NORMalize 2023), co-located with the ACM Conference on Recommender Systems 2023 (RecSys 2023) (2023)
22. Pariser, E.: The filter bubble: what the Internet is hiding from you. Penguin UK (2011)
23. Qi, T., Wu, F., Wu, C., Huang, Y.: News recommendation with candidate-aware user modeling. In: Proceedings of the 45th International ACM SIGIR Conference on Research and Development in Information Retrieval, pp. 1917–1921 (2022). https://doi.org/10.1145/3477495.3531778
24. de Souza Pereira Moreira, G., Ferreira, F., da Cunha, A.M.: News session-based recommendations using deep neural networks. In: Proceedings of the 3rd Workshop on Deep Learning for Recommender Systems, pp. 15–23 (2018). https://doi.org/10.1145/3270323.3270328
25. Touvron, H., et al.: Llama: open and efficient foundation language models. arXiv preprint arXiv:2302.13971 (2023). https://arxiv.org/abs/2302.13971
26. Wang, R., Wang, S., Lu, W., Peng, X.: News recommendation via multi-interest news sequence modelling. In: 2022 IEEE International Conference on Acoustics, Speech and Signal Processing (ICASSP), ICASSP 2022, pp. 7942–7946. IEEE (2022). https://doi.org/10.1109/ICASSP43922.2022.9747149
27. Wei, X., et al.: Polylm: an open source polyglot large language model. arXiv preprint arXiv:2307.06018 (2023). https://arxiv.org/abs/2307.06018
28. Wei, X., Weng, R., Hu, Y., Xing, L., Yu, H., Luo, W.: On learning universal representations across languages. In: International Conference on Learning Representations (2020)
29. Workshop, B., et al.: Bloom: a 176b-parameter open-access multilingual language model. arXiv preprint arXiv:2211.05100 (2022). https://arxiv.org/abs/2211.05100
30. Wu, C., Wu, F., An, M., Huang, J., Huang, Y., Xie, X.: Neural news recommendation with attentive multi-view learning. In: Proceedings of the 28th International Joint Conference on Artificial Intelligence, pp. 3863–3869 (2019). https://doi.org/10.24963/ijcai.2019/536
31. Wu, C., Wu, F., An, M., Huang, Y., Xie, X.: Neural news recommendation with topic-aware news representation. In: Proceedings of the 57th Annual Meeting of the Association for Computational Linguistics, pp. 1154–1159 (2019). https://doi.org/10.18653/v1/P19-1110
32. Wu, C., Wu, F., Huang, Y., Xie, X.: Personalized news recommendation: methods and challenges. ACM Trans. Inf. Syst. **41**(1), 1–50 (2023). https://doi.org/10.1145/3530257

33. Wu, C., Wu, F., Qi, T., Huang, Y.: Empowering news recommendation with pre-trained language models. In: Proceedings of the 44th International ACM SIGIR Conference on Research and Development in Information Retrieval, pp. 1652–1656 (2021). https://doi.org/10.1145/3404835.3463069
34. Wu, F., et al.: Mind: a large-scale dataset for news recommendation. In: Proceedings of the 58th Annual Meeting of the Association for Computational Linguistics, pp. 3597–3606 (2020). https://doi.org/10.18653/v1/2020.acl-main.331
35. Xue, L., et al.: mt5: a massively multilingual pre-trained text-to-text transformer. In: Proceedings of the 2021 Conference of the North American Chapter of the Association for Computational Linguistics: Human Language Technologies, pp. 483–498 (2021). https://doi.org/10.18653/v1/2021.naacl-main.41
36. Zuckerman, E.: The polyglot internet (2008). https://ethanzuckerman.com/the-polyglot-internet/

# A Human-in-the-Loop Tool for Annotating Passive Acoustic Monitoring Datasets (Extended Abstract)

Hannes Kath[1,2(✉)], Thiago S. Gouvêa[1], and Daniel Sonntag[1,2]

[1] German Research Centre for Artificial Intelligence (DFKI), Oldenburg, Germany
{hannes.kath,thiago.gouvea,daniel.sonntag}@dfki.de
[2] Applied Artificial Intelligence (AAI), University of Oldenburg, Oldenburg, Germany

**Abstract.** Passive Acoustic Monitoring (PAM) has become a key technology in wildlife monitoring, generating large amounts of acoustic data. However, the effective application of machine learning methods for sound event detection in PAM datasets is highly dependent on the availability of annotated data, which requires a labour-intensive effort to generate. This paper summarises two iterative, human-centred approaches that make efficient use of expert annotation time to accelerate understanding of the data: Combining transfer learning and active learning, we present an annotation tool that selects and annotates the most informative samples one at a time [11]. To annotate multiple samples simultaneously, we present a tool that allows annotation in the embedding space of a variational autoencoder manipulated by a classification head [10]. For both approaches, we provide no-code web applications for intuitive use by domain experts.

**Keywords:** Passive Acoustic Monitoring · Active Learning · Transfer Learning · Deep Generative Model · Annotation Tool

## 1 Introduction

Biodiversity loss is among the most pressing issues of our days [1]. Drivers of the negative change have been accelerating, and meeting internationally agreed conservation targets will require transformative change [7]. While machine learning methods have been increasingly brought to bear to support wildlife management, the tools available to those on the ecological front lines still lag the state-of-the-art in artificial intelligence research [6,21]. Passive acoustic monitoring (PAM) is a powerful technology for wildlife monitoring, allowing ecologists to gather extensive data on wildlife with minimal disturbance of habitats [18,19]. PAM systems can be used to continuously record sounds from various biomes, offering valuable insights into animal behaviour, species richness, and ecosystem health, with applications in ecosystem management and rapid biodiversity assessments [17].

However, effectively utilising this vast amount of data still poses significant management and analysis challenges. Sound event detection in particular (e.g.,


A. Hotho and S. Rudolph (Eds.): KI 2024, LNAI 14992, pp. 341–345, 2024.
https://doi.org/10.1007/978-3-031-70893-0_28

species identification) is limited by the need for annotated data to train supervised machine learning models. PAM data annotation is usually done manually, in a laborious and time-consuming process: domain experts listen to each audio file, annotating events by manually selecting time segments on a graphical representation of the sound (e.g., amplitude envelope or spectrogram) [2,16,20]. This approach is incompatible with the large volume of data generated by PAM. Seadash introduces a graphical implementation of data programming, but lacks evaluation on real life datasets [5]. Scikit-maad [22] and BamScape [14] are tools for large scale PAM data analysis by spectrogram segmentation and clustering; as command line tools, they lack interactivity. A promising, under-explored direction of research on interactive PAM annotation is the leveraging of small amounts of available labels (e.g., at early stages of annotation of a novel dataset) to improve the efficiency of the annotation process.

In the following, we present the core ideas of two human-centred approaches that efficiently annotate single or multiple samples of PAM datasets by iteratively incorporating previously annotated data.

## 2 Single Sample Annotation Tool

Active sampling significantly reduces the number of samples required to train a classifier compared to passive sampling. We use a combination of transfer learning and active learning for sample selection. The unlabelled dataset is embedded using BirdNet, a neural model trained on focal recordings of songbirds [8], which has shown superior performance as an embedding model for PAM data [3,12]. A number of uncertainty-based and diversity-based sampling strategies for multi-label active learning are then used to select segments for annotation (see [12] for details). While the user retains the power to choose which samples to annotate ((Fig. 1, left), the pipeline uses all available labels to recommend the most informative samples. Samples are presented as spectrograms and filterable audio ((Fig. 1, centre) and annotated with graphical interactions ((Fig. 1, right). Evaluations by a domain expert on a real life dataset show that using the interface results in a time improvement of a factor of 2–4 compared to conventional tools [11].

## 3 Multi Sample Annotation Tool

Semantic clustering of unlabelled data allows multiple samples to be annotated simultaneously. Our model architecture is inspired by [9,15] and extends a ariational autoencoder [13] with a classification head to adjust the latent space to represent events as outliers. The input data $X$ is processed by the encoder and mapped to the 2D-latent variable $Z$, which is presented to the user as a state-space and represents the input to the decoder and classifier. The decoder computes the reconstruction $\tilde{X}$ for the whole dataset. The classifier processes only annotated data by computing predicted labels $\tilde{Y}$ from $Z$ for all files with a label $Y$, grouping data points of the same category into $Z$.

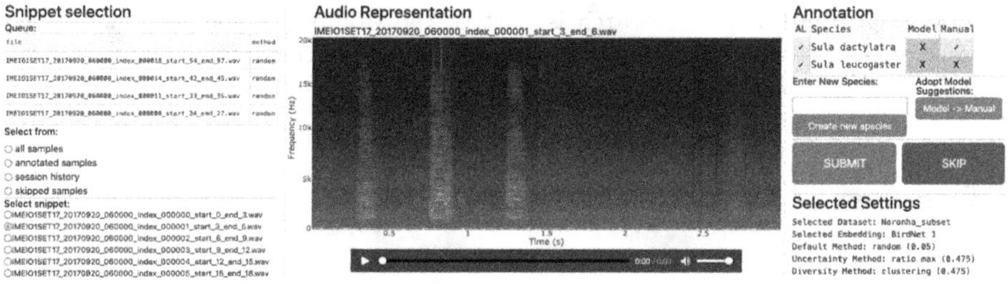


**Fig. 1.** Single sample annotation tool layout. Left: Sample selection column. Centre: Sample representation as spectrogram and audio. Right: Annotation column.

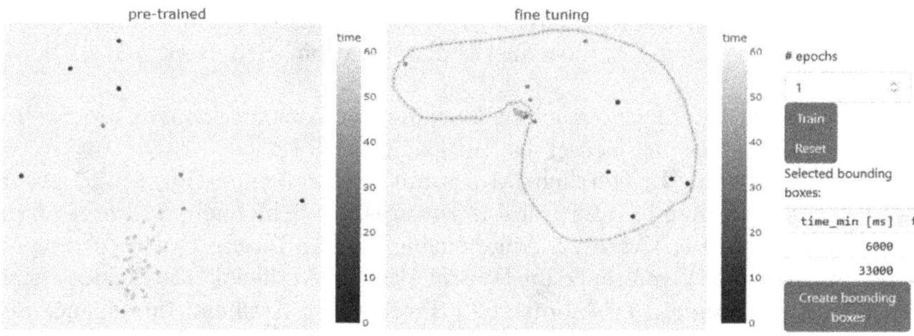


**Fig. 2.** Multi sample annotation tool layout. Left: Embedding space pre-trained. Centre: Embedding space fine-tuned. Right: Generated annotation table.

The VAE and classifier are jointly optimised by minimising the loss function $\mathcal{L} = \mathcal{L}_{\text{reconst}}(X, \tilde{X}) + D_{KL}\left(q_\phi(Z \mid X) \mid\mid p(Z)\right) + \mathcal{H}(Y, \tilde{Y})$. The first two terms are as in [13], and $\mathcal{H}$ is the cross entropy between $Y$ and $\tilde{Y}$. Figure 2 (left) shows the latent space of the pure VAE. Adding the classification head (Fig. 2, centre) distorts the latent space to represent events as outliers, which can be encircled by the user and further annotated using the generated table (Fig. 2, right). Evaluations on a synthetic dataset show that using the fine-tuned space for annotation outperforms the pre-trained model with an F-score of 94.2% compared to 77.0%.

## 4 Conclusion

Our work highlights the need for accessible machine learning-based annotation of PAM datasets and proposes two interactive, human-centred tools. For single sample annotation, we present an annotation tool that uses a combination of transfer learning and active learning. For multi sample annotation, we propose to modify the latent space of a VAE by adding a classifier head that generates an actionable, low-dimensional representation space of the input data.

Preliminary evaluations of the tools show promising results, and future work includes integrating our tool into real-world projects [4].

**Acknowledgements.** This research is part of the Computational Sustainability & Technology project area (https://cst.dfki.de/), and has been supported by the Ministry for Science and Culture of Lower Saxony (MWK), the Endowed Chair of Applied Artificial Intelligence, Oldenburg University, and DFKI.

## References

1. Cardinale, B.J., et al.: Biodiversity loss and its impact on humanity. Nature **486**, 59–67 (2012)
2. Cañas, J., Toro-Gómez, M., Sugai, L., et al.: A dataset for benchmarking neotropical anuran calls identification in passive acoustic monitoring. Sci. Data **10**(1), 771 (2023)
3. Ghani, B., Denton, T., Kahl, S., Klinck, H.: Global birdsong embeddings enable superior transfer learning for bioacoustic classification. Sci. Rep. **13**(1), 22876 (2023)
4. Gouvêa, T.S., et al.: Interactive machine learning solutions for acoustic monitoring of animal wildlife in biosphere reserves. In: IJCAI 2023 (2023)
5. Gouvêa, T.S., Troshani, I., Herrlich, M., Sonntag, D.: Annotating sound events through interactive design of interpretable features. In: Schlobach, S., Pérez-Ortiz, M., Tielman, M. (eds.) HHAI 2022: Augmenting Human Intellect - Proceedings of the First International Conference on Hybrid Human-Artificial Intelligence, Amsterdam, The Netherlands, 13–17 June 2022. Frontiers in Artificial Intelligence and Applications, vol. 354, pp. 305–306. IOS Press (2022). https://doi.org/10.3233/FAIA220225
6. Gouvêa, T., Kath, H., Troshani, I., et al.: Interactive machine learning solutions for acoustic monitoring of animal wildlife in biosphere reserves. In: Proceedings of the Thirty-Second International Joint Conference on Artificial Intelligence, Macau, SAR China, pp. 6405–6413 (2023)
7. IPBES: Summary for policymakers of the global assessment report on biodiversity and ecosystem services. Technical report, Zenodo (2019)
8. Kahl, S., Wood, C., Eibl, M., Klinck, H.: BirdNET: a deep learning solution for avian diversity monitoring. Eco. Inform. **61**, 101236 (2021). https://doi.org/10.1016/j.ecoinf.2021.101236
9. Kath, H., Gouvêa, T.S., Sonntag, D.: A deep generative model for interactive data annotation through direct manipulation in latent space. arXiv:2305.15337 (2023)
10. Kath, H., Gouvêa, T., Sonntag, D.: A human-in-the-loop tool for annotating passive acoustic monitoring datasets. In: Proceedings of the Thirty-Second International Joint Conference on Artificial Intelligence, Macau, SAR China, pp. 7140–7144 (2023). https://doi.org/10.24963/ijcai.2023/835
11. Kath, H., Serafini, P.P., Campos, I.B., Gouvêa, T.S., Sonntag, D.: Demo: enhancing wildlife acoustic data annotation efficiency through transfer and active learning. In: Proceedings of the 33nd International Joint Conference on Artificial Intelligence, IJCAI (2024)
12. Kath, H., Serafini, P.P., Campos, I.B., Gouvêa, T.S., Sonntag, D.: Leveraging transfer learning and active learning for sound event detection in passive acoustic monitoring of wildlife. In: 3rd Annual AAAI Workshop on AI to Accelerate Science and Engineering (AI2ASE), Vancouver, Canada (2024)
13. Kingma, D.P., Welling, M.: Auto-encoding variational bayes. In: Bengio, Y., LeCun, Y. (eds.) 2nd International Conference on Learning Representations, ICLR 2014, Banff, AB, Canada, 14–16 April 2014, Conference Track Proceedings (2014)

14. Michaud, F., Sueur, J., Cesne, M.L., Haupert, S.: Unsupervised classification to improve the quality of a bird song recording dataset. arXiv abs/2302.07560 (2022)
15. Paige, B., et al.: Learning disentangled representations with semi-supervised deep generative models. In: Guyon, I., et al. (eds.) Advances in Neural Information Processing Systems 30: Annual Conference on Neural Information Processing Systems 2017, Long Beach, CA, USA, 4–9 December 2017, pp. 5925–5935 (2017)
16. Perry, S., et al.: Pyrenote: a web-based, manual annotation tool for passive acoustic monitoring. In: IEEE 18th International Conference on Mobile Ad Hoc and Smart Systems, MASS 2021, Denver, CO, USA, 4–7 October 2021, pp. 633–638. IEEE (2021). https://doi.org/10.1109/MASS52906.2021.00091
17. Sueur, J., Pavoine, S., Hamerlynck, O., Duvail, S.: Rapid acoustic survey for biodiversity appraisal. PLoS ONE **3**(12), e4065 (2008). https://doi.org/10.1371/journal.pone.0004065
18. Sugai, L., Llusia, D.: Bioacoustic time capsules: using acoustic monitoring to document biodiversity. Ecol. Ind. **99**, 149–152 (2019). https://doi.org/10.1016/j.ecolind.2018.12.021
19. Sugai, L., Silva, T., Ribeiro, J., Llusia, D.: Terrestrial passive acoustic monitoring: review and perspectives. Bioscience **69**(1), 15–25 (2019). https://doi.org/10.1093/biosci/biy147
20. Tkachenko, M., Malyuk, M., Holmanyuk, A., Liubimov, N.: Label Studio: Data Labeling Software (2020)
21. Tuia, D., et al.: Perspectives in machine learning for wildlife conservation. Nat. Commun. **13**(1), 792 (2022)
22. Ulloa, J.S., Haupert, S., Latorre, J.F., Aubin, T., Sueur, J.: Scikit-maad: an opensource and modular toolbox for quantitative soundscape analysis in Python. Methods Ecol. Evol. 2041–210X.13711 (2021). https://doi.org/10.1111/2041-210X.13711

# Out-of-Distribution Detection with Logical Reasoning (Extended Abstract)

Konstantin Kirchheim(✉), Tim Gonschorek, and Frank Ortmeier

Otto-von-Guericke University, Magdeburg, Germany
{konstantin.kirchheim,tim.gonschorek,frank.ortmeier}@ovgu.de

**Abstract.** Machine learning models often only generalize reliably to samples from their training distribution which motivates out-of-distribution (OOD) detection in safety-critical applications. Current OOD detection methods, however, tend to be domain agnostic and are incapable of incorporating prior knowledge about the structure of the training distribution. To address this limitation, we introduce a novel, neuro-symbolic OOD detection algorithm that combines a deep learning-based perception system with a first-order logic-based knowledge representation. A reasoning system uses this knowledge base at run-time to infer whether inputs are consistent with prior knowledge about the training distribution. This not only enhances performance but also fosters a level of explainability that is particularly beneficial in safety-critical contexts.

**Keywords:** Out-of-Distribution · Deep Learning · Neuro-Symbolic

## 1 Introduction

In recent years, Deep Neural Networks (DNNs) outperformed classical machine learning models on virtually every task involving large amounts of high-dimensional data [6,10]. Current lines of research explore applications in potentially safety-critical areas, such as autonomous vehicles [2].

However, DNNs have been shown to only generalize well to new data points as long as they stem from the distribution for which the models have been optimized and tend to make egregiously wrong predictions with high confidence when applied to inputs that are out-of-distribution (OOD) [9]. For example, Fig. 1 shows an OOD example that is classified as *priority road* traffic sign with 100% confidence. Such examples motivate OOD detection methods.

'Priority Road' (100.00%)
Square | Red | Nosign
Consistent: No

**Fig. 1.** OOD Example

Current OOD detectors suffer from several limitations, particularly (1) the inability to incorporate prior domain knowledge about the structure of the data distribution (2) a lack


A. Hotho and S. Rudolph (Eds.): KI 2024, LNAI 14992, pp. 346–349, 2024.
https://doi.org/10.1007/978-3-031-70893-0_29

of explainability, in the sense that the raw outlier scores do not provide any human understandable justification. In this work, we aim to address these issues by using a neuro-symbolic approach.

## 2 Neuro-Symbolic OOD Detection

In many real-world settings, we possess extensive prior knowledge about what observations are plausible in some domains. However, constraints, such as *"All priority road signs are square shaped and yellow"*, often refer to abstract concepts that are difficult to correlate to raw observations, such as image pixels.

We propose a neuro-symbolic architecture, depicted in Fig. 2, that bridges the gap between abstract domain concepts and raw percepts by using several DNN-based concept detectors $f_i(\cdot)$ that map inputs $\mathbf{x}$ to human understandable concepts. At runtime, these models, which are trained in a supervised manner, are used to infer a *state* $z = F(\mathbf{x}) = \langle f_1(\mathbf{x}), ..., f_n(\mathbf{x})\rangle \in \mathcal{Z}$. We then use a First-order logic (FOL) knowledge base (KB) $\varphi$ to define constraints that this state must satisfy. We call the set of states that satisfy all constraints $\mathcal{V} = \{z \in \mathcal{Z} : z \models \varphi\}$ *valid states.*

**Naïve Logic.** A straightforward approach to performing OOD detection is to check if the state corresponding to some input is valid. Formally, we calculate outlier scores as follows:

$$\text{outlier}(\mathbf{x}) = \begin{cases} 0 & \text{if} \quad F(\mathbf{x}) \in \mathcal{V} \\ 1 & \text{else}\ . \end{cases}$$

This approach would reject the example in Fig. 1 since it is detected as *red*, which is inconsistent with prior knowledge about *priority road* signs. Since we can retrace decisions to violated rules, this approach is more explainable.

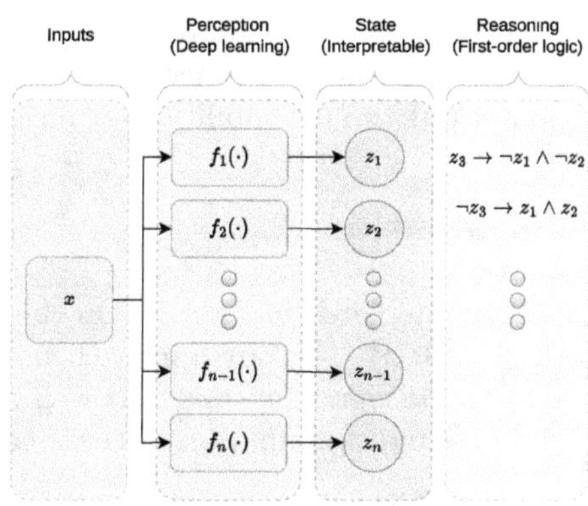


**Fig. 2.** Proposed neuro-symbolic architecture for OOD detection

**LogicOOD.** The naïve approach ignores the outlier scores of the individual concept detectors. We therefore propose to compute a refined outlier score as

$$\text{outlier}(\mathbf{x}) = \begin{cases} \sum_{i=1}^{N} \lambda_i D_{f_i}(\mathbf{x}) & \text{if} \quad F(\mathbf{x}) \in \mathcal{V} \\ 1 & \text{else} \end{cases}$$

where $\lambda$ are positive weighting coefficients, and $D_{f_i}$ is the Maximum Softmax Probability (MSP) baseline OOD detector [4].

**Table 1.** Mean performance and corresponding standard error of different OOD detection methods in our experiments. Results averaged over 10 experimental trials with different random seeds and different datasets

| DETECTOR | AUROC ↑ | AUPR-IN ↑ | AUPR-OUT ↑ | FPR95 ↓ |
|---|---|---|---|---|
| MSP [4] | 99.04 ± 0.07 | 98.35 ± 0.14 | 99.29 ± 0.05 | 2.54 ± 0.15 |
| EBO [8] | 99.03 ± 0.11 | 98.76 ± 0.14 | 99.08 ± 0.12 | 2.26 ± 0.27 |
| MaxLogit [3] | 99.01 ± 0.11 | 98.73 ± 0.14 | 99.07 ± 0.12 | 2.29 ± 0.27 |
| Entropy [1] | 99.15 ± 0.07 | 98.64 ± 0.13 | 99.33 ± 0.06 | 2.46 ± 0.15 |
| ReAct [12] | 99.04 ± 0.10 | 98.77 ± 0.13 | 99.08 ± 0.12 | 2.21 ± 0.24 |
| Mahalanobis [7] | 99.70 ± 0.02 | 99.40 ± 0.06 | 99.83 ± 0.01 | 1.11 ± 0.05 |
| ViM [13] | 96.96 ± 0.08 | 95.95 ± 0.10 | 99.75 ± 0.02 | 6.08 ± 0.16 |
| Ensemble [5] | 99.77 ± 0.03 | 99.58 ± 0.05 | 99.86 ± 0.01 | 0.99 ± 0.07 |
| Naïve Logic (ours) | 86.08 ± 0.91 | 91.76 ± 0.54 | 91.76 ± 0.45 | 100.00 ± 0.00 |
| LogicOOD (ours) | **99.85** ± 0.01 | **99.74** ± 0.02 | **99.92** ± 0.01 | **0.60** ± 0.04 |

The presented approach has some interesting theoretical properties. For instance, it can be shown that, for a KB of Horn clauses - a subset of FOL - enumerating $\mathcal{V}$ is #P-hard, while inference can be done in polynomial time. Also, adding additional concepts and constraints to the system can only increase (or maintain) the probability that an input is being rejected.

## 3 Evaluation

Results are depicted in Table 1. On the German Traffic Sign Recognition Benchmark (GTSRB) [11], our approach uses three concept detectors that infer the type of traffic sign, the shape, as well as the color. The KB encodes constraints on the possible combination of these concepts, such as

$$\forall z \quad \texttt{priority-road}(z) \rightarrow \texttt{yellow}(z) \wedge \texttt{square}(z) \ .$$

While the naïve approach already outperforms random guessing, LogicOOD achieves state-of-the-art results. Additional experiments with more complex constraints are provided in the main paper.

## References


1. Chan, R., Rottmann, M., Gottschalk, H.: Entropy maximization and meta classification for out-of-distribution detection in semantic segmentation. In: Proceedings of the IEEE/CVF International Conference on Computer Vision, pp. 5128–5137 (2021)
2. Grigorescu, S., Trasnea, B., Cocias, T., Macesanu, G.: A survey of deep learning techniques for autonomous driving. J. Field Robot. **37**(3), 362–386 (2020)

3. Hendrycks, D., Basart, S., Mazeika, M., Mostajabi, M., Steinhardt, J., Song, D.: Scaling out-of-distribution detection for real-world settings. In: International Conference on Machine Learning (2022)
4. Hendrycks, D., Gimpel, K.: A baseline for detecting misclassified and out-of-distribution examples in neural networks. In: International Conference on Learning Representations (2017)
5. Lakshminarayanan, B., Pritzel, A., Blundell, C.: Simple and scalable predictive uncertainty estimation using deep ensembles. In: Advances in neural information processing systems, pp. 6402–6413 (2017)
6. LeCun, Y., Bengio, Y., Hinton, G.: Deep learning. Nature **521**(7553), 436–444 (2015)
7. Lee, K., Lee, K., Lee, H., Shin, J.: A simple unified framework for detecting out-of-distribution samples and adversarial attacks. In: Advances in Neural Information Processing Systems, vol. 31 (2018)
8. Liu, W., Wang, X., Owens, J., Li, Y.: Energy-based out-of-distribution detection. In: Advances in Neural Information Processing Systems, vol. 33 (2020)
9. Nguyen, A., Yosinski, J., Clune, J.: Deep neural networks are easily fooled: high confidence predictions for unrecognizable images. In: Proceedings of the IEEE Conference on Computer Vision and Pattern Recognition, pp. 427–436 (2015)
10. Schmidhuber, J.: Deep learning in neural networks: an overview. Neural Netw. **61**, 85–117 (2015)
11. Stallkamp, J., Schlipsing, M., Salmen, J., Igel, C.: Man vs. computer: benchmarking machine learning algorithms for traffic sign recognition. Neural Netw. **32**, 323–332 (2012)
12. Sun, Y., Guo, C., Li, Y.: React: Out-of-distribution detection with rectified activations. Adv. Neural. Inf. Process. Syst. **34**, 144–157 (2021)
13. Wang, H., Li, Z., Feng, L., Zhang, W.: Vim: Out-of-distribution with virtual-logit matching. In: Proceedings of the IEEE/CVF Conference on Computer Vision and Pattern Recognition, pp. 4921–4930 (2022)

# A Brief Systematization of Explanation-Aware Attacks

Maximilian Noppel(✉) and Christian Wressnegger

KASTEL Security Research Labs, Karlsruhe Institute of Technology, Karlsruhe, Germany
maximilian.noppel@kit.edu

**Abstract.** Due to the overabundance of trained parameters modern machine learning models are largely considered black boxes. Explanation methods aim to shed light on the inner working of such models, and, thus can serve as debugging tools. However, recent research has demonstrated that carefully crafted manipulations at the input or the model can successfully fool the model *and* the explanation method. In this work, we briefly present our systematization of such explanation-aware attacks. We categorize them according to three distinct attack types, three types of scopes, and three different capabilities an adversary can have. In our full paper [12], we further present a hierarchy of robustness notion and various defensive techniques tailored toward explanation-aware attacks.

**Keywords:** XAI · Explanation-Aware Attacks · Adversarial ML

## 1 Introduction

The research field of explainable machine learning has evolved to shed light on the inner working of ML classifiers. Thus, explanation methods can be used to debug models during development and to build trust in the reasonability of a classifier's decision-making process. In their core the developed explanation methods are necessarily based on heuristics. They compress a classifier's scattered and highly-complex decision surface into a relatively low-dimensional explanation space, e.g., the space of pixels in the input resolution. The resulting explanations can be visualized as heatmaps overlaid on the input indicating "where the model looks" (cf. Fig. 1). We denote methods that attribute importance scores to input features as *feature attribution* methods and focus on such methods for this work.

Unfortunately, recent research has demonstrated that many of the proposed explanation methods are unreliable in adversarial environments [4,5,7,10–12,14]. The classifier *and* the explanation method can be fooled through attacks at inference time or at training time, and in a vast number of different attack scenarios. These adversarial corner cases raise concerns on the trustworthiness of explainable machine learning in general. In this short paper, we present a


A. Hotho and S. Rudolph (Eds.): KI 2024, LNAI 14992, pp. 350–354, 2024.
https://doi.org/10.1007/978-3-031-70893-0_30

brief systematization of such attacks against explanations, so-called explanation-aware attacks. For the full paper we refer to Noppel and Wressnegger, *SoK: Explainable Machine Learning in Adversarial Environments* [12].

## 2 Systematization

We systematize explanation-aware attackers according to their goals and their capabilities. The goals are specified by the attack types and the attack scopes, as introduced below. The attacker's capabilities to reach the respective goals are specified thereafter. Lastly, we outline the further contributions in the full paper.

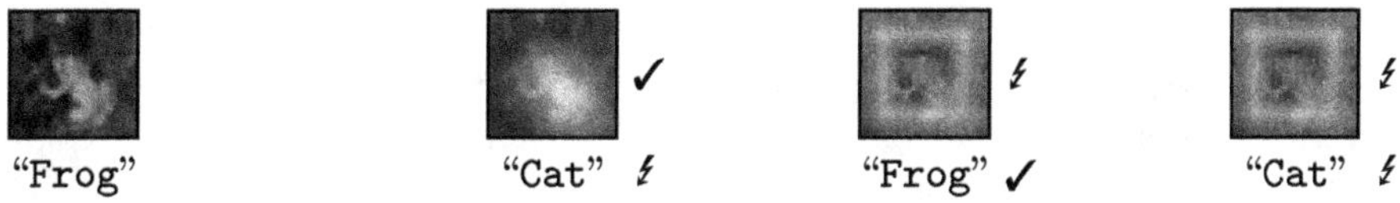


(a) Explanation Preserving (b) Prediction Preserving (c) Dual

**Fig. 1.** The three attack types with a square representing an altered explanation. A ✓ symbol indictates correctness, while ↯ indicates something being wrong.

**Attack Types.** Above, in Fig. 1, we depict the three attack types. Each type requires the prediction and the explanation to be either preserved or altered:

(a) *Explanation-Preserving Attack.* The aim of an explanation-preserving attack is to change the prediction but to preserve the explanation in comparison to a benign case. Note that the benign case might be specified by a clean in-distribution input and/or a benignly trained model, depending on the threat model.
(b) *Prediction-Preservering Attack.* In a prediction-preserving attack the prediction should be correct, but the explanation should deviate from the benign case. We provide details how this deviation could look like in the next paragraph.
(c) *Dual Attack.* Lastly, in a dual attack the predictions and the explanations differ from the benign case. This type of attack allows the greatest flexibility.

We have found legitimate argumentation for each type in the literature, hence, the relevant attack type depends heavily on the concrete application scenario.

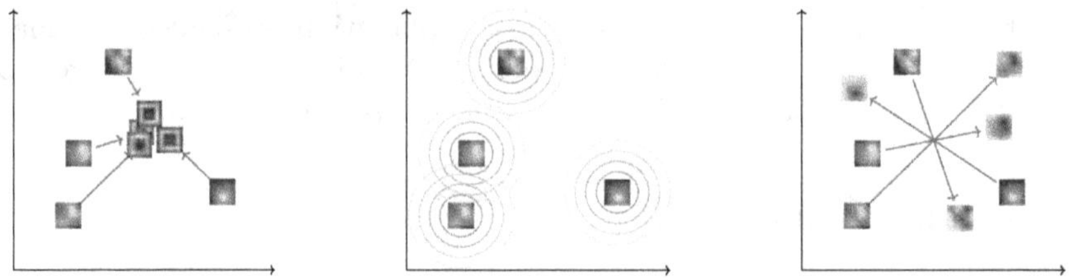

(a) Targeted (b) Untargeted (c) Semi-Targeted

**Fig. 2.** The three scopes in a simplified 2D explanation space, but with the displayed explanation being heatmaps. The importance scores of the first and the second feature are shown on the x-axis, and the y-axis respectively.

**Attack Scopes.** Next, we identify three attack scopes, which further shape our understanding of an altered explanation. We provide a simplified and explanatory depiction of each scope in depiction of each scope in Fig. 2, using a 2D explanation space.

(a) *Targeted.* In targeted attacks, the adversary aims to show a certain fixed target explanation, e.g., a square, as depicted in Fig. 2a.
(b) *Untargeted.* In untargeted attacks the adversarial goal is to produce a maximally different explanation compared to the benign scenario. The difference is measured by a certain norm in explanation space, visualized as circles in Fig. 2b.
(c) *Semi-Targeted.* Lastly, there are scenarios where the sample-specific target explanation is defined by a function in explanation space. We denote such attacks as semi-targeted, and depict an inverting function in Fig. 2c as an example.

**The Attacker's Capabilities.** We broadly distinguish the attacker's capabilities in three major categories: Firstly, *input manipulators* can perturb the inputs at inference time to show the wished effects on the explanations and the predictions [5,8,14]. The manipulations are restricted to be within a certain norm ball as a proxy for the imperceptibility of the manipulation. Secondly, *model manipulators* can manipulate indirectly via data poisoning or code poisoning, or directly via model poisoning [3,6,7,10]. Lastly, a powerful *system manipulator* can manipulate the whole explainable system [1,2,9,13]. This capability explicitly includes using multiple models or entirely different explanation methods. Anything that generates predictions and explanations in the correct format is allowed. Note that the system manipulator in this case is the system operator, who wants to hide a certain property of the system, e.g., an unfair decision-making process. Importantly, the first two, input manipulators and model manipulators, are also present in classical attacks against predictions, while the capability to manipulate the whole explainable system is only reasonable with explanations.

**Further Contributions.** In the full paper, we provide further details on the above aspects, and categorize related work according to the introduced schema. Additionally, we present a hierarchy of explanation-aware robustness notions. Moreover, we discuss defensive techniques at training time and operational defenses at inference time from the viewpoint of explanation-aware attacks.

## 3 Conclusion

In conclusion, we systematize attacks against explainable systems according to three primary dimensions. Our systematization enables the community to identify potential attacks in the concrete application scenario at hand. We intend to raise awareness for the unreliability of explainable machine learning in adversarial environments. In particular, we raise doubt on the effectiveness of defensive techniques utilizing explanations. Our work should not be seen as concluding the research field but as laying the groundwork to support more focused investigations toward a secure, reliable, and trustworthy AI utility.

**Acknowledgement.** The authors gratefully acknowledge funding from the German Federal Ministry of Education and Research (BMBF) under the project DataChain-Sec (FKZ 16KIS1700) and by the Helmholtz Association (HGF) within topic "46.23 Engineering Secure Systems."

## References

1. Aïvodji, U., Arai, H., Fortineau, O., Gambs, S., Hara, S., Tapp, A.: Fairwashing: the risk of rationalization. In: Proceedings of the International Conference on Machine Learning (ICML), Proceedings of Machine Learning Research, vol. 97 (2019)
2. Aïvodji, U., Arai, H., Gambs, S., Hara, S.: Characterizing the risk of fairwashing. In: Proceedings of the Annual Conference on Neural Information Processing Systems (NeurIPS) (2021)
3. Anders, C.J., Pasliev, P., Dombrowski, A.K., Müller, K.R., Kessel, P.: Fairwashing explanations with off-manifold detergent. In: Proceedings of the International Conference on Machine Learning (ICML), Proceedings of Machine Learning Research, vol. 119 (2020)
4. Baniecki, H., Biecek, P.: Adversarial attacks and defenses in explainable artificial intelligence: a survey. In: Proceedings of the IJCAI Workshop of Explainable AI (XAI) (2023)
5. Dombrowski, A.K., Alber, M., Anders, C., Ackermann, M., Müller, K.R., Kessel, P.: Explanations can be manipulated and geometry is to blame. In: Proceedings of the Annual Conference on Neural Information Processing Systems (NeurIPS) (2019)
6. Fang, S., Choromanska, A.: Backdoor attacks on the DNN interpretation system. In: Proceedings of the National Conference on Artificial Intelligence (AAAI) (2022)
7. Heo, J., Joo, S., Moon, T.: Fooling neural network interpretations via adversarial model manipulation. In: Proceedings of the Annual Conference on Neural Information Processing Systems (NeurIPS) (2019)

8. Ivankay, A., Girardi, I., Frossard, P., Marchiori, C.: Fooling explanations in text classifiers. In: Proceedings of the International Conference on Learning Representations (ICLR) (2022)
9. Lakkaraju, H., Bastani, O.: "How do I fool you?": manipulating user trust via misleading black box explanations. In: Proceedings of the AAAI/ACM Conference on AI, Ethics, and Society (AIES) (2020)
10. Noppel, M., Peter, L., Wressnegger, C.: Disguising attacks with explanation-aware backdoors. In: Proceedings of the IEEE Symposium on Security and Privacy (S&P) (2023)
11. Noppel, M., Wressnegger, C.: Explanation-aware backdoors in a nutshell. In: Proceedings of the German Conference on Artificial Intelligence (KI) (2023)
12. Noppel, M., Wressnegger, C.: SoK: explainable machine learning in adversarial environments. In: Proceedings of the IEEE Symposium on Security and Privacy (S&P) (2024)
13. Slack, D., Hilgard, S., Jia, E., Singh, S., Lakkaraju, H.: Fooling LIME and SHAP: adversarial attacks on post hoc explanation methods. In: Proceedings of the AAAI/ACM Conference on AI, Ethics, and Society (AIES) (2020)
14. Zhang, X., Wang, N., Shen, H., Ji, S., Luo, X., Wang, T.: Interpretable deep learning under fire. In: Proceedings of the USENIX Security Symposium (2020)

# Star-Shaped Denoising Diffusion Probabilistic Models (Extended Abstract)

Andrey Okhotin[1,2](✉), Dmitry Molchanov[7](✉), Vladimir Arkhipkin[3], Grigory Bartosh[4], Viktor Ohanesian[8], Aibek Alanov[1,5], and Dmitry Vetrov[6](✉)

[1] HSE University, Moscow, Russia
[2] Lomonosov Moscow State University, Moscow, Russia
[3] Sber Artificial Intelligence Laboratory (Sber AI), Moscow, Russia
[4] AMLab, Informatics Institute University of Amsterdam, Amsterdam, The Netherlands
[5] Artificial Intelligence Research Institute (AIRI), Moscow, Russia
[6] Constructor University, Bremen, Germany
dvetrov@constructor.university
[7] London, United Kingdom
[8] Belgrade, Serbia

**Keywords:** Artificial Intelligence · Generative Modeling · Diffusion Models

## 1 Introduction

Deep generative models have shown outstanding sample quality in a wide variety of modalities. Generative Adversarial Networks (GANs) [4,9], autoregressive models [14], Variational Autoencoders [10,15], Normalizing Flows [2,5] and energy-based models [17] show impressive abilities to synthesize objects. Recently, diffusion probabilistic models [7,16] have received a lot of attention. These models generate samples using a trained Markov process that starts with white noise and iteratively removes noise from the sample. The main principle of diffusion models is to destroy information during the forward process and then restore it during the reverse process. In conventional diffusion models like denoising diffusion probabilistic models (DDPM) destruction of information occurs through the injection of Gaussian noise, which is reasonable for some types of data, such as images. However, for data distributed on manifolds, bounded volumes, or with other features, the injection of Gaussian noise can be unnatural, breaking the data structure. Unfortunately, it is not clear how to replace the noise distribution within traditional diffusion models. The problem is that we have to maintain a connection between the distributions defining the Markov noising process that gradually destroys information and its marginal distributions. While some papers explore other distributions, such as delta functions [1] or Gamma distribution [12], they provide ad hoc solutions for special cases that are not easily generalized.

D. Molchanov and V. Ohanesian—Independent Researcher


A. Hotho and S. Rudolph (Eds.): KI 2024, LNAI 14992, pp. 355–359, 2024.
https://doi.org/10.1007/978-3-031-70893-0_31

## 2 Model Definition

We define star-shaped diffusion as a *non-Markovian* forward process $q^{\mathrm{ss}}(x_{0:T})$ that has the following structure:

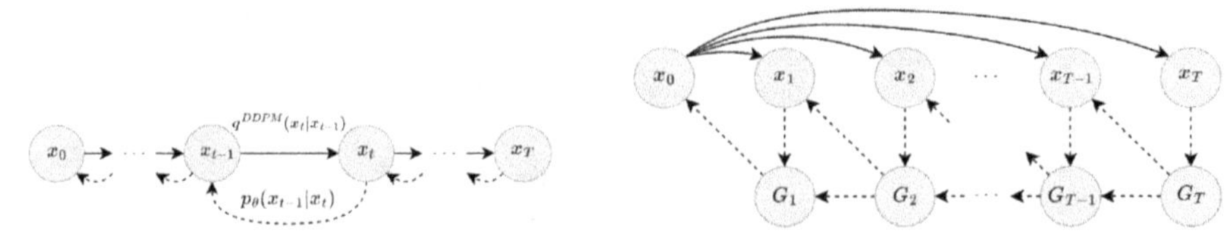

(a) Denoising Diffusion Probabilistic Models

(b) Star-Shaped Denoising Diffusion Probabilistic Models

**Fig. 1.** Model structure of DDPM and SS-DDPM

$$q^{\mathrm{ss}}(x_{0:T}) = q(x_0)\prod_{t=1}^{T} q^{\mathrm{ss}}(x_t|x_0), \tag{1}$$

where $q(x_0)$ is the data distribution. In contrary with DDPM, for the star-shaped diffusion, however, the Markovian assumption about the structure of the reverse process breaks, so it can be defined as follows:

$$p_\theta^{\mathrm{ss}}(x_{0:T}) = p_\theta^{\mathrm{ss}}(x_T)\prod_{t=1}^{T} p_\theta^{\mathrm{ss}}(x_{t-1}|x_{t:T}), \tag{2}$$

where $p_\theta^{\mathrm{ss}}$ is a trainable model with parameters $\theta$. Base on the provided structure of forward and reverse processes, we can write down the variational lower bound as follows:

$$\mathcal{L}^{\mathrm{ss}}(\theta) = \mathbb{E}_{q^{\mathrm{ss}}}\left[\log p_\theta(x_0|x_{1:T}) - \sum_{t=2}^{T} D_{KL}\left(q^{\mathrm{ss}}(x_{t-1}|x_0)\,\|\,p_\theta^{\mathrm{ss}}(x_{t-1}|x_{t:T}\right)\right] \tag{3}$$

With this VLB, we only need the marginal distributions $q(x_{t-1}|x_0)$ to define and train the model, which allows us to use a wider variety of noising distributions. Since conditioning the predictive model $p_\theta(x_{t-1}|x_{t:T})$ on the whole tail $x_{t:T}$ is typically impractical, we propose a more efficient way to implement the reverse process next. Instead of using the full tail $x_{t:T}$, we would like to define some statistic $G_t = \mathcal{G}_t(x_{t:T})$ that would extract all information about $x_0$ from the tail $x_{t:T}$. Formally speaking, we call $G_t$ a *sufficient tail statistic* if the following equality holds:

$$q^{\mathrm{ss}}(x_{t-1}|x_{t:T}) = q^{\mathrm{ss}}(x_{t-1}|G_t). \tag{4}$$

This condition could be satisfied for a specific subset of the exponential family, which we call an *exponential family with linear parameterization.* Moreover, we can rewrite the probabilistic model in terms of $G_t$ and see that variables $(x_0, G_{1:T})$ form a (not necessarily Gaussian) DDPM. In the case of Gaussian distributions, this demonstrates equivalence between SS-DDPM and DDPM. The resulting model, illustrated in the Fig. 1, is similar to DDPM in spirit.

## 3 Experiments

We evaluate SS-DDPM with different families of noising distributions. Our implementation is available at https://github.com/andrey-okhotin/star-shaped (Table 1).

**Table 1.** We consider two examples of star-shaped diffusion processes with Dirichlet and Wishart noise to generate data from the probabilistic simplex and from the manifold of p.d. matrices respectively. We apply SS-DDPM to a geodesic dataset of fires on the Earth's surface [3] using a three-dimensional von Mises–Fisher distribution. The first row is real data and the second is generated

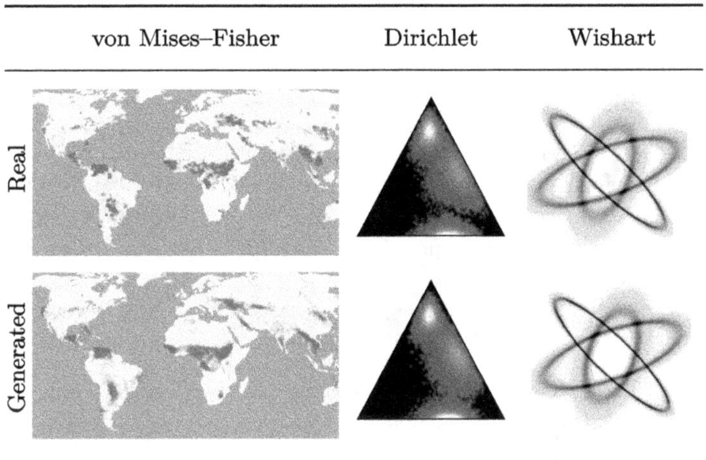


*Discrete Data.* Categorical SS-DDPM is similar to Multinomial Text Diffusion (MTD, [8]). We follow a similar setup to MTD and apply Categorical SS-DDPM to unconditional text generation on the `text8` dataset [11]. As shown in Table 2, SS-DDPM achieves similar results to MTD, allowing to use different distributions in a unified manner.

**Table 2.** Comparison of Categorical SS-DDPM and Multinomial Text Diffusion on `text8`

| Model | NLL (bits/char) |
|---|---|
| MTD | $\leq 1.72$ |
| SS-DDPM | $\leq \mathbf{1.69}$ |

*Image Data.* Finally, we evaluate SS-DDPM on CIFAR-10. Since the training data is constrained to a [0, 1] segment, we use Beta distributions. We evaluate our model with various numbers of generation steps and report the resulting FID [6] in Fig. 2. Beta SS-DDPM achieves comparable quality with the Improved DDPM [13] and is slightly better on lower numbers of steps, challenging the optimality of Gaussian noise in this setting.

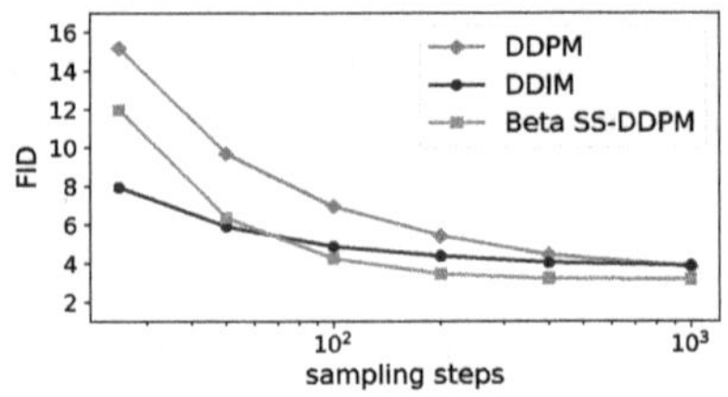


**Fig. 2.** Quality of Beta SS-DDPM, DDPM and DDIM on CIFAR10

## 4 Conclusion

Conducted experiments show the ability of SS-DDPM to work with different distributions from a general subset of the exponential family and generate data of decent quality from various domains. The star-shaped formulation opens new applications of diffusion-like probabilistic models, especially for data from exotic domains where domain-specific non-Gaussian diffusion is more appropriate.

## References

1. Bansal, A., et al.: Cold diffusion: inverting arbitrary image transforms without noise. arXiv preprint arXiv:2208.09392 (2022)
2. Chen, R.T., Behrmann, J., Duvenaud, D.K., Jacobsen, J.H.: Residual flows for invertible generative modeling. In: Advances in Neural Information Processing Systems, vol. 32 (2019)
3. EOSDIS: land, atmosphere near real-time capability for eos (lance) system operated by nasas earth science data and information system (esdis) (2020). https://earthdata.nasa.gov/earth-observation-data/near-real-time/firms/active-fire-data
4. Goodfellow, I., et al.: Generative adversarial nets. In: Advances in Neural Information Processing Systems, vol. 27 (2014)
5. Grathwohl, W., Chen, R.T., Bettencourt, J., Sutskever, I., Duvenaud, D.: Ffjord: free-form continuous dynamics for scalable reversible generative models. arXiv preprint arXiv:1810.01367 (2018)
6. Heusel, M., Ramsauer, H., Unterthiner, T., Nessler, B., Hochreiter, S.: Gans trained by a two time-scale update rule converge to a local nash equilibrium. In: Advances in Neural Information Processing Systems, vol. 30 (2017)
7. Ho, J., Jain, A., Abbeel, P.: Denoising diffusion probabilistic models. arXiv preprint arXiv:2006.11239 (2020)
8. Hoogeboom, E., Nielsen, D., Jaini, P., Forré, P., Welling, M.: Argmax flows and multinomial diffusion: learning categorical distributions. Adv. Neural. Inf. Process. Syst. **34**, 12454–12465 (2021)
9. Karras, T., et al.: Alias-free generative adversarial networks. In: Advances in Neural Information Processing Systems, vol. 34 (2021)

10. Kingma, D.P., Welling, M.: Auto-encoding variational bayes. arXiv preprint arXiv:1312.6114 (2013)
11. Mahoney, M.: Large text compression benchmark (2011). http://www.mattmahoney.net/dc/text.html
12. Nachmani, E., Roman, R.S., Wolf, L.: Denoising diffusion gamma models. arXiv preprint arXiv:2110.05948 (2021)
13. Nichol, A., Dhariwal, P.: Improved denoising diffusion probabilistic models. arXiv preprint arXiv:2102.09672 (2021)
14. Ramesh, A., et al.: Zero-shot text-to-image generation. In: International Conference on Machine Learning, pp. 8821–8831. PMLR (2021)
15. Rezende, D.J., Mohamed, S., Wierstra, D.: Stochastic backpropagation and approximate inference in deep generative models. In: International Conference on Machine Learning, pp. 1278–1286. PMLR (2014)
16. Sohl-Dickstein, J., Weiss, E., Maheswaranathan, N., Ganguli, S.: Deep unsupervised learning using nonequilibrium thermodynamics. In: International Conference on Machine Learning, pp. 2256–2265. PMLR (2015)
17. Xiao, Z., Kreis, K., Kautz, J., Vahdat, A.: Vaebm: a symbiosis between variational autoencoders and energy-based models. arXiv preprint arXiv:2010.00654 (2020)

# Author Index


**A**
Alanov, Aibek 355
Albayrak, Sahin 308
Amalraj, Anandraj 275
Arkhipkin, Vladimir 355
Asif, Haris 292

**B**
Bartosh, Grigory 355
Bayerkuhnlein, Moritz 3
Beecks, Christian 275
Bertram, Timo 17
Bonenberger, Christopher 30, 300
Borst, Vanessa 282

**D**
D'Eramo, Carlo 316
Dang, Xuzhe 45
Dasanayake, Buddhika 232
De Candido, Oliver 190
Dittus, Timo 282
Diyanananda, Yasas 232

**E**
Edelkamp, Stefan 45, 60, 73
Ertel, Wolfgang 30

**F**
Filter, Björn 86
Freund, Michael 99
Fries, Justus 99
Fürnkranz, Johannes 17

**G**
Glavaš, Goran 335
Gonschorek, Tim 346
Göttl, Quirin 292
Götz, Julia 316
Gouvêa, Thiago S. 114, 260, 341
Graß, Alexander 275
Großmann, Jürgen 147

**H**
Haasis, Julian 300
Hammam, Ahmed 190
Harth, Andreas 99
Hartwig, Mattis 175
Hotho, Andreas 246

**I**
Iana, Andreea 335

**J**
Jakobs, Oliver 218
Jentsch, Marc 275

**K**
Karjauv, Aray 308
Kath, Hannes 114, 341
Kemmerzell, Nils 128
Kharma, Sami 147
Kirchheim, Konstantin 346
Kitschke, Felix 275
Kounev, Samuel 282
Krause, Anna 246
Kruspe, Anna 160

**L**
Leuschel, Michael 190
Lugrin, Birgit 316
Luttermann, Malte 175

**M**
Manzoor, Aqsa 232
Marzilger, Robert 292
Mattick, Alexander 292
Molchanov, Dmitry 355
Möller, Ralf 86, 175




A. Hotho and S. Rudolph (Eds.): KI 2024, LNAI 14992, pp. 361–362, 2024.
https://doi.org/10.1007/978-3-031-70893-0

Müller, Konstantin 282
Müller, Martin 17

**N**
Noppel, Maximilian 350
Norz, Maximilian 275

**O**
Ohanesian, Viktor 355
Okhotin, Andrey 355
Ortmeier, Frank 346
Özçep, Özgür Lütfü 86

**P**
Paulheim, Heiko 335
Plinge, Axel 292

**R**
Riedmann, Anna 316
Roßbach, Jan 190
Ruttmann, Julia 204

**S**
Saha, Sreyashi 232
Schäffer, Patric 275
Scheele, Stephan 324
Schiendorfer, Alexander 204
Schlör, Daniel 246
Schmid, Ute 324
Schneider, Markus 30, 300
Schon, Claudia 218
Schreiner, Annika 128
Schwartz, Tobias 3
Schwenker, Friedhelm 30
Sentanoe, Wilson 232
Slany, Emanuel 324
Sonntag, Daniel 114, 260, 341
Stillman, Mila 160

**T**
Thyssens, Daniela 232
Tritscher, Julian 246
Troshani, Ilira 260

**V**
Vetrov, Dmitry 355

**W**
Wolf, Maximilian 246
Wolter, Diedrich 3
Wressnegger, Christian 350

Printed in the USA
CPSIA information can be obtained
at www.ICGtesting.com
CBHW071148070924
14227CB00009B/577

9 783031 708923